AF377671

Socio-Hydrology: The New Paradigm in Resilient Water Management

Socio-Hydrology: The New Paradigm in Resilient Water Management

Editors

Tamim Younos
Tammy E. Parece
Juneseok Lee
Jason Giovannettone
Alaina J. Armel

MDPI • Basel • Beijing • Wuhan • Barcelona • Belgrade • Manchester • Tokyo • Cluj • Tianjin

Editors

Tamim Younos
Green Water-Infrastructure Academy
USA

Tammy E. Parece
Colorado Mesa University
USA

Juneseok Lee
Manhattan College
USA

Jason Giovannettone
Dewberry
USA

Alaina J. Armel
AECOM
USA

Editorial Office
MDPI
St. Alban-Anlage 66
4052 Basel, Switzerland

This is a reprint of articles from the Special Issue published online in the open access journal *Hydrology* (ISSN 2306-5338) (available at: https://www.mdpi.com/journal/hydrology/special_issues/SH).

For citation purposes, cite each article independently as indicated on the article page online and as indicated below:

LastName, A.A.; LastName, B.B.; LastName, C.C. Article Title. *Journal Name* **Year**, *Volume Number*, Page Range.

ISBN 978-3-0365-2203-6 (Hbk)
ISBN 978-3-0365-2204-3 (PDF)

Cover image courtesy of Tammy E. Parece

Contents

About the Editors

Tamim Younos is the Founder & President of the Green Water-Infrastructure Academy, a nonprofit organization, and a former Research Professor of Water Resources at Virginia Tech. Dr. Younos is a water scientist with research and educational interests in urban environmental sustainability topics and watershed assessment. He has authored/co-authored more than 150 research/technical publications and edited seven books on water science topics. Dr. Younos has offered numerous seminars/lectures and organized regional, national/international symposia & workshops. His professional recognitions include American Water Resources Association (AWRA) Icko Iben Award, AWRA Fellow Award, Fulbright Scholar Award, and Japan Society for Promotion of Science (JSPS) Award.

Tammy E. Parece is the Assistant Professor of Geography in the Department of Social & Behavioral Sciences at Colorado Mesa University. She manages Colorado Mesa University's Geography Education and Outreach Program (GEO_PRO). In 2016, she received her PhD from Virginia Tech in Geospatial and Environmental Analysis. Her dissertation evaluated the physical and social aspects of urban agriculture, including urban agriculture's ability to address food insecurity and rainwater harvesting for irrigation. She has published several book chapters, journal articles, technical manuals and GIS laboratory manuals. She has also edited numerous books and special journal editions.

Junseok Lee is Associate Professor of Civil and Environmental Engineering at Manhattan College, N.Y. Dr. Lee has about 70+ technical publications, including articles in highly respected journals and conference proceedings. In addition, he has delivered numerous presentations and invited talks at national and international conferences on water, environment, and infrastructure engineering. In 2018 & 2020, he won the Best Paper Awards from the *American Water Works Association* (AWWA)'s Distribution & Plant Operations Division. Dr. Lee, who has a Ph.D. in Civil and Environmental Engineering from Virginia Tech, is a registered Professional Engineer of Civil Engineering in California and a board-certified Diplomate, Water Resources Engineer (D.WRE) from the American Academy of Water Resources Engineer. He is the recipient of the 2021 *Environmental Water Resources Institute's* (*EWRI*) Service to the Profession Award.

Jason Giovannettone is a multidisciplinary civil and environmental engineering professional/ atmospheric scientist specializing in climate change and environmental sustainability. He has worked on an array of projects within the government and private sectors related to climate variability, climate change adaptation, flood susceptibility mapping, and green water-infrastructure. As the Director, Climate and Sustainability at the Sisters of Mercy of the Americas, he provides leadership in helping them attain their goal of becoming carbon neutral prior to the year 2040. With this goal in mind, Dr. Giovannettone is leading the planning and implementation of a large solar array to be installed at an initial pilot location. He is also leading efforts to reduce emissions due to travel, facility and campus operations, and plastic consumption, through software tools that allow the organization to optimize meeting locations based on attendee travel distances and to maintain a comprehensive record of all direct and indirect sources of greenhouse gas emissions.

Alaina J. Armel is a multidisciplinary civil and environmental engineering professional specializing in water resources management. She has worked on an array of corrective action, watershed management, and stormwater permitting and planning projects for military, government, and private sector clients. As task manager and lead client liason for water and environmental services projects, she provides leadership in the areas of TMDLs, water resources, and water quality design. Ms. Armel applies expertise in federal and state regulations to ensure compliance with the Chesapeake Bay TMDL and Virginia Small MS4 General Permit through the recommendation of structural and non-structural best management practice technologies that primarily result in reductions in nitrogen, phosphorous, and sediment loads.

 **MDPI**

Editorial

Introduction to the Special Issue "Socio-Hydrology: The New Paradigm in Resilient Water Management"

Tamim Younos [1,*], Tammy E. Parece [2], Juneseok Lee [3], Jason Giovannettone [4] and Alaina J. Armel [5]

[1] Green Water-Infrastructure Academy, Blacksburg, VA 24060, USA
[2] Department of Social and Behavioral Sciences, Colorado Mesa University, Grand Junction, CO 81501, USA; tparece@coloradomesa.edu
[3] Department of Civil & Environmental Engineering, Manhattan College, Riverdale, NY 10471, USA; juneseok.lee@manhattan.edu
[4] Dewberry, Fairfax, VA 22031, USA; jpgiovannettone@gmail.com
[5] AECOM, Arlington, VA 22201, USA; alaina.armel@aecom.com
* Correspondence: tamim.younos@gwiacademy.org

Citation: Younos, T.; Parece, T.E.; Lee, J.; Giovannettone, J.; Armel, A.J. Introduction to the Special Issue "Socio-Hydrology: The New Paradigm in Resilient Water Management". *Hydrology* **2021**, *8*, 138. https://doi.org/10.3390/hydrology8030138

Received: 31 August 2021
Accepted: 8 September 2021
Published: 11 September 2021

Publisher's Note: MDPI stays neutral with regard to jurisdictional claims in published maps and institutional affiliations.

Water is life! Ancient human communities were established in close proximity of natural water systems, i.e., streams and rivers, lakes, springs and oceans, where water was readily available for human consumption. Science and technology, environmental and social values related to water source development and use have evolved over generations. In modern times, the 18th century Industrial Revolution can be considered as a benchmark for human intervention in natural systems. Energy production and use, in the form of fossil fuels (i.e., coal and petroleum), enabled water transport from source to distant consumers and caused the emergence of high-density population (urban) centers, accelerated industrial activity, and modernized the agricultural and food production systems. It is important to note that the environmental and social values of the 18th century were dictated by the 18th century state-of-knowledge and values and did not foresee the unintended consequences of accelerated human intervention on natural systems and environment degradation throughout the 20th century, and even up until today in some parts of the world.

On a positive note, the 20th century witnessed significant modernization in the arena of water resources development and water infrastructure. The technological accomplishments of the 20th century include building dams and reservoirs, using powered pumps to extract deep groundwater and surface water resources, construction of centralized water treatment plants and water delivery pumps and pipelines to transport clean drinking water to homes and public and commercial buildings, and installing sewer and stormwater pipes to move wastewater and stormwater runoff away from population centers. However, the technologies and water management strategies that were implemented in the 20th century did not fully integrate environmental impact assessments and social and anthropogenic factors in the planning and design of sustainable water management systems [1].

During the third decade of the 21st century, human societies across the world are being challenged with significant water-related problems such as ecosystem degradation, groundwater depletion, natural and anthropogenic drought and floods, water borne health issues, and deforestation. These problems are exacerbated by climate change, a phenomenon that is largely accelerated due to human intervention in natural systems since the Industrial Revolution. The impacts of climate change demonstrate how the overall benefit of economic development during the past two centuries can be negated by neglecting the influence of anthropogenic factors in the planning and design of water management systems. The impact of climate change on water resources includes, but is not limited to, increases in regional and average global temperatures, resulting in changes in precipitation patterns and intensity; the severity and length of droughts; sea level rise and its associated consequences (e.g., the flooding of coastal cities and the encroachment of saline waters into freshwater aquifers); and increased acidity of the oceans acidity and its consequences, resulting in a potentially significant impact on marine ecosystems, as well as the health of coral reefs, shellfish and fisheries [2].

Furthermore, due to the accelerated interconnectedness of the global economy, the adverse impact on global hydrologic conditions in recent decades has become a critical factor that impedes the sustainable management of water resources across the world. There is a significant need to better understand the interaction of hydrological systems, such as climate variability and anthropogenic factors, that contribute to the dynamics and resilience of coupled human–water systems and effective risk management in the arena of water resources management. Consequently, there is significant demand for a paradigm shift toward seeking resilient solutions for complex water management problems.

The emerging field of socio-hydrology intends to address these complex water management issues. Socio-hydrology is an interdisciplinary field that integrates the natural and social sciences and aims to study the long-term dynamics of bidirectional feedback in coupled human–water systems [3–5]. The science of coupled human–water systems is based on the interconnectedness of global biophysical and social processes and aims to examine tradeoffs and synergies in order to provide scientific feedback for defining resilient solutions that resolve complex water management problems. The feedback approach in socio-hydrologic studies further advances the concept of Integrated Water Resources Management (IWRM), a practice that promotes the coordinated development and management of water, land and related resources in order to maximize economic and social benefits.

This Special Issue on socio-hydrology was planned to compile interdisciplinary scientific endeavors and innovations on coupled human–water systems research development, education, and applications. Articles were sought to explore and discuss the following topics and research case studies within the context of socio-hydrology: (1) methods applied in socio-hydrology research and education; (2) socio-hydrologic studies across interdisciplinary boundaries; (3) analysis of the spatial dimension of socio-hydrologic factors; (4) resilience assessment and risk management in coupled human–water systems; (5) socio-hydrologic modeling techniques and applications; and (6) other topics relevant to socio-hydrology.

This Special Issue is by no means a comprehensive coverage of the socio-hydrology issues but attempts to open a window to observe and understand emerging topics of socio-hydrology through various applications and case studies. The articles published in this Special Issue represent diverse and broad aspects of water management in the context of socio-hydrology systems around the globe. The topics discussed include hydrologic and economic implications of irrigation efficiency policy; the role of public trust, risk perceptions and social salience in selecting drinking water sources; landslide susceptibility to anthropogenic influences, which can cause profound social and economic impacts; the assessment of the influence of anthropogenic factors on watershed scale water quality; the assessment of the societal readiness and resilience against water-related crises such as floods, dam failure, drinking water disruption and excessive precipitation; the application of the common pool resource principle to foster groundwater resilience and avoid water source exploitation; social barriers and the hiatus from the successful implementation of green stormwater infrastructure; the application of a socio-hydrology model in agrarian communities focused on knowledge sharing among scientists, local knowledge-holders and students; and a case study of the impact of regional climate change on coastal tourism.

The articles and ideas presented in this Special Issue can be significant reference sources for interdisciplinary water science programs and an excellent guide for experts involved with the futuristic planning and management of water resources.

References

1. Younos, T.; Lee, J.; Parece, T.E. Twenty-first century urban water management: The imperative for holistic and cross-disciplinary approach. *J. Environ. Stud. Sci.* **2019**, *9*, 90–95. [CrossRef]
2. Younos, T.; Grady, C.A. (Eds.) Climate change and water resources. In *The Handbook of Environmental Chemistry*; Springer: Berlin/Heidelberg, Germany, 2013; Volume 25, 221p.
3. Vogel, R.M.; Lall, U.; Cai, X.; Rajagopalan, B.; Weiskel, P.K.; Hooper, R.P.; Matalas, N.C. Hydrology: The interdisciplinary science of water. *Water Resour. Res.* **2015**, *51*, 4409–4430. [CrossRef]
4. Pande, S.; Sivapalan, M. Progress in socio-hydrology: A meta-analysis of challenges and opportunities. *Wiley Interdiscip. Rev. Water* **2017**, *4*, e1193. [CrossRef]
5. Konar, M.; Garcia, M.; Sanderson, M.R.; Yu, D.J.; Sivapalan, M. Expanding the scope and foundation of sociohydrology as the science of coupled human-water systems. *Water Resour. Res.* **2019**, *55*, 874–887. [CrossRef]

Article

System Dynamics Modeling for Evaluating Regional Hydrologic and Economic Effects of Irrigation Efficiency Policy

Yining Bai [1,*], Saeed P. Langarudi [1] and Alexander G. Fernald [1,2]

1   College of Agricultural, and Environmental Sciences, New Mexico State University,
     Las Cruces, NM 88003, USA; lang@nmsu.edu (S.P.L.); afernald@nmsu.edu (A.G.F.)
2   New Mexico Water Resources Research Institute, New Mexico State University, Las Cruces, NM 88003, USA
*   Correspondence: ynb@nmsu.edu

Abstract: Exploring the dynamic mechanisms of coupled sociohydrologic systems is necessary to solve future water sustainability issues. This paper employs system dynamics modeling to determine hydrologic and economic implications of an irrigation efficiency (IE) policy (increased conveyance efficiency and field efficiency) in a coupled sociohydrologic system with three climate scenarios. Simulations are conducted within the lower Rio Grande region (LRG) of New Mexico for the years 1969 to 2099, including water, land, capital, and population modules. Quadrant analysis is utilized to compare the IE policy outcomes with the base case and to categorize results of simulations according to hydrologic and economic sustainability. The four categories are beneficial, unacceptable, unsustainable agricultural development, and unsustainable hydrology. Simulation results for the IE policy analyzed here fall into the categories of unsustainable agricultural development or unacceptable, suggesting there are long-term negative effects to regional economies in all scenarios with mixed results for hydrologic variables. IE policy can yield water for redistribution as increased unit water supply in the field produces more deep percolation; however, IE policy sacrifices regional connectivity. Specifically, simulation results show that the policy increases abundance by 4.7–74.5% and return flow by −3.0–9.9%. These positive results, however, come at the cost of decreased hydrologic connectivity (−31.5 to −25.1%) and negative economic impacts (−32.7 to −5.7%). Long-term net depletions in groundwater are also observed from loss of hydrologic connectivity and increased agricultural water demand from projections of increased consumptive use of crops. Adaptive water management that limits water use in drought years and replenishes groundwater in abundant years as well as economic incentives to offset the costs of infrastructure improvements will be necessary for the IE policy to result in sustainable agriculture and water resources.

Keywords: socio-hydrology; irrigation efficiency; surface water-groundwater interactions; sustainability

Citation: Bai, Y.; Langarudi, S.P.; Fernald, A.G. System Dynamics Modeling for Evaluating Regional Hydrologic and Economic Effects of Irrigation Efficiency Policy. *Hydrology* **2021**, *8*, 61. https://doi.org/10.3390/hydrology8020061

Academic Editor: Tamim Younos

Received: 4 March 2021
Accepted: 30 March 2021
Published: 2 April 2021

1. Introduction

The inclusion of the social component in hydrologic modeling is necessary for considering the implications of water management decisions in coupled socio-hydrologic systems. Sivapalan et al. [1] presented the importance of socio-hydrology to illustrate the evolution of coupled socio-hydrologic systems, potential co-evolution tracks, and emerging or even unexpected patterns that depart from the previous framework setting for scenario analyses. System dynamics modeling is a particularly useful method to simulate the dynamic mechanisms within a coupled system [2]. By properly modeling the interactions within a coupled socio-hydrologic system, proposed changes such as policy implementation may be included to model future impacts on varying aspects of both systems. In this way, simulations can offer insights as to which components of the system may benefit or be negatively impacted, which may contribute to water policy decision-making processes.

5

1.1. System Dynamics and Its Application for Sociohydrology

Communicating a coupled socio-hydrologic system's complexity is difficult when considering the dynamics of the system, varied perspectives of water stakeholders, and potential conflicts among water users. Local changes to a system result in unexpected changes to the broader system and interconnected systems such as economies [3]. A change in the system can yield nonlinear and indirect effects [4–6]. The causes for an outcome with a management strategy are varied due to delays between actions and effects, making it challenging to identify policy options.

System dynamics modeling is one approach to represent the key feedback structures of a coupled system with a social component. Key feedback structures in socio-hydrologic modeling track systematic impacts that arise due to anthropogenic factors. Liu et al. [7] used one exploratory and simplified conceptual model to depict the socio-hydrologic system's co-evolution processes. Gober et al. [8] stated that system dynamics view human activities as internal driving factors and include the interaction between social and hydrologic processes that threatens current water systems' viability through feedbacks and unintended consequences.

System dynamics modeling has long been recognized as an approach to realize socio-hydrology research since Forrester developed the concept [9,10]. System dynamics modeling can long-term reflecting the impacts of social components. Then influence of social components may not have been obtained by modeling of the individual parts of the system separately. In this method, systems are represented through feedback loops, stocks, and flows. Feedback loops dictate the behavior of the system based upon physical processes and thresholds. Stocks depict the state of the system and maintain stepwise trends. Flows affect the stocks as inflow and outflow and interlink the stocks within a system [11].

System dynamics modeling is categorized as either qualitative and quantitative or conceptual and numerical [12,13]. Feedback loops improve the understanding of a system qualitatively [14]. Stocks and flows visualize the effects of system behavior through simulation quantitatively. Meanwhile, system dynamics modeling aggregates a wide range of input parameters as key factors in a meaningful way [15]. In particular, it can incorporate different forms of human decision-making processes and behavioral rules [16]. These characteristics can guide water management strategies responding to the crucial changes in an adaptive way [17,18]. System dynamics modeling can dynamically simulate the consequences of evolutionary systems as a decision support tool for strategic policy testing [19,20].

1.2. Regional Irrigation Efficiency

Precautionary measures for ensuring future supply often focus on reducing future demand by increasing IE. IE's formal definitions are explained in Appendix A. IE policy aims to maximize the consumptive use portion of water withdrawals to obtain "more crop per drop" [21]. The essential assumption about irrigation efficiency is that by reducing the portion of non-consumptive use (e.g., surface water diverted that percolates back to groundwater) of total diversions, water is being conserved. The investment in advanced irrigation systems leads IE's increase, which keeps the root zone saturated with limited irrigation. The saving of water at the field scale impacts the whole socio-hydrology system. The water saved results from IE policy implementation; however, it is not aligned with regional sustainability in agriculture and water. Furthermore, IE's investment and its impacts regionally are particularly of interest to agricultural development in dry regions with limited available water supplies. Agricultural water demand is affected by various factors. IE policy achieved in one study performed in New Mexico led to a reduction in water applied per hectare but increasing water depletion based on various basin data analyses [22].

Achieving real water savings requires understanding institutional, technical, and accounting measures that accurately track and economically reward reduced water depletions. Conservation programs that target reduced water diversions or applications provide no guarantee of saving water. The systematic understanding leads to cooperation on regional sustainability goals from diverse water users.

Implications of IE policy should be calculated in a broader context reflecting a mathematical perspective as well as a systematic perspective that includes uses that are traditionally classified as nonbeneficial. The mathematical perspective reflects IE policy's performance based on IE's definition; systematic perspective reflects the changes in the system's components affected by IE policy. Encompassing water management calls for understanding the interconnections between potential solutions and systematic consequences [17].

This paper aims to examine the utility of system dynamics modeling by analyzing the costs and benefits of implementing an IE policy. To address the overall effect of IE policy on agriculture sustainability and water sustainability, this study couples the dynamics of the agricultural hydrologic cycle, irrigation management, population, and economic development with interconnected feedbacks through system dynamics modeling. IE policy's impact is investigated with three climate scenarios through system dynamics simulations within New Mexico's LRG region from 1969 to 2099. It is hypothesized that system dynamics modeling would provide useful insights for informing IE policy. Specifically, it is hypothesized that unintended consequences of IE policy would be illuminated, such as decreased hydrologic connectivity, groundwater depletion, and negative economic impacts. Simulation results report the model's pertinent performance measures of irrigated agriculture and local economies in response to the policy scenario.

2. Overview of the Research Area

The Rio Grande River flows through three states (Colorado, New Mexico, and Texas) and forms the border between two countries (the Republic of Mexico and the United States of America). The Rio Grande has its headwaters in the San Juan Mountains of Colorado and terminates in the Gulf of Mexico. The river forms a valley through New Mexico; a large majority of the land use in the valley consists of irrigated agriculture [23]. The valleys surrounding the Rio Grande in New Mexico are categorized into three regions: upper Rio Grande, middle Rio Grande, and lower Rio Grande (LRG). The latter is focused on here (Figure 1). The LRG planning region includes all of Doña Ana County, and the total area of the planning region is 3814 square miles (9878 square km). Agriculture is the predominant land use adjacent to the Rio Grande in this area [23]. The climate is semiarid, and annual precipitation ranges from 8 to 20 inches, depending on topography. The majority of precipitation falls as rainfall during the monsoon season. Geology surrounding the LRG consists of 150 to 400 feet of alluvium in which unconfined aquifers are highly connected with the river [24]. Fuchs et al. [25] stated that groundwater storage change is positively correlated with surface water use in the LRG region. The LRG region is located downstream of the Elephant Butte Reservoir, which supplies surface water to irrigators and happens to be experiencing a megadrought (drought persisting more than 20 years) [26].

The majority of water diverted from the Elephant Butte Reservoir is for agricultural use (87%), with only a small amount diverted for residential use [27]. Groundwater pumping is an indispensable supplement to surface irrigation supply impacting Rio Grande's management operation. The current operating agreement since 2008 among the Elephant Butte Irrigation District, El Paso County Water Improvement District, and the U.S. Bureau of Reclamation has been to release 150,000 additional acre-ft from the reservoir in full-supply years to meet downstream delivery requirements of Texas and Mexico [28]. Because agriculture in this region plays a central role in the local economy, stakeholders' perception and the economic value of crop yields jointly determine the water demand resulting in a low elasticity for water demand [29]. Water delivery requirements not met by surface water are supplied by groundwater. During drought, groundwater is presumably the crucial alternative water source.

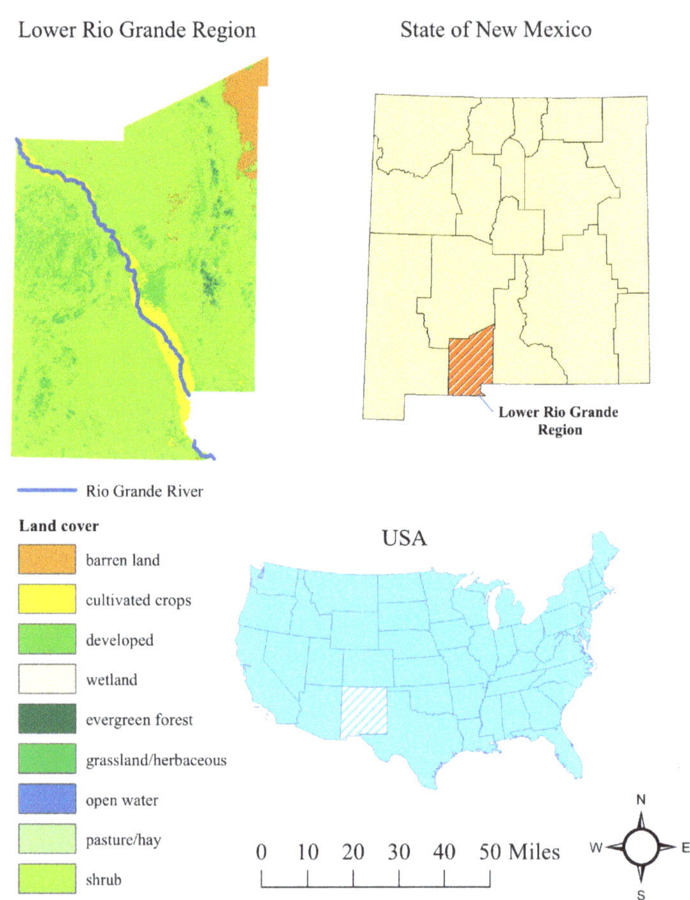

Figure 1. Generalized map of the lower Rio Grande region, its location in the US, and land cover.

The LRG region's recorded population in 2010 was 209,000, which is continuing to increase [30]. The rising municipal water demand to match population growth creates the risk that water rights will be transferred from agriculture to urban areas, compressing the available agricultural water supply. New Mexico has been leading national pecan production after 2018. Hurricane Michael severely impacted the Georgia pecan industry [18]. Of the state's 92 million pounds of pecans in 2017, Doña Ana county produced 66.9 million pounds. Pecan orchards comprise over 30% of the LRG region. During an irrigation season, pecan orchards need 3.6–6.6 ft irrigation in the southwestern United States [19,20]. Doña Ana also leads statewide pasture production [21]. A sufficient water supply is required to ensure profitable agricultural production, preventing substantial reallocation of water from agriculture to other sectors [31].

3. Methods

3.1. Model Structure

The model details are available from the accompanying Supplementary Material and Langarudi et al. [3]. This paper's system dynamics model consists of 15 stocks and 33 flows and simulates from 1969 to 2099. The model is structured into four modules: water, capital, land, and population. All the details of the assumptions and modeling choices, including equations, parameter values, estimations and measurements, and data sources, are reported in the Supplementary Material. These modules interact with each other within a complex feedback system. The relationships among modules are constructed based on the literature, previous models, or empirical studies (black and blue linkages in Figure 2). The water module consists of surface water, soil moisture of irrigated and non-irrigated land, and groundwater as stocks. The primary physical processes are integrated into the model building, including surface inflow/outflow, river leakage, surface water withdrawal, canal leakage, field percolation, gaining flow from groundwater, and groundwater withdrawal. Income, capital development, irrigated land, and population growth contribute to water demand (Figure 2). Agricultural parameters, such as crop type and soil properties, are aggregated into the model design, addressing their impacts on the hydrologic cycle dynamics and water demand in socio-economic processes. Water availability is driven by precipitation, surface inflow, and groundwater storage. Water availability impacts income, capital development, irrigated land and population in turn. The model successfully passes the confidence building (validation) tests presented in the Supplementary Materials, supporting that this model can provide insights into regional water issues.

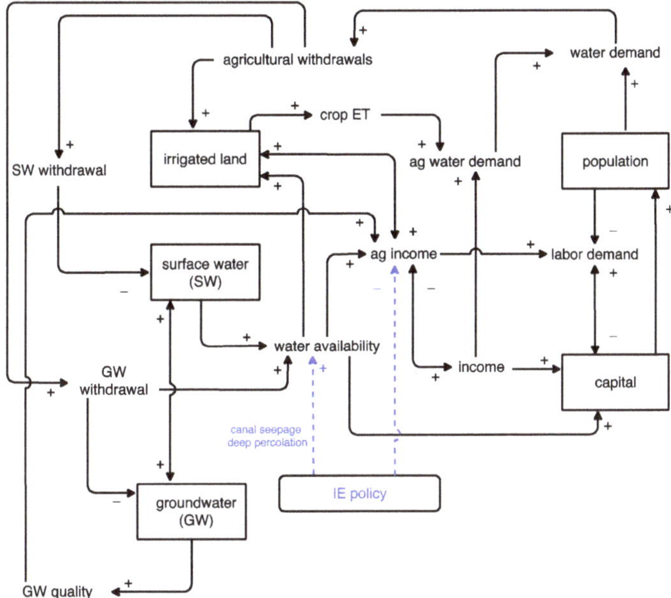

Figure 2. Causal structure of the model: positive or negative causality is marked as plus or minus; double sides arrows explains that two variables have mutual feedback; blue dash line represents where the policy implements.

3.2. Scenarios

3.2.1. Climate Scenarios

Hydroclimate scenarios are introduced in a changing future. Model inputs (precipitation, temperature, and surface water inflow) are acquired from the New Mexico Dynamic Statewide Water Budget (DSWB) model [32]. The DSWB model generates data by using climate projections (Global Circulation Model), including GFDL (Geophysical Fluid Dynamics Laboratory), UKMO (United Kingdom Met Office), and NCAR (National Center for Atmospheric Research). Based on different greenhouse gas emission scenarios, each climate projection offers different drought conditions [26].

The GFDL, UKMO, and NCAR projections are used as inputs because they represent low, moderate and, high emission scenarios. The scenarios are listed in Table 1, where the prediction part of the simulation is divided between 2017–2050 and 2051–2099. The simulation interval is divided to illustrate the short and long-term results of the climate scenarios.

Table 1. Climate inputs for scenario tests with average values of periods 2017–2050 and 2051–2099 (Units of surface inflow are in thousands of acre-feet, KAF).

	Precipitation (in)	Temperature (°F)	Surface Inflow (KAF/Year)
Historical	10.0 ± 2.7	61.3 ± 1.1	675.9 ± 208.6
GFDL 2017–2050	9.5 ± 2.7	62.2 ± 1.2	679.1 ± 159.8
GFDL 2051–2099	9.2 ± 3.0	66.4 ± 2.0	508.5 ± 178.6
UKMO 2017–2050	10.1 ± 2.7	62.6 ± 1.1	775.4 ± 175.8
UKMO 2051–2099	10.9 ± 2.4	65.7 ± 1.3	660.3 ± 150.1
NCAR 2017–2050	10.2 ± 2.3	62.1 ± 0.9	818.7 ± 137.8
NCAR 2051–2099	10.1 ± 2.4	62.6 ± 0.7	804.0 ± 139.3

3.2.2. Climate Scenarios as Input

Precipitation has relatively high variability during 2017–2050 and 2051–2099 in the LRG planning region. Compared to historical average annual precipitation, the GFDL projection shows a decreasing trend, UKMO shows an increasing trend, and NCAR is similar to historical conditions.

Projections of temperature have relatively low variability in projections but show a potential increase of 5.1 °F in the long-term. All projections show temperature increasing to some degree as time progresses. Compared with historical average annual temperature (61.3 °F), GFDL shows a greatly increasing trend, UKMO shows a moderately increasing trend, and NCAR shows a mildly increasing trend.

Projections of surface inflow had high variability and no clear trends over time as all projections showed similar or increased values in the near-term which all decreased in the long-term. Compared with historical average annual surface inflow, the GFDL has similar then decreased flow, UKMO shows increased then slightly decreased flow, and NCAR shows greatly increased then moderately increased.

3.2.3. IE Policy

Canal lining and precision irrigation are strategies implemented here to represent IE policy. The scenario analysis is conducted with the cost of infrastructure development deducted from agricultural income. The deduction of cost counts as the impact of investment on agriculture or non-agriculture, which does not impact the cost of capital development. This means that agricultural income in the paper, in effect, represents a "net income" that reflects the impact of technological investment costs.

Canal lining aims to improve conveyance efficiency. Conveyance efficiency increases through investment in irrigation infrastructure that reduces canal seepage and increases water delivery to the fields relative to water released from the reservoir upstream. Improvement of conveyance efficiency necessitates irrigation infrastructure and involves deduction from profit. Canal lining cost varies depending on lining material, canal cross-section design, and installation approaches. Based on the Bureau of Reclamation's historical record, the canal lining cost ranges from 14.9 to 37 dollars per square yard [33]. The canal lining cost ranges from up to 2000 dollars per acre-foot of water saved [34]. Multiplying the unit cost of canal lining and surface water divisions yields the total cost. More of surface water withdrawals will be delivered to the field with canal lining, and less water will be "lost" as canal seepage. In the model, as conveyance efficiency increases, the total cost for canal lining will be deducted from agriculture income. The regional canal lining project is a long-term process, so canal lining takes ten years to plan and ten more years to implement. The service life of the canal lining is deemed to be forty years [34]. The canal lining process continues until the total irrigation supply reaches the largest yearly water supply of surface water. The variables involved in the policy test are listed in Table 2.

Table 2. Variables setting for IE policy.

Policy	Affected Parameters	Setting
	IE$_c$	+20%
	Canal lining cost	$100 per acre-ft
IE policy	Deep percolation fraction	−50%
	Precision irrigation cost	$800 per acre

Precision irrigation aims to improve field irrigation efficiency. Irrigation systems could be altered to improve field efficiency (e.g., through drip and sprinkler irrigation). High IE irrigation systems typically maintain the same productivity with less water applied or increase productivity with the same water application rate due to more water delivered to the fields being available to crops [35,36]. Field irrigation efficiency refers to irrigation technology and practices that attempt to apply only the water necessary to replenish the amount of water lost through ET and reduce percolation past the root zone. Costs associated with irrigation conservation have been cited for having varied installation costs per acre and operation costs [37]. As summarized in a report of drip irrigation experiments, the cost for design, materials, and installation of a drip irrigation system for the case of a farm in Rincon, New Mexico, which was a 26-acre farm, was 52,000 dollars [38]. The variables involved in the policy test are listed in Table 2.

3.3. Evaluation Framework

The simulated results are examined based on performance indicators such as water abundance, groundwater dependency, field IE, irrigation return, and hydrologic connectivity. These measures are described below.

- Abundance: the difference between available water (sum of water supply from surface water, recharge or leakage from river channels and canals as well as percolation from the land surface) and total withdrawals (sum of withdrawals from surface water and groundwater for all uses);
- Irrigation return: a proportion of irrigation drainage, which is excess water left in the root zone after soil saturation;
- Connectivity: a measure for recharge connectivity of surface water and groundwater defined as the sum of river leakage, canal leakage, and deep percolation.
- Groundwater dependency: the portion of agricultural groundwater withdrawals in total agricultural water withdrawals;
- Agricultural water demand: water needed to be withdrawn from surface water and groundwater for sustaining the desired level of agricultural yield from irrigated land.

The policy performance is analyzed to measure water sustainability while investigating water economics. To understand the tradeoffs between the selected scenarios from both economical and hydrological perspectives, the above-described measures were employed. The percentage deviation is given by Equation (1):

$$\Delta y_{ij} = \frac{\sum y_{ijt} - \sum y_{ojt}}{\sum y_{ojt}} \times 100 \tag{1}$$

The percentage deviation is given by Equation (1), where y_{ij} represents the value of measure j with scenario i. Note that $i = 0$ indicates the base case simulation whereas $i = 1$ represents irrigation efficiency. The values are then summed for t (time) during the periods 2017–2050 and 2051–2099. Policy notations are listed in Table 3. For each of these measures, we calculate their relative deviation from base run values. For each comparison, the climate scenario remains unchanged. For example, we do not compare a UKMO policy scenario (Ui) with a GFDL base scenario. These simulations provide three alternative future possibilities as benchmarks that the policy scenarios could be compared with.

Table 3. Notation of scenarios.

Notation	GFDL	UKMO	NCAR
Default	GFDL	UKMO	NCAR
IE policy	G1	U1	N1

The Δy_{ij}'s value of hydrology performance indicators are plotted against percentage change in agriculture income on two-dimensional quadrant grids. On the presented quadrants, the x-axis always represents agricultural income, while the y-axis represents one of the hydrologic measures (IE, water abundance, irrigation return, and hydrologic connectivity). The title of each row in the figure indicates which hydrologic measures should be on the y-axis for the graphs in that column. Each column of the graphs is for a specific climate scenario. The first column indicates the results for GFDL, the second for UKMO, and the third for NCAR. Each point on the grid represents the discrepancy between a policy simulation run and a corresponding base run (no policy applied). Therefore, for each row of graphs, the origin of a grid represents an average steady-state, which is equivalent to a comparison of a base case and itself. Because there is no discrepancy between a simulation run and itself, such comparisons yield $x = 0$ and $y = 0$. The analysis is broken into two periods being the years 2017–2050 and 2051–2099. The impact of policy implementation could be prolonged and yield unexpected consequences. The comparison between the two periods can illustrate the expected and unexpected performance.

A quadrant analysis is applied to interpret the hydrology performance indicators and agricultural income. Each quadrant represents a potential outcome: beneficial, unacceptable, unsustainable agriculture development, and unsustainable hydrology. The upper right quadrant (beneficial) represents policies that are beneficial both economically and hydrologically, which would be the most favorable state. Thus, results that are located in the lower-left quadrant (unacceptable) are neither economically beneficial nor sustainable, thus unacceptable. The upper left quadrant (unsustainable agriculture development) represents results with a positive effect on water system performance but a negative influence on agricultural income. The lower right quadrant (unsustainable hydrology) indicates results that would be beneficial economically but would need additional water management policies to offset associated negative hydrologic impacts. The following sections detail each grid presented in Figure 3 and demonstrate different performance measures of irrigated agriculture and how they deviate from the base runs due to the policy applications.

3.4. Model Suitability

The purpose of this modeling exercise is to determine long-term hydrologic and economic trends. Simulated results reveal implications of a specific water management policy. Only limited regional hydrologic and economic data are available in the literature from which model inputs and relationships are acquired. System dynamics modeling is specifically chosen for this application because it has the benefit over other modeling methods of being able to proficiently model long-term trends while having limited data as inputs [13].

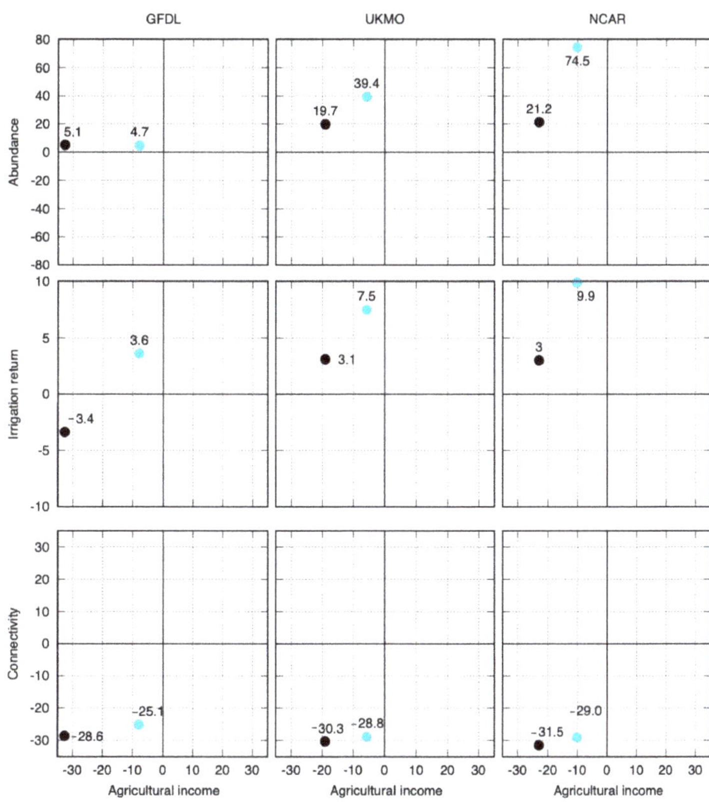

Figure 3. Tradeoff between selected performance measures and agriculture income; *x*-axis in all plots indicates agriculture income while the *y*-axis for each row of graphs from top to bottom are abundance, irrigation return, and connectivity; graph columns from left to right indicate climate scenarios GFDL, UKMO, and NCAR; value of hydrology indicators is labeled.

4. Results

Basic behavior-over-time diagrams of the model's outputs showed the strong relationships between independently measured variables (see Appendix A Figures A1 and A2). These diagrams, along with formal statistical methods guided by practical examples [39], are used to validate the model outputs, including seven water variables and seven socioeconomic variables. These variables displaying the model's outputs differ from the variables used for calibration that can be adjusted manually. Chapter 9 of the Supplementary Material offers a complete methodology. Results-over-time of base run and scenario tests are graphed in Figure A3. Results are interpreted in the quadrant analysis.

4.1. Policy Scenarios

The performance of water abundance, irrigation return, and connectivity are graphed in Figure 3. Black dots represent results during 2017–2050; blue dots represent results during 2051–2099.

4.1.1. Agricultural Income

Agricultural income is negatively affected by the IE policy. The results of simulations are constrained to the categories of unsustainable agricultural development or unacceptable in the quadrant analysis. Agriculture income reduces by 32.7% in G1, 19.1% in U1, 23.0% in N1 during 2017–2050, and 7.8% in G1, 5.7% in U1, and 10.0% in N1 during 2051–2099. The average reductions in agricultural income are -24.9% in the short-term and -7.8% in the long-term. The proposed infrastructure related to the IE policy is a long-term investment on a large scale, so reducing agricultural income is not surprising. It should be noted that the negative impacts to agricultural income are greatly reduced in the long-term but still have a consistently negative impact compared to the base case.

4.1.2. Abundance

Positive effects on water abundance in all scenarios are observed, and results fall within the category of unsustainable agricultural development. Abundance increases during 2017–2050 (5.1% in G1, 19.7% in U1, and 21.2% in N1) and 2051–2099 (4.7% in G1, 39.4% in U1, and 74.5% in N1). The average increase of abundance is 15.3% in the short-term and 39.5% in the long-term. The water "saved" positively affects water availability. An interesting result is that in the drought scenario (G1), the benefits of abundance show a slight decrease from the short-term to the long-term. The other two scenarios show a large increase from the short-term to the long-term. Copious precipitation in UKMO and surface inflow of NCAR during the long-term period contribute to a total agricultural water supply.

4.1.3. Irrigation Return

Irrigation return is affected positively for the majority of simulations, and results fall within the category of unsustainable agricultural development. An exception is the G1 scenario in the short-term, which falls into the category of unacceptable. The evaluated IE policy yields irrigation return changes (-3.4% in G1, 3.1% in U1, and 3.0% in N1) during 2017–2050. Irrigation return increases (3.6% in G1, 7.5% in U1, and 9.9% in N1) during 2051–2099. The average increase in irrigation return is 0.9% in the short-term and 7.0% in the long-term. The increasing irrigation return over the simulation period shows that the IE policy benefits the long-term irrigation supply.

4.1.4. Connectivity

The IE policy has consistently poor results regarding connectivity, and all simulations fell into the category of unacceptable. Decreases (over 25.1%) in connectivity (Figure 3) are due to a nearly complete reduction of canal seepage resulting from canal lining and the reduction of deep percolation in the field due to precision irrigation technologies. From 2017–2050 to 2051–2099, connectivity changes from -28.6% in G1, -30.3% in U1, and -31.5% in N1, to -25.1% in G1, -28.8% in U1, and -29.0% in N1. The average decrease in connectivity is -30.1% in the short-term and -27.6% in the long-term. These changes show that canal seepage and deep percolation are important sources of recharge to groundwater.

4.2. Water Resilience

In this paper, water resilience represents stable groundwater storage and agricultural water demand. Groundwater (GW) dependency and agricultural water demand are analyzed for exploring water resilience within efficiency-oriented management. Table 4 summarizes GW dependency and agricultural water demand in simulations.

IE policy lowered GW dependency under all three climate scenarios throughout the simulation. GW dependency reduction ranges from 10.7% to 39.1% during 2017–2050 and 1.7% to 14.1% during 2051–2099. The average change in GW dependency was -22.1% in the short-term and -6.2% in the long-term. GW dependency declined initially; however, after 2050, the benefits were consistently diminished.

Table 4 also suggests that IE policy failed to regulate agricultural water demand. Agricultural water demand changes are positive, which means more pressure would be put on the agricultural water supply. Agricultural water demand exhibited rising trends with precision irrigation policy in all three climate scenarios. The increase in agricultural demand ranges from 4.8% to 9.3% during 2017–2050 and 1.8% to 5.6% during 2051–2099.

The average change in agricultural water demand was 6.8% in the short-term and 2.7% in the long-term.

Table 4. Relative changes of groundwater dependency and agricultural water demand between scenarios and base runs (percent).

	IE Policy	
	GW Dependency	**Agricultural Water Demand**
GFDL 2017–2050	−16.5	9.3
GFDL 2051–2099	−1.7	2.7
UKMO 2017–2050	−10.7	4.8
UKMO 2051–2099	−2.7	1.8
NCAR 2017–2050	−39.1	6.3
NCAR 2051–2099	−14.1	3.6

5. Discussion

5.1. Regional Water Reuse

Water "lost" to irrigation return and groundwater through connectivity can be beneficial on the regional level for other users [40–43]. Results (Figure 3) show positive irrigation return (except G1 during 2017–2050) and negative connectivity. This suggests that, from one perspective, IE policy can yield water for redistribution as increasing unit water supply in the field. From another perspective, IE policy sacrifices regional connectivity. In general, groundwater at the regional scale is not sufficiently recharged. Since the 1960s, IE policy impacts of upstream areas upon downstream resources have been reported in the literature [44,45]. However, real savings at a regional scale require encompassing water management for downstream re-allocation. Simons et al. [45] showed that the recoverable flow from a water user could be reused multiple times by downstream water users. The "saved" amount in water conveyance impacts downstream uses that may rely on this portion of water.

Connectivity reductions indicate inadequacies of recharge from reduced irrigated agriculture on a regional scale. Similar to the irrigation return flow, the connectivity exhibits the time lag of mass movement in the water cycle. It buffers the disturbance of cumulative consequences such as water quality deterioration, soil contamination, edaphon alternation, and groundwater table depression. Declining connectivity and irrigation return reduce water availability for other users, as these non-consumptive flows play a vital role in instream water supplies [46,47].

In a broad context, connectivity contributes to groundwater resilience. In simulations, a loss of connectivity between surface water and groundwater is one of the unintended consequences of IE policy. Pringle [48] argued that human alterations of hydrologic processes that eliminate hydrologic connectivity have already influenced ecological patterns regionally and globally. As Gleick et al. [49] demonstrated, traditional IE improvement failed to depict the co-benefits, including water quality, reductions in water-related energy costs, ecosystem health, and improved crop quality. Although valuable, these benefits are not always as tangible as direct and immediate benefits because they do not yield "new water." As Lexartza-Artza and Wainwright [50] suggested, practical studies should consider the opposite structural and functional components of connectivity and system boundaries. In water balance closure, any water use changes in one part of the region or basin will impact another water use. The consideration that water is "lost" when not applied directly to anthropogenic uses is a misleading perception. Applying these labels presents a lack of quantification of non-consumptive values. The definitions of natural and anthropogenic flows should be embedded in water management to represent multiple water users' interdependency. IE may maximize water supply locally yet concurrently need synergies in a broader context. Synergistic practices such as aquifer storage and recovery should be included in future strategic water management alongside the efficiency practice to guarantee ecological benefits.

5.2. Groundwater Resilience

The evaluation of simulated scenarios indicates mixed results for groundwater resilience related to the policy tested. Folke et al. [51] defined resilience as both the capacity to undertake continuous changes and the ability to develop unceasingly. In light of this definition, groundwater resilience is not increased. Results suggest that IE policy could reduce groundwater dependency in the short-term, but this effect is diminished as time progresses. In simulations, net increases in agricultural water demand and reduced surface water supply and precipitation eventually lead to net groundwater mining and even reduced aquifer storage. Even though the GW dependency decreases, the total amount of groundwater withdrawals may increase because of increasing agricultural water demand. The incentives to grow higher water use crops lead to agricultural water demand increase in simulations for ensuring agricultural profits with increasing ET caused by higher temperature. The LRG region's growing season is also longer than the surface water irrigation season. Regardless of surface water supply, irrigation in the LRG region is dependent upon groundwater at times. Increases in agricultural water demand (Table 4) show that IE policy fails to reduce total water withdrawals in times of drought without fundamental changes to current water delivery mechanisms.

5.3. Economic Implications

Reliance upon groundwater is costly in multiple aspects. Groundwater depletion, as a supplement to surface water supply, raises the cost of energy for pumping, the need for drilling new wells, and the need for deepening or repairing wells. Groundwater withdrawals need adequate energy facilities for pumping. Dehghanipour et al. [52] concluded that a reduction in surface water supply raises a challenge to groundwater pumping's sustainable capacity. The sum of direct and indirect losses to agriculture in California was estimated at 2.2 billion dollars in 2014 [53] as the costs associated with groundwater pumping were an estimated 454 million dollars [54]. A farmer in the LRG Valley with a 325-acre farm had to spend about 150,000 dollars to drill a new well capable of producing sufficient quantities for irrigation [55]. The decreases in simulated agricultural income in this paper are in line with these observations.

The simulated initial costs for the IE policy are large during 2017–2050, as would be expected for infrastructure development. As such, the IE policy is a long-term policy that likely requires an up-front investment from the government [56]. Increased IE can support immediate irrigated agriculture requirements. Initial water supply security may encourage farmers to plant permanent crops such as orchards that yield higher profits. This may lead to inflexible production systems and expose the investment to increased risk [57]. In the long-term, irrigators have to cover the costs of maintaining the investment and cope with the returning drought. The investment for IE policy requires irrigation water productivity improvement to cover the costs in the long-term. Irrigated agriculture is not profitable for the majority of farmers when productivity remains low [58]. Enhancements in agricultural structure, such as cropland allotment and cropping intensity, promotion of resource conservation technologies, and farm mechanization, will ensure irrigated agriculture's benefits. Turral et al. [59] reviewed global irrigation development and concluded that agricultural water use productivity is needed to achieve a higher value of agriculture. Constant adaptation and flexibility in water strategies and assorted economic incentives will enable regional agriculture development.

5.4. Management Strategies

Results in Figure 3 explain the sustainability stage in multiple aspects. In terms of abundance and irrigation return, results point to unsustainable agriculture development. Even though abundance and irrigation return increase, profits are decreased. From a connectivity perspective, the results show an unacceptable stage of sustainability where both connectivity and agricultural income suffer. Together these indicate that economic measures and alternative water management strategies are both lacking.

IE policy lays a foundation for ensuring irrigation supply as it maximizes the possible use of surface water. The investment in IE policy (loss in agricultural income in Figure 3) indicates that the next step is to increase agriculture benefits through various means, such as subsidies and technologies ensuring production and product quality to offset costs. Further, flexible water management strategies that focus on compensating for lost natural and historic flows during years with abundant water supply are necessary to prevent long-term consequences from the lost connectivity. If there are no assorted water management strategies to replenish the aquifer in wet years, recurring drought will still have an adverse impact on irrigation supply due to predicted increased and inelastic agricultural water demand. Scherberg et al. [60] applied modeling to verify that canal piping combined with an aquifer recharge program stabilized groundwater storage. Results suggested that canal piping alone failed to keep summer flows and could be detrimental to the local hydrology.

Economic incentives to offset costs of infrastructure investment such as subsidies or technologies that increase agricultural profits are necessary for the policy tested to be economically viable, particularly in the short-term. Maintaining a flexible cropping pattern could prevent groundwater reliance in times of drought, forgoing costs associated with well installation, repair, and pumping.

6. Summary and Conclusions

This study determined the systematic behavior of an arid region's irrigation system within socio-hydrologic dynamics. A system dynamics model was developed for the lower Rio Grande region to determine the impacts of increased irrigation efficiency upon local hydrology and economics under different climate scenarios through 2099. The model simulated the variables' dynamics with three climate projections and surface inflow that corresponded to the three climate projections as inputs. Regional IE policy was tested as a model scenario. Results showed that regional IE policy could yield water for redistribution as increasing unit water supply in the field; however, it may also have unintended consequences such as decreased hydrologic connectivity, groundwater depletion, and negative economic impacts. Regional IE policy fails to reverse the declining trends of groundwater and regulate agricultural water demand. Specifically, irrigation return flow and abundance are increased in the majority of simulations. These benefits come at the cost of decreased connectivity and overall economic losses in all scenarios.

In these scenarios, losses to agricultural economies are observed, particularly in the short-term. This suggests that subsidies, technologies, or management strategies that increase agricultural profits or decrease costs of IE infrastructure, particularly through the year 2050, are necessary for this IE policy to be economically viable. Water use in drought years and replenishment of groundwater in abundant years as well as economic incentives to offset the costs of infrastructure improvements will be necessary for the IE policy to result in sustainable agriculture and water resources.

Research on management strategies for increasing water sustainability is a prerequisite for future development. System dynamics modeling provides insights into interdependent water systems and how they may enable social and economic development. Further efforts may increase understanding of water management strategies, uncertainties of policy outcomes, and potential synergies. Successful implementation of this policy would include adaptive water management that limits water use in drought years and replenishes groundwater in abundant years. Economic policies or incentives to mitigate investment costs, particularly through 2050, are a necessary addition for this IE policy to be viable.

Supplementary Materials: The following are available online at https://www.mdpi.com/article/10.3390/hydrology8020061/s1.

Author Contributions: Conceptualization, Y.B. and S.P.L.; methodology, Y.B., S.P.L. and A.G.F.; software, Y.B. and S.P.L.; validation, S.P.L.; formal analysis, Y.B.; resources, A.G.F.; writing and original draft preparation, Y.B.; writing review and editing, Y.B., S.P.L. and A.G.F.; visualization, Y.B;

supervision, A.G.F.; project administration, A.G.F.; funding acquisition, A.G.F. All authors have read and agreed to the published version of the manuscript.

Funding: This research was provided by the U.S. Bureau of Reclamation/New Mexico State University Cooperative Agreement R16AC00002 and the New Mexico Agricultural Experiment Station, New Mexico State University.

Acknowledgments: We thank Ian Hewitt for his constructive revisions. The authors wish to acknowledge the anonymous reviewers for their insightful comments and helpful suggestions which improved the earlier version of the manuscript.

Conflicts of Interest: The authors declare no conflict of interest.

Appendix A

Appendix A.1 Irrigation Efficiency Equations

We define the regional irrigation efficiency as a product of two components: conveyance efficiency (IE_c) and field irrigation efficiency (IE_f). Conveyance efficiency is defined as

$$IE_c = \frac{SD - CD}{SD} \tag{A1}$$

where SD is surface water diversion and CD is the amount of water that is lost through conveyance. Field irrigation efficiency is defined as follows:

$$IE_f = \frac{ET - P_{ag}}{SD - CD + GWD} \tag{A2}$$

where ET is evapotranspiration; P_{ag} is effective precipitation during the growing season; GWD is groundwater diversion. Thus, the following may be deduced:

$$IE = \frac{ET}{SD + GWD} \tag{A3}$$

where IE is total irrigation efficiency.

Appendix A.2 Behavior over Time Diagrams of System Dynamics Model Outputs

The figures below show basic behavior-over-time diagrams of the model's outputs results with complete methodology and full descriptions in the Supplementary Materials, pp. 81–82. Black lines represent the results simulated by the model applied in this paper. These results are plotted against the Dynamic Statewide Water Budget data (blue dots) and observations (red dots).

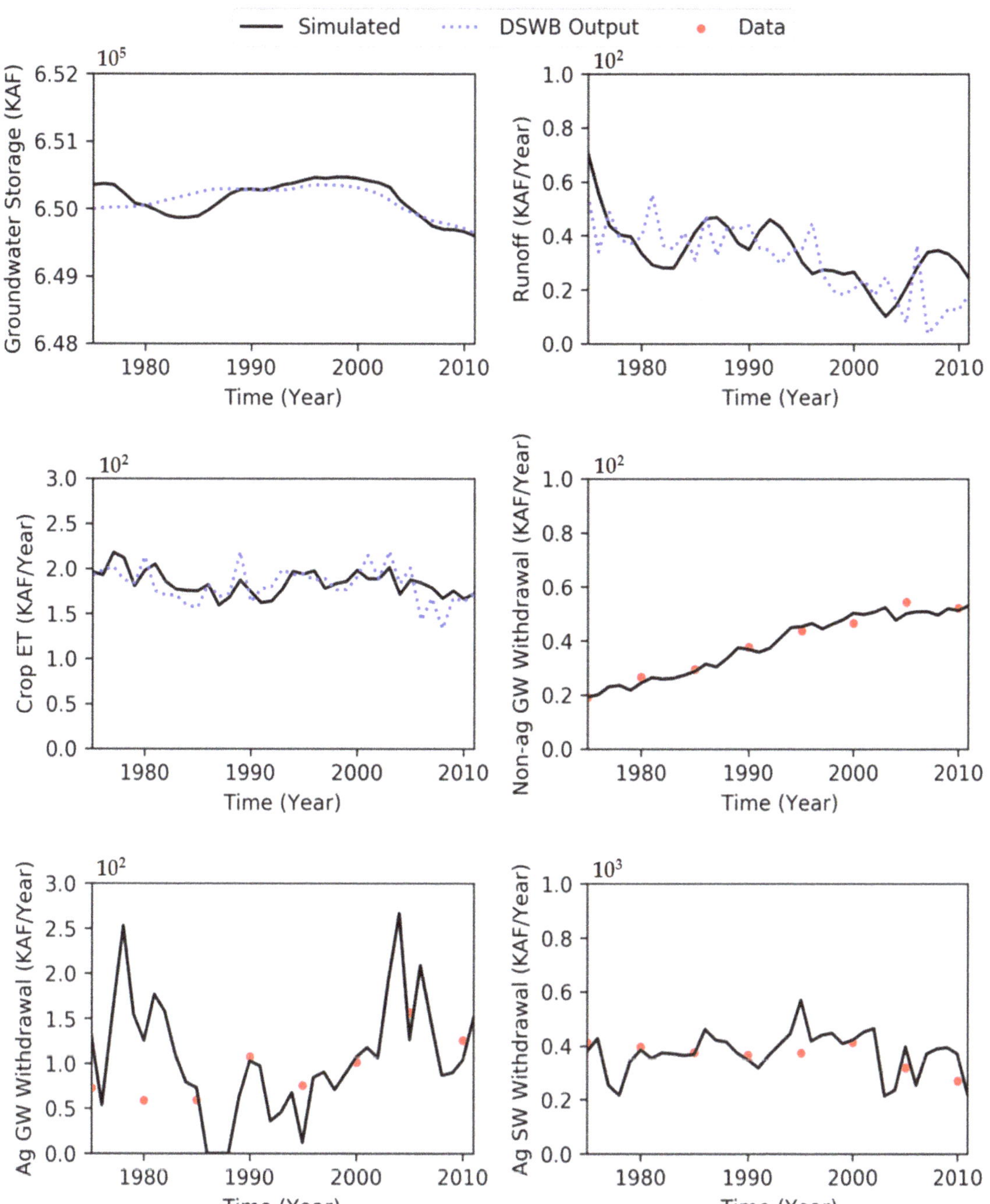

Figure A1. Behavior reproduction results (water variables).

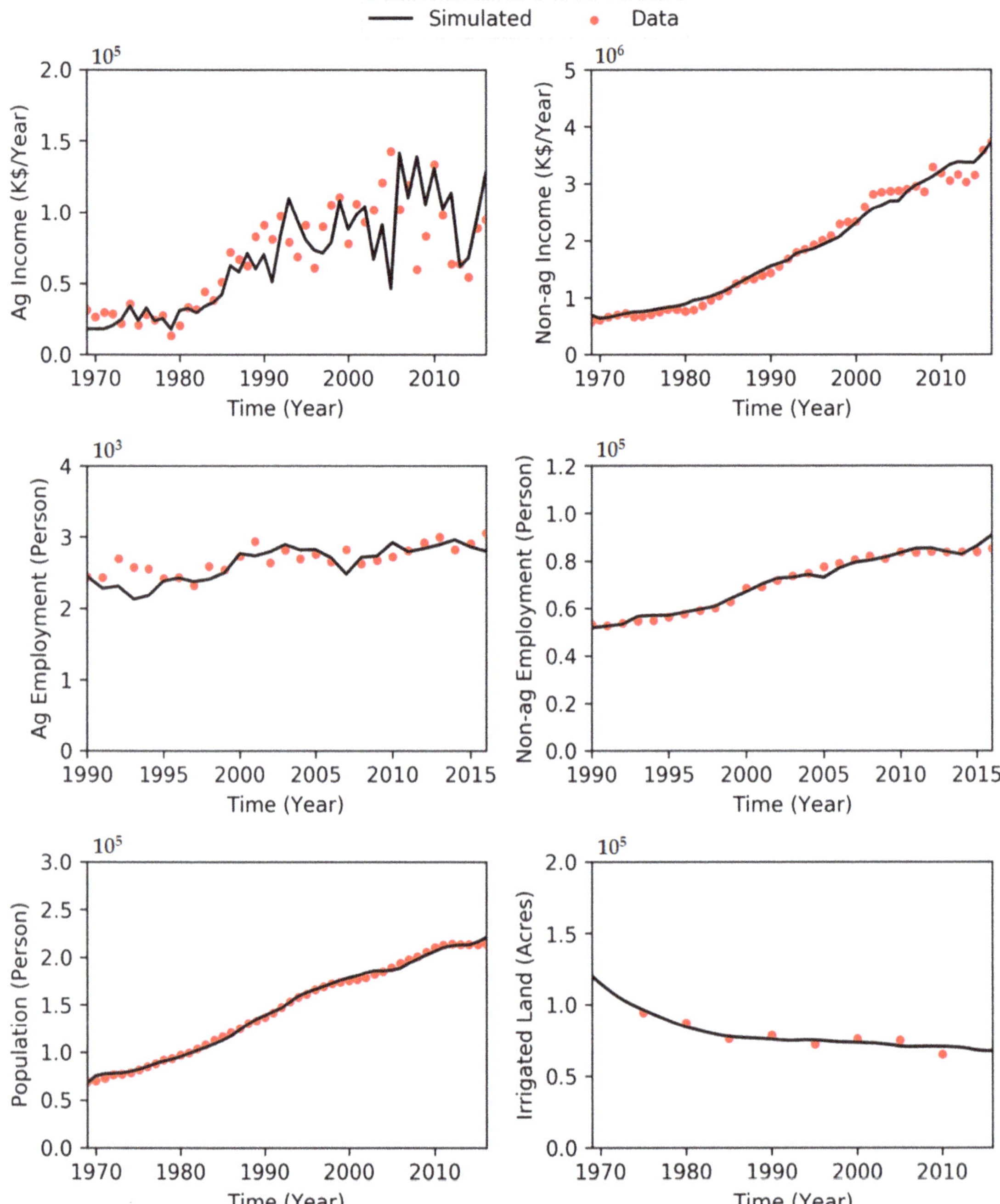

Figure A2. Behavior reproduction results (socioeconomic variables).

Appendix A.3 Result over Time Diagrams of Simulations

The figures below show simulation results-over-time of base runs and scenario tests of variables chosen in this paper. Simulation results show varied trends as climate inputs and model's dynamics correspondingly.

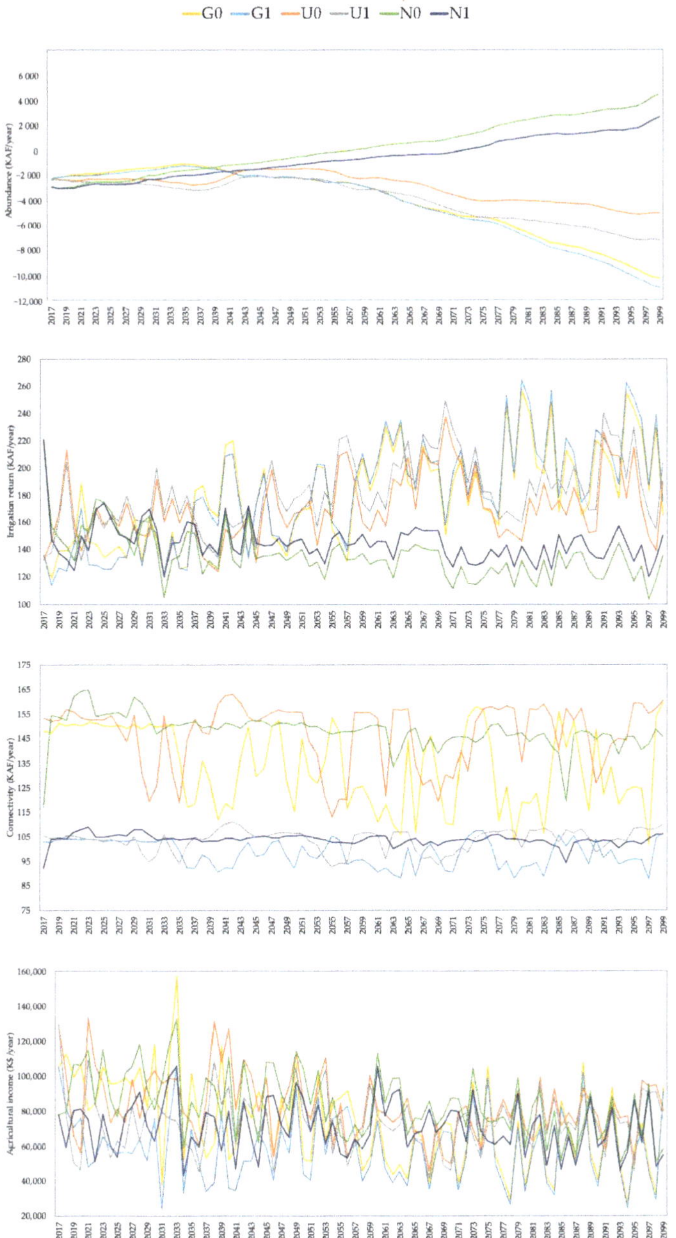

Figure A3. Simulation results for selected variables with three climate projections.

References

1. Sivapalan, M.; Savenije, H.H.G.; Blöschl, G. Socio-Hydrology: A New Science of People and Water. *Hydrol. Process.* **2012**, *26*, 1270–1276. [CrossRef]
2. Mashaly, A.F.; Fernald, A.G. Identifying Capabilities and Potentials of System Dynamics in Hydrology and Water Resources as a Promising Modeling Approach for Water Management. *Water* **2020**, *12*, 1432. [CrossRef]
3. Langarudi, S.P.; Maxwell, C.M.; Bai, Y.; Hanson, A.; Fernald, A. Does Socioeconomic Feedback Matter for Water Models? *Ecol. Econ.* **2019**, *159*, 35–45. [CrossRef]
4. Sterman, J.D. All Models Are Wrong: Reflections on Becoming a Systems Scientist. *Syst. Dyn. Rev.* **2002**, *18*, 501–531. [CrossRef]
5. House-Peters, L.A.; Chang, H. Urban Water Demand Modeling: Review of Concepts, Methods, and Organizing Principles. *Water Resour. Res.* **2011**, *47*, W05401. [CrossRef]
6. An, L. Modeling Human Decisions in Coupled Human and Natural Systems: Review of Agent-Based Models. *Ecol. Model.* **2012**, *229*, 25–36. [CrossRef]
7. Liu, D.; Tian, F.; Lin, M.; Sivapalan, M. A Conceptual Socio-Hydrological Model of the Co-Evolution of Humans and Water: Case Study of the Tarim River Basin, Western China. *Hydrol. Earth Syst. Sci.* **2015**, *19*, 1035–1054. [CrossRef]
8. Gober, P.; Wheater, H.S. Debates—Perspectives on Socio-Hydrology: Modeling Flood Risk as a Public Policy Problem. *Water Resour. Res.* **2015**, *51*, 4782–4788. [CrossRef]
9. Forrester, J.W. *Industrial Dynamics*; Productivity Press: Cambridge, MA, USA, 1961; ISBN 0-915299-88-7.
10. Forrester, J.W. System Dynamics, Systems Thinking, and Soft OR. *Syst. Dyn. Rev.* **1994**, *10*, 245–256. [CrossRef]
11. Sterman, J.D. System Dynamics Modeling: Tools for Learning in a Complex World. *Calif. Manag. Rev.* **2001**, *43*, 8–25. [CrossRef]
12. Wolstenholme, E.F. Qualitative vs Quantitative Modelling: The Evolving Balance. *Null* **1999**, *50*, 422–428. [CrossRef]
13. Coyle, G. Qualitative and Quantitative Modelling in System Dynamics: Some Research Questions. *Syst. Dyn. Rev.* **2000**, *16*, 225–244. [CrossRef]
14. Clancy, T. Systems Thinking: Three System Archetypes Every Manager Should Know. *IEEE Eng. Manag. Rev.* **2018**, *46*, 32–41. [CrossRef]
15. Forrester, J.W. Information Sources for Modeling the National Economy. *J. Am. Stat. Assoc.* **1980**, *75*, 555–566. [CrossRef]
16. Langarudi, S.; Bar-On, I. Utility Perception in System Dynamics Models. *Systems* **2018**, *6*, 37. [CrossRef]
17. Stave, K.A. A System Dynamics Model to Facilitate Public Understanding of Water Management Options in Las Vegas, Nevada. *J. Environ. Manag.* **2003**, *67*, 303–313. [CrossRef]
18. Maxwell, C.M.; Langarudi, S.P.; Fernald, A.G. Simulating a Watershed-Scale Strategy to Mitigate Drought, Flooding, and Sediment Transport in Drylands. *Systems* **2019**, *7*, 53. [CrossRef]
19. Konar, M.; Garcia, M.; Sanderson, M.R.; Yu, D.J.; Sivapalan, M. Expanding the Scope and Foundation of Sociohydrology as the Science of Coupled Human-Water Systems. *Water Resour. Res.* **2019**, *55*, 874–887. [CrossRef]
20. Page, A.; Langarudi, S.P.; Forster-Cox, S.; Fernald, A. A Dynamic Hydro-Socio-Technical Policy Analysis of Transboundary Desalination Development. *J. Environ. Account. Manag.* **2019**, *7*, 87–114. [CrossRef]
21. Luquet, D.; Vidal, A.; Smith, M.; Dauzat, J. 'More Crop per Drop': How to Make It Acceptable for Farmers? *Agric. Water Manag.* **2005**, *76*, 108–119. [CrossRef]
22. Ward, F.A.; Pulido-Velazquez, M. Water Conservation in Irrigation Can Increase Water Use. *Proc. Natl. Acad. Sci. USA* **2008**, *105*, 18215–18220. [CrossRef] [PubMed]
23. Ackerly, N. *Irrigation Systems in the Mesilla Valley: An Historical Overview*; Center for Anthropological Research New Mexico State University: Las Cruces, NM, USA, 1992.
24. New Mexico Bureau of Geology and Mineral Resources. *Geologic Map of New Mexico*; Geologic Map; New Mexico Bureau of Geology and Mineral Resources: Socorro, NM, USA, 2003.
25. Fuchs, E.H.; Carroll, K.C.; King, J.P. Quantifying Groundwater Resilience through Conjunctive Use for Irrigated Agriculture in a Constrained Aquifer System. *J. Hydrol.* **2018**, *565*, 747–759. [CrossRef]
26. NMWRRI. *The New Mexico Dynamic Statewide Water Budget*; New Mexico Water Resources and Research Institute, New Mexico State University: Las Cruces, NM, USA, 2018.
27. Contor, B.A.; Taylor, R.G. Why Improving Irrigation Efficiency Increases Total Volume of Consumptive Use. *Irrig. Drain.* **2013**, *62*, 273–280. [CrossRef]
28. Benson, M.H.; Morrison, R.R.; Llewellyn, D.; Stone, M. Governing the Rio Grande: Challenges and Opportunities for New Mexico's Water Supply. In *Practical Panarchy for Adaptive Water Governance: Linking Law to Social-Ecological Resilience*; Cosens, B., Gunderson, L., Eds.; Springer International Publishing: Cham, Switzerland, 2018; pp. 99–114. ISBN 978-3-319-72472-0.
29. Levidow, L.; Zaccaria, D.; Maia, R.; Vivas, E.; Todorovic, M.; Scardigno, A. Improving Water-Efficient Irrigation: Prospects and Difficulties of Innovative Practices. *Agric. Water Manag.* **2014**, *146*, 84–94. [CrossRef]
30. OSE. *Region 11-NM Regional Water Planning*; State of New Mexico, Interstate Stream Commission, Office of the State Engineer: Albuquerque, NM, USA, 2017.
31. Hooke, J. Coarse Sediment Connectivity in River Channel Systems: A Conceptual Framework and Methodology. *Geomorphology* **2003**, *56*, 79–94. [CrossRef]

32. Peterson, K.; Hanson, A.; Roach, J.; Randall, J.; Thomson, B. *A Dynamic Statewide Water Budget for New Mexico: Phase III—Future Scenario Implementation*; New Mexico Water Resources Research Institute, New Mexico State University: Las Cruces, NM, USA, 2019.
33. Hill, R. How Well Does Your Irrigation Canal Hold Water? Does It Need Lining? *ENGR/BIE/WM* **2000**, *3*, 1.
34. King, P.; Hawley, J.; Hernandez, J.; Kennedy, J.; Martinez, E. Study of Potential Water Salvage on the Tucumcari Project Arch Hurley Conservancy District Phase I: A Pre-Appraisal-Level Study of the Potential Amount of Water that May Be Saved, and the Costs of Alternative Methods of Reducing Carriage Losses from District Canals. 2006. Available online: https://nmwrri.nmsu.edu/wp-content/uploads/2015/technical-reports/tr335.pdf (accessed on 17 September 2019).
35. Burt, C.M.; O'Neill, B.P. Drip and Furrow on Processing Tomato-Field Performance. In Proceedings of the 28th Annual Irrigation Association Technical Conference, San Diego, CA, USA, 9 December 2007; p. 10.
36. Perry, C.; Steduto, P.; Allen, R.G.; Burt, C.M. Increasing Productivity in Irrigated Agriculture: Agronomic Constraints and Hydrological Realities. *Agric. Water Manag.* **2009**, *96*, 1517–1524. [CrossRef]
37. Wells, L. *Cost of Pecan Production*; UGA Pecan Extension: Athens, GA, USA, 2014.
38. Bevacqua, R.F. *Drip Irrigation for Row Crops*; New Mexico State University: Las Cruces, NM, USA, 2001.
39. Langarudi, S.P.; Radzicki, M.J. Resurrecting a Forgotten Model: Updating Mashayekhi's Model of Iranian Economic Development. In *Energy Policy Modeling in the 21st Century*; Qudrat-Ullah, H., Ed.; Understanding Complex Systems; Springer: New York, NY, USA, 2013; pp. 197–233. ISBN 978-1-4614-8605-3.
40. Perry, C. Efficient Irrigation; Inefficient Communication; Flawed Recommendations. *Irrig. Drain.* **2007**, *56*, 367–378. [CrossRef]
41. Seckler, D.W. *The New Era of Water Resources Management: From "Dry" to "Wet" Water Savings*; IWMI: Gujarat, India, 1996; ISBN 978-92-9090-325-3.
42. Burt, C.M.; Clemmens, A.J.; Strelkoff, T.S.; Solomon, K.H.; Bliesner, R.D.; Hardy, L.A.; Howell, T.A.; Eisenhauer, D.E. Irrigation Performance Measures: Efficiency and Uniformity. *J. Irrig. Drain. Eng.* **1997**, *123*, 423–442. [CrossRef]
43. Guillet, D. Rethinking Irrigation Efficiency: Chain Irrigation in Northwestern Spain. *Hum. Ecol.* **2006**, *34*, 305–329. [CrossRef]
44. Perry, C.J. The IWMI Water Resources Paradigm–Definitions and Implications. *Agric. Water Manag.* **1999**, *40*, 45–50. [CrossRef]
45. Simons, G.W.H.; Bastiaanssen, W.G.M.; Immerzeel, W.W. Water Reuse in River Basins with Multiple Users: A Literature Review. *J. Hydrol.* **2015**, *522*, 558–571. [CrossRef]
46. Conover, C.S. *Ground-Water Conditions in the Rincon and Mesilla Valleys and Adjacent Areas in New Mexico*; US Government Printing Office: Washington, DC, USA, 1950.
47. Whittlesey, N. Improving irrigation efficiency through technology adoption: When will it Conserve water? In *Developments in Water Science*; Alsharhan, A.S., Wood, W.W., Eds.; Water Resources Perspectives: Evaluation, Management and Policy; Elsevier: Amsterdam, The Netherlands, 2003; Volume 50, pp. 53–62.
48. Pringle, C. What Is Hydrologic Connectivity and Why Is It Ecologically Important? *Hydrol. Process.* **2003**, *17*, 2685–2689. [CrossRef]
49. Gleick, P.H.; Christian-Smith, J.; Cooley, H. Water-Use Efficiency and Productivity: Rethinking the Basin Approach. *Water Int.* **2011**, *36*, 784–798. [CrossRef]
50. Lexartza-Artza, I.; Wainwright, J. Hydrological Connectivity: Linking Concepts with Practical Implications. *CATENA* **2009**, *79*, 146–152. [CrossRef]
51. Folke, C.; Carpenter, S.R.; Walker, B.; Scheffer, M.; Chapin, T.; Rockström, J. Resilience Thinking: Integrating Resilience, Adaptability and Transformability. *Ecol. Soc.* **2010**, *15*. [CrossRef]
52. Dehghanipour, A.H.; Zahabiyoun, B.; Schoups, G.; Babazadeh, H. A WEAP-MODFLOW Surface Water-Groundwater Model for the Irrigated Miyandoab Plain, Urmia Lake Basin, Iran: Multi-Objective Calibration and Quantification of Historical Drought Impacts. *Agric. Water Manag.* **2019**, *223*, 105704. [CrossRef]
53. Kerlin, K. Drought Impact Study: California Agriculture Faces Greatest Water Loss Ever Seen. Available online: https://www.ucdavis.edu/news/drought-impact-study-california-agriculture-faces-greatest-water-loss-ever-seen (accessed on 17 September 2019).
54. McGhee, G. Understanding California's Groundwater | Water in the West 2014. Available online: https://waterinthewest.stanford.edu/programs/groundwater/California-groundwater-overdraft (accessed on 17 September 2019).
55. Guido, Z. The Costs of Drought on the Rio Grande | NOAA Climate.Gov 2012. Available online: https://www.climate.gov/news-features/features/costs-drought-rio-grandeseen (accessed on 17 September 2019).
56. Inam, A.; Adamowski, J.; Prasher, S.; Halbe, J.; Malard, J.; Albano, R. Coupling of a Distributed Stakeholder-Built System Dynamics Socio-Economic Model with SAHYSMOD for Sustainable Soil Salinity Management. Part 2: Model Coupling and Application. *J. Hydrol.* **2017**, *551*, 278–299. [CrossRef]
57. Adamson, D.; Loch, A. Possible Negative Feedbacks from 'Gold-Plating' Irrigation Infrastructure. *Agric. Water Manag.* **2014**, *145*, 134–144. [CrossRef]
58. Hasan, F.u.; Fatima, B.; Heaney-Mustafa, S. A Critique of Successful Elements of Existing On-Farm Irrigation Water Management Initiatives in Pakistan. *Agric. Water Manag.* **2021**, *244*, 106598. [CrossRef]

59. Turral, H.; Svendsen, M.; Faures, J.M. Investing in Irrigation: Reviewing the Past and Looking to the Future. *Agric. Water Manag.* **2010**, *97*, 551–560. [CrossRef]
60. Scherberg, J.; Keller, J.; Patten, S.; Baker, T.; Milczarek, M. Modeling the Impact of Aquifer Recharge, in-Stream Water Savings, and Canal Lining on Water Resources in the Walla Walla Basin. *Sustain. Water Resour. Manag.* **2018**, *4*, 275–289. [CrossRef]

hydrology

Article

Landslide Susceptibility Analysis: A Logistic Regression Model Case Study in Coonoor, India

Evangelin Ramani Sujatha [1] and Venkataramana Sridhar [2,*]

[1] Centre for Advanced Research on Environment, School of Civil Engineering, SASTRA Deemed to be University, Thanjavur 613401, India; sujatha@civil.sastra.edu

[2] Biological Systems Engineering Department, Virginia Tech, Blacksburg, VA 24061, USA

* Correspondence: vsri@vt.edu

Abstract: Landslides are a common geologic hazard that disrupts the social and economic balance of the affected society. Therefore, identifying zones prone to landslides is necessary for safe living and the minimal disruption of economic activities in the event of the hazard. The factors causing landslides are often a function of the local geo-environmental set-up and need a region-specific study. This study evaluates the site characteristics primarily altered by anthropogenic activities to understand and identify the various factors causing landslides in Coonoor Taluk of Uthagamandalam District in Tamil Nadu, India. Studies on landslide susceptibility show that slope gradient, aspect, relative relief, topographic wetness index, soil type, and land use of the region influence slope instability. Rainfall characteristics have also played a significant role in causing landslides. Logistic Regression, a popular statistical tool used for predictive analysis, is employed to assess the various selected factors' impact on landslide susceptibility. The factors are weighted and combined in a GIS platform to develop the region's landslide susceptibility map. This region has a direct link between natural physical systems, hydrology, and humans from the socio-hydrological perspective. The landslide susceptibility map derived using the watershed's physical and environmental conditions offers the best tool for planning the developmental activities and prioritizing areas for mitigation activities in the region. The Coonoor region's tourism and agriculture sectors can significantly benefit from identifying zones prone to landslides for their economic stability and growth.

Keywords: landslides; logistic regression; slope gradient; land use; soil; Coonoor

check for **updates**

Citation: Sujatha, E.R.; Sridhar, V. Landslide Susceptibility Analysis: A Logistic Regression Model Case Study in Coonoor, India. *Hydrology* **2021**, *8*, 41. https://doi.org/10.3390/ hydrology8010041

Academic Editor: Tamim Younos

Received: 29 January 2021
Accepted: 3 March 2021
Published: 5 March 2021

Publisher's Note: MDPI stays neutral with regard to jurisdictional claims in published maps and institutional affiliations.

1. Introduction

Landslides are a common geologic hazard in the hill and mountain terrains of the world. The landslides impact the society and livelihoods of the affected communities. Landslides can result in loss of lives and cause potential damage to infrastructure facilities, agricultural land, public and private assets. UNESCO has also recognized landslides as a significant geohazard globally and attributes 14% of total casualties from various natural hazards like earthquakes, floods, etc. to landslides [1,2]. Global landslides cause nearly 1000 fatalities and a loss of approximately 4 million USD in a year [3]. They can impede the region's economic growth and development and hamper the region's social set-up by isolating the hill communities for long periods from the rest of the surrounding areas. Landslides also lead to environmental degradation by the removal of soil and tree cover. They have significant economic value and affect the environment adversely and hence are a severe concern in mountainous terrains. Assessment of the regions prone to landslides is therefore mandatory for any developmental, land use, and mitigation planning in the hill and mountain communities.

Landslides have profound social and economic impacts. Landslides affect public and private properties and cause both direct and indirect losses that can have either consequential or inconsequential economic impacts [4]. Linear infrastructure like roads or railroads are often severely damaged by landslides causing disruption to normal traffic or completely cutting off the access to the affected areas. This directly affects the tourism industry in the region. The direct economic impact which is consequential involves the repair of damaged infrastructure which includes property and installations or its replacement and clean-up activities. Fatal and non-fatal injuries or accident costs caused by landslide also fall under this category. In cases of remote hamlets or villages, whose economy depends on transporting their raw material or manufactured goods using the roads affected by landslides suffer economic losses due to traffic disruption caused by landslides, though indirect losses are consequential [4]. Moreover, the decrease in tourist activity due to landslides or even landslide vulnerability is an indirect consequential loss. Indirect losses also include reduced real estate values, devaluation of tax revenues, loss of industrial/agriculture revenue, loss of productivity of labor force due to injury, death, or trauma caused by landslides, and the capital spent on prevention/mitigation measures [5]. Landslides can significantly reduce the revenue of the affected regions causing a social set-back [6]. They impose severe constraints to the affected population in terms of economic loss and social set-back [7]. This significant impact of landslides on the socio-economic system of the affected region mandates a thorough understanding of factors causing landslides in the specific geo-environment.

Landslides are caused by several topographical, environmental, geological, hydrological, and geotechnical factors such as terrain features, slope morphometry, drainage pattern, land use and land cover in the region, geomorphological set-up, etc. These factors are usually termed as causative factors [8,9]. Landslides are triggered by extreme rainfall events or snowmelt, seismic activity, and/or anthropogenic activities [2,10]. Climate change and extreme rainfall events trigger landslides more frequently, causing considerable losses to the society, particularly in areas with a large settlement [11–14]. Haque et al. [15] investigated the human cost of global warming, focusing on deadly landslides and triggers between 1995–2014. They reported that there was a significant increase in the number of fatal landslides in the said period. Haque et al. [15] also demonstrated the linkage between catastrophic landslides and extreme rainfall events in their study, particularly in densely populated areas. The effect of various factors contributing to landslides in a region can be perceived in a landslide susceptibility map that describes the spatial propensity of landslide vulnerability in a selected geographic or geomorphic boundary.

Landslide susceptibility assessment is a complex process and involves determining the spatial association between various factors causing landslides and its location. Several statistical, deterministic, and heuristic methods are employed to evaluate landslide susceptibility [13,15–24]. Data-driven statistical models are widely favored for their simplicity and ease of application, while the limitations can come from a lack of local data including temperature and precipitation [25]. Popular statistical methods used to assess landslide susceptibility are bivariate methods [26,27], multivariate regression [28,29], and logistic regression [19,20,30–32]. Bivariate models like frequency ratio, weights of evidence, information value, and yule coefficient assess the spatial association between landslide occurrence and each causative factor using a set of observations. Bivariate models are simple and straightforward, but the relative importance of the factors influencing landslides cannot be determined using bivariate methods. Multiple regression models attempt to evaluate the relationship between landslides and numerous causative factors. They also estimate the importance of these factors in causing landslides and can also identify outliers. However, the multiple regression model's success depends on the data used, and the results are too difficult to interpret. Logistic Regression is a statistical modelling approach used to parameterize a non-linear relationship between dependent and independent variables [31,33,34], particularly when the dependent variable has a binary or dichotomous output. Logistic regression, like linear regression, evaluates the relationship between several predictor variables and the dependent variable. Unlike linear regression, which requires continuous variables, logistic regression can use any type of independent variables—continuous and categorical. It is also not mandatory for independent variables to have a normal distribution and evaluate multiple independent variables. The logistic regression model features make it an ideal choice for modelling landslide susceptibility in this study. However, it should be noted that this method requires a large dataset and is sensitive to the large variance in the dataset used.

Landslide susceptibility models are region-specific and are, to a large extent, dictated by the local geo-environmental set-up.

Coonoor is a popular hill-station in Tamil Nadu, India and is located in the western ghats, a zone prone to intense slope stability problems. Tourism and tourism-related activities such as flower shows, vegetable shows, hiking, special events in botanical gardens, eco-tourism, etc. are prevalent throughout the year. People witness landslides every year, particularly in the months between October and December due to intense and prolonged monsoonal rainfall. These landslides cause severe distress to the hill community in terms of social and economic losses [10]. Therefore, it is necessary to study the factors causing landslides in the Coonoor Taluk to map the regions susceptible to landslides.

This study evaluates the factors that contribute to landslide occurrences, understands their spatial association with the landslide, and map landslide susceptibility using a logistic regression model for Coonoor Taluk, India. The objective of the study is to throw light on the relation between landslide susceptibility and anthropogenic activities in this region. Geo-environmental factors are used to build the landslide susceptibility map and are compared with the most significant anthropogenic activities that have modified the natural setting in the region.

2. Methodology

2.1. Study Area: Coonoor Taluk, Tamil Nadu

Coonoor, a popular hill station in the Western Ghats, is a sensitive eco-system in South India (Figure 1). It covers approximately 230 square kilometers and is bound on the southeastern flank by Doddabetta ranges, Hulikal ravine on the southwest, and Kothagiri ridges on the northeast. It has a well-connected road and rail network through Mettupalayam. The region falls under a tropical zone on account of its elevation and experiences a subtropical highland climate. Its altitude ranges between 394 m and 2033 m, and the average altitude can be described as 1800 to 2000 m above mean sea level. The annual average temperature is 17 °C, with the highest temperature records in May and the lowest temperature in January. The average yearly rainfall is 1335 mm [35]. Precipitation is minimum in the winter months between January and March and maximum in the north-east monsoon season, between October and December. The relative humidity is high almost throughout the year. Bedrock geology consists predominantly of the charnockite rock group with Satyamangalam schist enclaves [36,37]. They are acidic, deeply weathered, and capped with aluminous laterite in several places forming an irregular soil horizon [37].

The most common landforms observed in the region are gentle mounds with thick soil cover and high peaks with steep escarpment. The area is characterized by crests, valleys, deep gorges, cascades, and high-velocity streams [37]. It is drained by numerous streams of the first order that originate from the peaks, and the drainage pattern is predominantly dendritic.

Coonoor Taluk has a population of 157,754, according to the 2011 census. The rural and urban population is 27,128 and 130,626 respectively, i.e., nearly 82.8% of the population are urban dwellers. The metropolitan region is densely populated. It is also home to native tribes like Badugas and Todas, who constitute about 1.5% of the Taluk population. The literacy rate in the Coonoor Taluk is nearly 88%. The economy is dependent on tourism-related activities and the tea industry. Almost 24% of the tea plantations and factories of Nilgiris District, Tamil Nadu, are located in Coonoor. Coonoor being a hill station, has a massive tourist influx all through the year. In 2016, the tourist population visiting Coonoor was 2,463,779, of which nearly 98% were domestic tourists. While this study does not consider the human footprint on landslides, the factors considered in identifying the susceptible zones, including the inhabitants of those regions, will have a direct bearing on broader conclusions in making them more resilient to these landslide disasters.

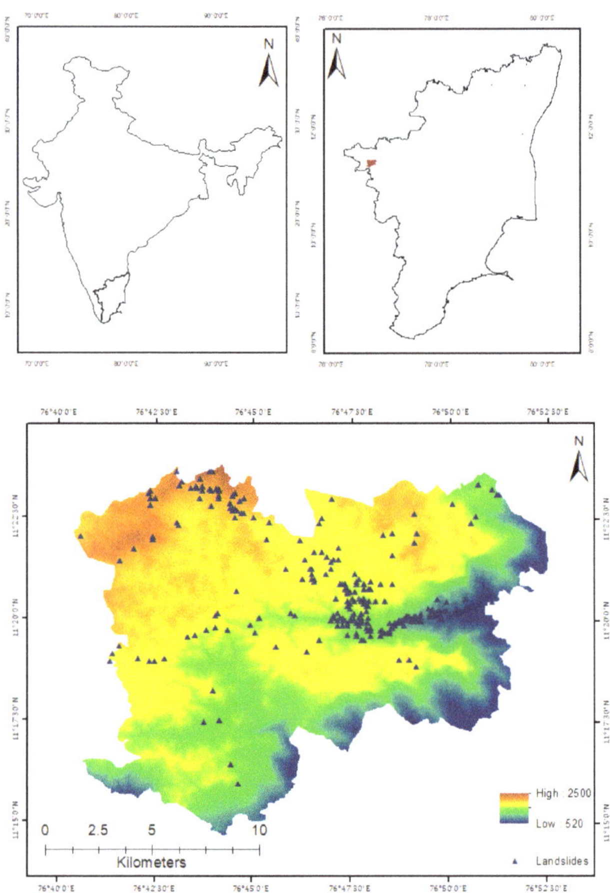

Figure 1. Location map of the study area showing the Digital Elevation Model of Coonoor Taluk, Tamil Nadu, India.

2.2. Data Sources

Data sources include both analogue and digital formats of data. Landslide inventory was collected from the National Highway Authority, Southern Railways Coonoor Division, State Highway Department, Coonoor, Office of the District Collector, Nilgiris, and from various literature sources [10,37,38]. Freely downloadable digital data products like ASTERDEM (30 m × 30 m), LANDSAT 8 OLI, and analogue maps like Survey of India (SOI) topographic maps 58 A11 and 58 A15 of 1:50,000 scale obtained from GSI, India were used to derive the various thematic layers. The soil map was adopted from the Tamil Nadu Agriculture University (TNAU), Coimbatore, India. Limited soil samples were also collected from selected locations for geotechnical analysis and cross-verified with the TNAU soil map. Daily rainfall data for the various rainfall stations in Coonoor Taluk and Uthagamandalam were obtained from Tamil Nadu Statistical Department, Chennai, India, for the years 2007–2017.

2.3. Landslide Characterization

Landslide inventory is a vital dataset required for modelling landslide susceptibility. Scientific records of landslide incidences, including their spatial and temporal attributes, are rare in the region. The inventory has been constructed for the period between 1992 and 2018 based on available records from Government departments, literature, and field survey of known locations. Landslide data collected from the records of the Geological Society of India (GSI) shows that nearly 367 landslides have occurred in the period of study (1992–2018) and includes landslides of small, medium, and large volumes. The information on the length, width, and depth of the recorded landslides is available for only 270 landslides. The volume of these landslides ranges from 3.925 m^3 to 2,512,000 m^3. Nearly 38% of the landslides have a volume less than 100 m^3 and 31% above 1000 m^3. Landslides have primarily occurred due to cut slopes' failure, toe erosion in road cuts or natural slopes, removal of material from the toe, and failure of steep cuts. Approximately 74% of the landslides have occurred due to road cutting activity. The slope failures are shallow in nature, and rainfall is observed to be the triggering factor most often. Hence the landslides triggered by rains alone are considered for this study. Nearly 57% of the landslides recorded are earth slides, slumps, and soil slides, and about 40% of the landslides fall in the category of debris and debris slides. Slope failures like subsidence, boulder fall, and rock toppling are also observed in the region but rare. Figure 2 shows the location of the landslides used for the training of the logistic regression model.

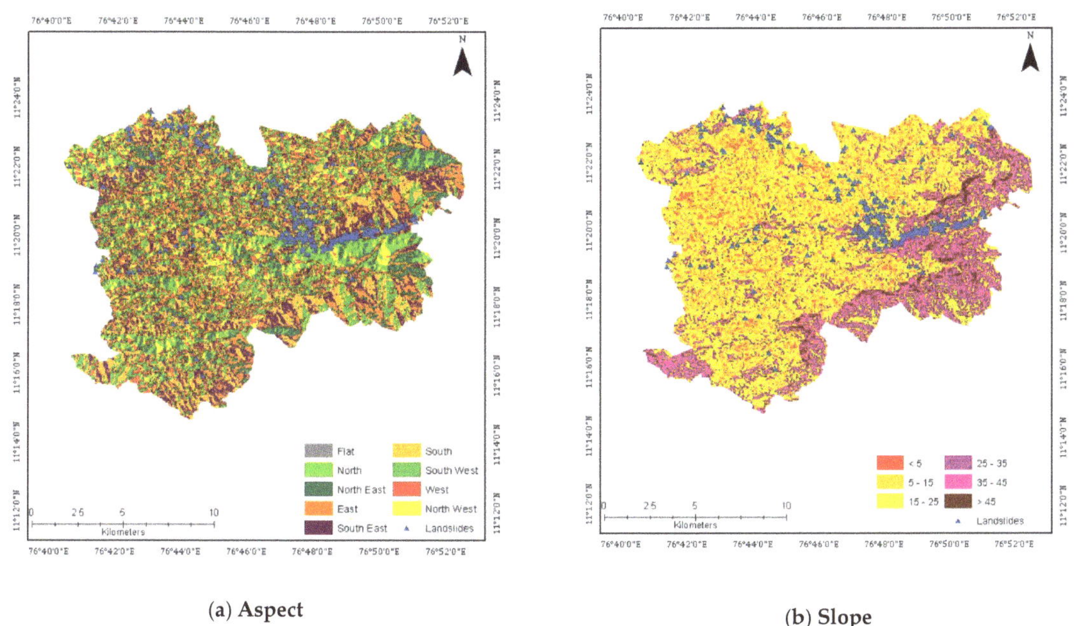

(a) **Aspect** (b) **Slope**

Figure 2. *Cont.*

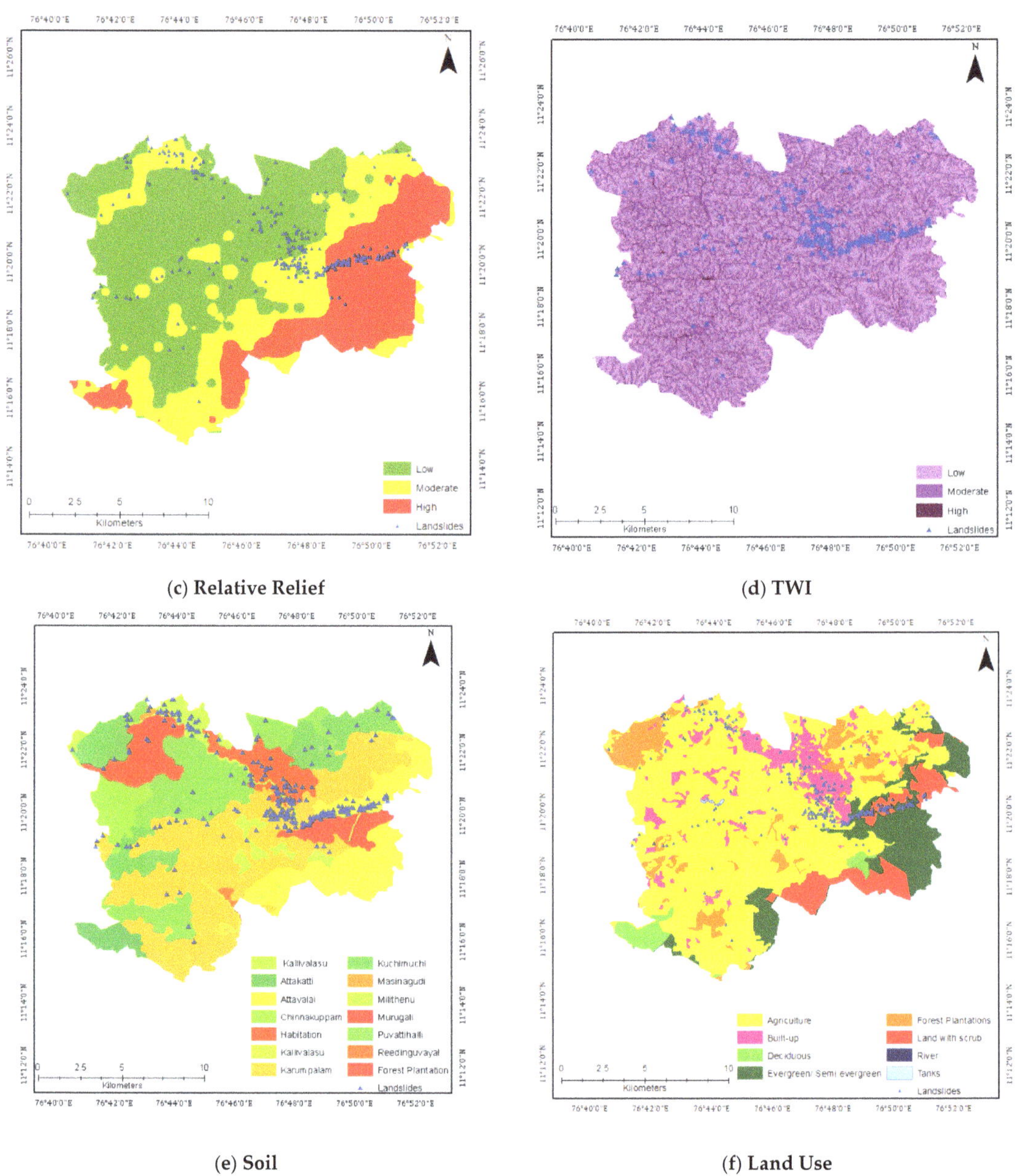

Figure 2. *Cont.*

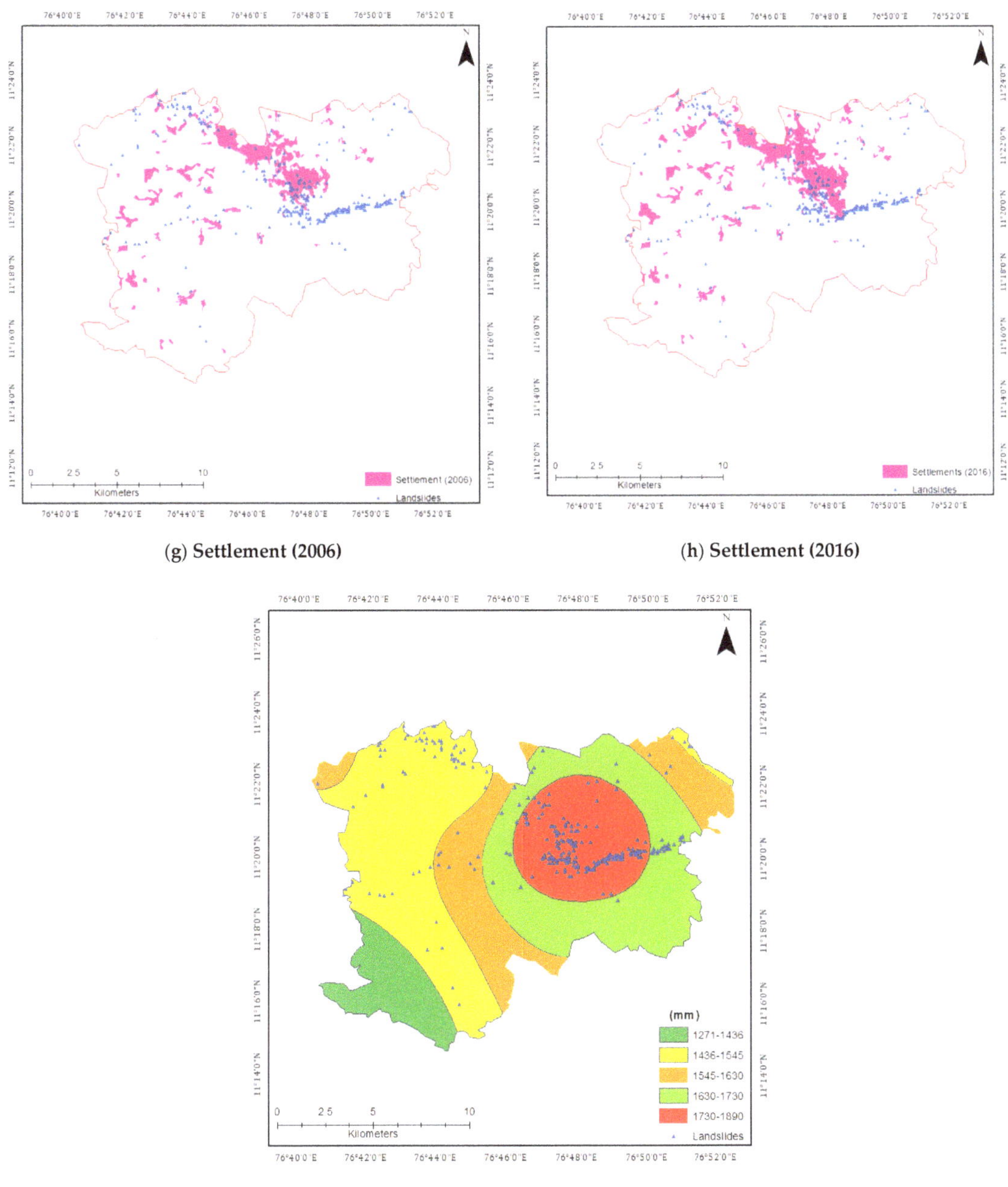

Figure 2. Spatial Databases of Factors Causing Landslides in Coonoor, Tamil Nadu. (**a**) Aspect (**b**) Slope (**c**) Relative Relief (**d**) Topographic Wetness Index-TWI (**e**) Soil (**f**) Land Use (**g**) Settlement-2006 (**h**) Settlement-2016 (**i**) Average Annual Precipitation.

2.4. Spatial Database of Causative Factors

Numerous factors, including topographical, geotechnical, geological, environmental, hydrologic, and climatic factors, contribute to landslides occurrences. The factors are often a function of the local geo-environmental set-up. Therefore, it is necessary to assess each factor's influence with respect to its local geo-environment based on the landslide occurrences. Table 1 presents the various factors commonly used to map landslide susceptibility in literature. The factors that influence landslides in this area were selected based on their correlation to causing landslides. Factors such as aspect, slope gradient, curvature, relative relief, land use, soil, topographic wetness index, distance from lineaments, distance from streams, and average annual rainfall were considered for assessing landslide susceptibility in this region [13,15,19,31,32,34,39]. A Pearson's correlation analysis was carried out for each of the factors listed above to assess their relationship to landslides in this specific geo-environment. The factors that showed a correlation greater than 0.4 were selected to model landslide susceptibility using logistic regression. The factors selected for the study based on Pearson's correlation were aspect, slope, relative relief, TWI, soil, land use and annual precipitation.

Table 1. Common Physical and Environmental Factors causing Landslides used in Literature.

Factor	Reference
Aspect	Sujatha and Rajamanickam (2011) [40]; Akgun (2012) [32]; Eker and Aydin (2014) [19]; Talaei (2014) [34]; Lee et al. (2017) [13]; Pourghasemi and Rahmati (2018) [39]
Slope	Sujatha and Rajamanickam (2011) [40]; Akgun (2012) [32]; Eker and Aydin (2014) [19]; Talaei (2014) [34]; Lee et al. (2017) [13]; Basu and Pal (2018) [41]; Youssef (2015) [42]; Pourghasemi and Rahmati 2018 [39]
Relief	Sujatha and Rajamanickam (2011) [40]; Eker and Aydin (2014) [19]; Talaei (2014) [34]; Youssef (2015) [42]; Pourghasemi and Rahmati 2018 [39]
Relative Relief	Qui et al., 2018
Curvature	Sujatha and Rajamanickam (2011) [40]; Eker and Aydin (2014) [19]; Talaei (2014) [34]; Youssef (2015) [42]; Lee et al. (2017) [13]; Pourghasemi and Rahmati 2018 [39]
Soil	Sujatha and Rajamanickam (2011) [40]; Lee et al. (2017) [13]
Geology	Eker and Aydin (2014) [19]; Talaei (2014) [34]; Youssef (2015) [42]; Lee et al. (2017) [13]; Pourghasemi and Rahmati 2018 [39]
Distance from Fault/Lineament	Sujatha and Rajamanickam (2011) [40]; Akgun (2012) [32]; Talaei (2014) [34]; Youssef (2015) [42]; Lee et al. (2017) [13]
Distance from Streams	Akgun (2012) [32]; Talaei (2014) [34]; Youssef (2015) [42]; Pourghasemi and Rahmati 2018 [39]
Drainage Density	Pourghasemi and Rahmati 2018 [39]
Topographic Wetness Index (TWI)	Sujatha and Rajamanickam (2011) [40]; Lee et al. (2017) [13]
Stream Power Index (SPI)	Lee et al. (2017) [13]
Land use	Sujatha and Rajamanickam (2011) [40]; Eker and Aydin (2014) [19]; Talaei (2014) [34]; Lee et al. (2017) [13]; Pourghasemi and Rahmati 2018 [39]; Haque et al. (2019) [15]
NDVI	Youssef (2015) [42]
Distance from Roads	Sujatha and Rajamanickam (2011) [40]; Akgun (2012) [32]; Talaei (2014) [34]; Youssef (2015) [42]; Pourghasemi and Rahmati 2018 [39]
Peak Ground Acceleration	Talaei (2014) [34]
Rainfall	Talaei (2014) [34]; Yousef (2015); Haque et al. (2019) [15]

Thematic layers aspect, slope gradient, relative relief, and topographic wetness index were extracted from ASTER GDEM of 30 m × 30 m resolution using the spatial analyst tool of ArcMap. Aspect represents slope direction and was divided into eight cardinal directions and a category flat (Figure 2a). Slope gradient was reclassified into five classes as 0°–5°, 5°–15°, 15°–25°, 25°–35°, 35°–45°, and greater than 45° (Figure 2b). Relative relief represents the difference in maximum and minimum elevation within a pixel. Relative relief was classified into three categories low (92–250), moderate (250–495), and high (495–922) and is presented in Figure 2c. The spatial distribution of topographic wetness index (TWI) was generated using flow accumulation and slope datasets. Flow accumulation was also extracted from ASTERDEM of 30 m resolution using flow direction raster. It was reclassified as low (5.77–9.44), moderate (9.44–12.64), and high (12.64–22.78) and is shown in Figure 2d.

The soil was classified based on the region of occurrence as Kallivalasu, Attakatti, Attavalai, Chinnakupam, Karumpalam, Kuchimuchi, Masinagudi, Milithenu, Murugali, Puvattihalli, and Reedinguvayalu soil series (Figure 2e). The soil's textural classification in the region indicates that loam, loamy sand, sandy clay loam, sandy clay, and clay were present as stratified layers. Rock outcrops were also noticed in the north-western part of the study area. They were converted into plantations taking advantage of the terrain. The thickness of the soil varies between 51 cm in Murugali and 7 m in Attavalai. Land use was extracted from BHUVAN data provided by the National Remote Sensing Agency, India. The land use map is of scale 1: 50,000. Figure 2f shows the land use map for the year 2016. The major land use categories are agriculture, forests, land with scrub, settlements, and water bodies. Tea cultivation and agriculture occupy nearly 78.4% of the area in the region. Settlements are dense in the northern part of the study area. An analysis of the spatial spread of settlements in a decade between 2006 and 2016 indicates an increase of nearly 28% (Figure 2g,h).

Daily rainfall data were used to calculate the average annual rainfall. Rainfall data from six rainfall stations—Coonoor, Ketti, Kothagiri, Runnymede, Kundah, and Uthagamandalam, was used to map the spatial variations of average annual rainfall in the study area using spatial kriging (Figure 2i). It was classified into four classes based on natural breaks— 1275 mm–1436 mm, 1436 mm–1545 mm, 1545 mm–1603 mm, and 1603 mm–1890 mm.

2.5. Landslide Susceptibility Assessment

A binary logistic regression model was used to map the spatial variability of the zones prone to landslides in Coonoor. The spatial variation of the factors causing landslides is shown in Figure 2.

Logistic function f(z) describes the probability of occurrence of a landslide event and is defined as

$$f(z) = \frac{e^z}{1 + e^z} = \frac{1}{1 + e^{-z}} \tag{1}$$

and it varies from zero to one. "z" is expressed as the linear combination of predictors i.e., independent variables that cause landslides and respective coefficients. The model is expressed by

$$z = b_0 + b_1 X_1 + b_2 X_2 + b_3 X_3 + \ldots \ldots + b_n X_n \tag{2}$$

b_0 represents model coefficient i.e., the intercept or constant; $b_1 \ldots \ldots .b_n$ are coefficients representing the measure of the contribution of predictor variables $X_1 \ldots .. \ldots \ldots X_n$ in causing landslides. The terms b_0 to b_n are unknown and are determined based on the relationship between the independent variables and landslide conditions and are estimated by the maximum likelihood approach, which is a derivative of the probability distribution of landslides, the dependent variable [34]. The independent variables are spatially represented as thematic layers and illustrate each factor causing a landslide. "z" varies between $-\infty$ and $+\infty$ and is an index that allows the user to combine the various independent variables responsible for landslide occurrence. Sample observations are used to fit a multiple logistic regression model. The coefficients b_0, b_1, b_2, $b_3 \ldots \ldots .b_n$ are estimated and used to ascertain landslide probability.

Logistic regression (LR) model is built by (i) selection of independent variables based on its association with landslide occurrence (ii) checking the statistical significance of the selected variables using *p*-value significance test (iii) verifying the lack of inter-dependency between the selected independent variable using collinearity statistics—tolerance and VIF (iv) modelling landslide probability through logistic regression model and (v) validation through Area Under Curve (AUC) and landslide density function using the validation dataset. In this study, the landslide density function, computed for each class of a thematic layer is used to transform nominal variables into numerical variables, and is used as input variables for determination of the LR model. This helps prevent the creation of a large number of dummy variables. Moreover, it incorporates the knowledge of landslide history into the model. The landslide density function is defined as

$$ LDF = \frac{\frac{Area\ of\ Landslide\ pixels\ in\ a\ particular\ class}{Total\ Area\ of\ landslide\ pixels}}{\frac{Area\ of\ pixels\ in\ a\ particular\ class}{Total\ Area}} \tag{3} $$

Landslides cover nearly 1.1% of the total area, which is many times smaller than the area in which landslides are not present, and hence, it can be considered a rare event [43,44]. The ratio of landslide to non-landslide pixels used for developing the training dataset of the model is based on sensitivity analysis conducted on a different ratio of 1:1, 1:2, 1:2.5 and 1:5 based on various literature [31,44,45]. Seed cells of 100 m radius surrounding a landslide location were considered to extract the independent variable's feature in a landslide affected region. Similarly, random locations not affected by landslides were also selected to represent zones not prone to landslides. The landslide pixels' ratio to non-landslide pixels was maintained as 1:1, 1:2, 1:2.5 and 1:5 to generate the training dataset. It was observed that the ratio of 1:2.5 performed better consistently in these trials, and hence it was selected for the study. Different random sets of pixels with no landslides were selected to verify the consistency of the results. A binary variable to indicate the absence (0) or presence (1) of the landslide was added to the dataset.

2.5.1. Multicollinearity Analysis

Collinearity among the selected independent variables profoundly affects the model performance [39,46]. Tolerance and variance inflation factor (VIF) is used to measure multicollinearity in selected variables. Tolerance values less than 0.2 indicate marginal multicollinearity among selected independent variables, while tolerance less than 0.1 advocates multicollinearity to a great extent. Similarly, a variable with VIF greater than 2 indicates serious multicollinearity [44,47]. All the variables selected have a tolerance greater than 0.2 and VIF less than two, which indicates that the variables are not unduly correlated with each other (Table 2). Hence, all the selected variables were used to build the model.

Table 2. Multi-Collinearity Analysis of the selected Predictor Variables.

Predictor Variable	Aspect	Slope	Relative Relief	TWI	Soil Type	Land Use	Average Annual Rainfall
Tolerance	0.852	0.899	0.951	0.569	0.815	0.624	0.617
VIF	1.173	1.112	1.052	1.758	1.228	1.603	1.409

Statistical Package for Social Sciences (SPSS) was used to build the logistic regression model. The logistic regression method based on the forward likelihood ratio was selected to assess the effect of the predictor variables of landslide occurrences. The statistical significance of the chosen variable was evaluated using the χ^2 score. The Wald χ^2 score's significance level for a predictor variable to enter the model was set at 0.1. The training sets were evaluated based on the χ^2 value of Hosmer–Lemeshow, Nagelkerke R^2, and Cox and Snell R^2.

2.5.2. Landslide Susceptibility Map and Validation

The model was trained using the landslides that occurred between 1992 and 2009. The pixel size of the raster dataset used for the landslide model was 30 m $\times$ 30 m. The total area of Coonoor region was represented with 255,700 pixels and the landslide data used for training the model consisted of 7859 pixels. Random selection of landslide and non-landslide pixels was adopted to build the logistic regression model. Validation of the model was carried out using the landslides that happened between the years 2010–2018. The coefficients calculated using the logistic regression model are assigned as weights for the thematic layer. The weighted thematic layers are combined in a GIS environment. The landslide probability is determined from the logistic function "z". The spatial distribution of landslide probability represents the landslide susceptibility of the region. Hence, the spatial variation of probability is reclassified into five categories: very low, low, moderate, high, and very high, using quantile classification to represent landslide susceptibility. The landslide susceptibility map is validated using the landslide density index and area under the curve (AUC) to envisage the prediction and success rate of the landslide model built. Cumulative landslide percentage and area were plotted, and the area under these curves was calculated using both the training dataset and validation dataset of landslides. These represent the prediction and success rate of the model developed.

3. Results and Discussion

3.1. Logistic Regression Model for Mapping Landslide Susceptibility

The variables—aspect, slope, relative relief, TWI, soil, land use, and average annual rainfall were used to build the landslide susceptibility model. The model included all the selected variables. The null hypothesis for the test is set as the coefficient is zero. The estimated coefficient of the selected factors was statistically different from zero. The logistic regression model employed to assess the impact of predictor variables on landslides' occurrence showed that the goodness of fit was acceptable as the significance of χ^2 was greater than 0.05.

Similarly, Cox and Snell R^2 and Nagelkerke R^2 of 0.589 and 0.838, respectively, which are greater than 0.2 [48] indicate that selected independent i.e., predictor variables explained the dependent variable successfully. Table 3 shows the coefficient for the factors influencing landslide susceptibility in the region, and the model's summary of classification is presented in Table 4. The model indicates that aspect (slope direction), relative relief, and TWI are negatively correlated, while all other factors are positively correlated. The influence of the parameters on increasing the susceptibility to landslides can be understood by exponentiating these factors' coefficient, which expresses their odds. Table 3 shows that this study's most influential parameters are average annual rainfall and land use, followed by slope and soil, indicating that climate and anthropogenic interference are very significant in causing landslides in this region.

Table 3. Factors selected for modelling landslide susceptibility and their estimated coefficient.

Variables	βi	SE	Wald	df	Sig.	Exp(βi)
Aspect	−2.542	0.762	11.119	1	0.001	0.079
Slope	1.204	0.954	1.593	1	0.207	3.334
Relative Relief	−6.288	1.298	23.453	1	0.000	0.002
TWI	−3.044	1.198	6.458	1	0.011	0.048
Soil	0.995	0.153	42.556	1	0.000	2.705
Land use	1.885	0.534	12.078	1	0.001	6.391
Average Annual Rainfall	2.081	0.170	150.038	1	0.000	8.014
Constant	2.765	2.317	1.425	1	0.233	15.886

Table 4. Classification Summary.

Observed		Predicted		Percentage Correct
		Landslides		
		Non-Occurrence	Occurrence	
Landslides	Non-Occurrence	605	30	95.3
	Occurrence	26	239	90.2
Overall Percentage				93.4

Cut-off value = 0.50.

The landslide susceptibility index based on the logistic regression model is

$$z = 2.765 - 2.542 \text{ (Aspect)} + 1.204 \text{ (slope)} - 6.288 \text{ (Relative Relief)} - 3.044 \text{ (TWI)} + 0.995 \text{ (Soil)} + 1.885 \text{ (Land Use)} + 2.081 \text{ (Average Annual Rainfall)} \tag{4}$$

The classification summary of the model shows that the model has a 93.4% successful prediction rate. The model's capability to delineate areas not prone to landslides is higher (95.3%) than the ability to identify zones prone to landslides (90.2%).

3.2. Spatial Variation of Landslide Susceptibility

Landslide susceptibility map (Figure 3) of the region has been reclassified into five zones for better understanding: very low, low, moderate, high, and very high using quantile classification. Quantile classification is used as the distance between categories is not known. The upper limit of probability of landslide susceptibility in the very low, low, moderate, and high zones result to be 7%, 29%, 59%, and 84%, respectively. Zones with a probability of landslide susceptibility greater than 84% are classified as very high hazard zones. The zones demarcated as high and very high susceptible constitute 18.5% and 17.6% of the total area, but nearly 34% and 48.6% of the landslides have occurred in these zones, respectively. Settlements fall in these high and very high susceptible zones. The major road network that connects hill town to the plains and further to the district center Udhagamandalam falls in these high and very high susceptible zones. A part of the Nilgiri Mountain Railway, a UNESCO world heritage site, falls in these zones. The high susceptible region is densely populated, intensely modified for agriculture, and has a high linear infrastructure density. The southern part of the study area with intense agriculture is not much affected by landslides. Therefore, slope modifications for development may enhance landslide susceptibility.

Simple biological stabilization techniques like turfing the slopes with plants like hedge grass-like vetiver, asparagus can be popularized among high-density settlements. The roots will act as reinforcement and improve the shear strength of the slopes. The water outlets and drainages can also be regulated along the slopes to avoid the slopes' saturation.

The performance of the logistic regression model is assessed using landslide occurrences between the years 2010 and 2018. A buffer of 100 m radius was used to delineate the landslide area. The landslide density function calculated with the validation dataset indicates that landslide density function increases exponentially with the susceptibility class (Table 5).

It increases from 0.17 for low susceptible areas to 2.76 for very highly susceptible areas. Area Under the Curve (AUC) is a standard indicator used to assess the susceptibility model's spatial forecasting capacity [49]. The AUC of the landslide susceptibility map generated using the logistic regression model is portrayed in Figure 4. The AUC for the prediction and success rates are 79% and 83%, respectively (Figure 4), indicating that the model built assesses landslide susceptibility in Coonoor Taluk satisfactorily. The model shows a better success rate, indicating that this model's susceptibility map can be used for planning schemes for hazard preparedness and land use planning with greater accuracy.

Table 5. Landslide Density Function for the Susceptibility Classes using Validation Landslide Dataset.

Susceptibility Class	Area Pixels	Landslide Pixels	Area Ratio	Landslide Ratio	Landslide Density Function
Very Low	45,240	0	0.000	0.177	0.00
Low	61,739	14	0.040	0.241	0.17
Moderate	56,364	48	0.138	0.220	0.63
High	47,406	117	0.336	0.185	1.81
Very High	44,951	169	0.486	0.176	2.76

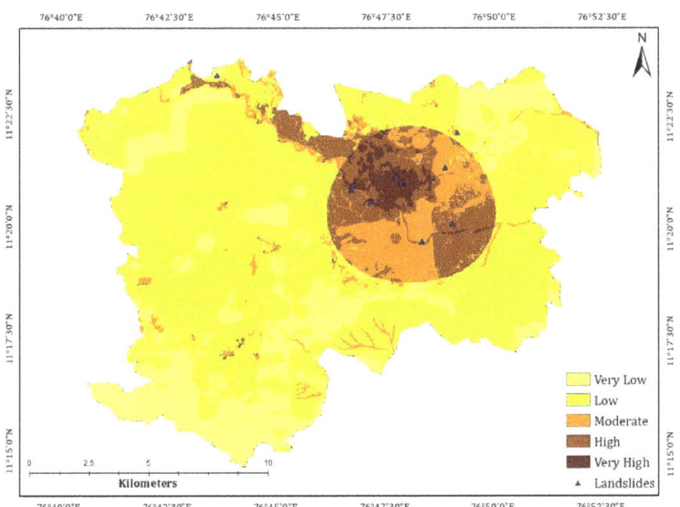

Figure 3. Landslide Susceptibility Map of Coonoor Taluk of Nilgiris District, India using Logistic Regression Model.

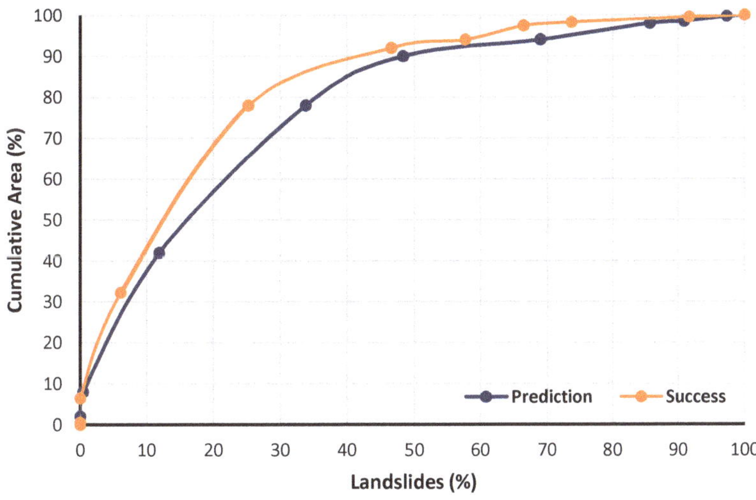

Figure 4. Prediction and Success Rate of the Logistic Regression Model using Area Under the Curve Method.

3.3. Effect of Local Geo-Environment on Landslides

Average annual rainfall is observed as the most influential parameter that contributes to landslide susceptibility in this region. Bisht et al. [50] reported that the study area region had witnessed both 95th and 99th percentile extreme precipitation events between 1971–2015. An analysis of the relationship between the past landslide occurrence and rainfall indicates that antecedent rainfall plays a vital role in initiating landslides. A minimum of five days antecedent rainfall of 132 mm is required to cause small and medium volume landslides. Landslides are more prevalent in the zones where the average annual rainfall ranges between 1730 mm–1890 mm, which is the highest recorded in this region i.e., nearly 55% of the landslides have been registered in this zone. It is also noted that around 26% of the landslides are observed in the zones where average annual rainfall ranges between 1436–1545 mm. This region is intensely cultivated, and a high number of landslide incidences may be due to the land modification for agriculture and related agriculture practices.

The descriptive statistics of slope gradient indicates that most slopes fall in the gentle—moderate category where landslides are most likely to occur [51–54]. The majority of landslides have occurred in slopes less than 28°. Gentle slopes appear to be more prone to landslides [13,19]. Nearly 40% of the landslides have occurred in the slopes with a gradient between 15°–25° that cover 34% of the total area. Slopes with steeper angles have significantly less overburden as the material tends to erode faster due to its gradient. The overburden covering the slope is highly resistant to movement. Slope direction often dictates the flow direction, and the amount of rainfall received. Nearly 45% of the study area's slopes face the southeast, south and southwest directions, and 59% of the landslides have occurred in the slopes facing these directions. These slopes are frequently affected by landslides by virtue of their slope morphometry.

Soil is ranked as the fourth factor that influences landslide susceptibility. Masinagudi series and Kallivalasu series are more prone to landslides. Nearly 22% of the total slides fall in 3% of the Masinagudi series area, and 9% of landslides fall in 19% of the total area occupied by the Kallivalasu series. Both the series have soil in the category sandy clay loam. The average hydraulic conductivities of Kallivalasu and Masinagudi series are 2.29×10^{-4} cm/s and 6.31×10^{-5} cm/s, respectively, with an average thickness of 3 m each. These deposits being moderately permeable and lesser in thickness, allow the water to reach the nearly impermeable bedrock and shear resistance at the soil's interface overburden and rock bed reduces to an insignificant amount causing the entire overburden to fail.

Topographic wetness index (TWI), a steady-state wetness index quantifies the topographic control on the hydrologic processes. While it considers the slope morphometry and upstream contributing area per unit width perpendicular to the direction of flow [55], a suite of soil moisture indices is widely used for predicting hydrologic extremes [56,57]. It is a more relevant metric for hill terrains than flat areas. It also explains the distribution of soil moisture [58]. It quantifies the tendency to distribute soil water, which is influenced by topography [59] and is often used in vegetation studies. Landslides are more prevalent in the high TWI zones indicating that soil moisture is an essential factor that causes landslide susceptibility. It is rather challenging to be spatially mapped. Hence, TWI can be effectively used in place of soil moisture despite its inability to consider the humidity, heterogeneity of soil, and vegetation cover [58].

Relative relief helps to characterize the relief characteristics without taking into account the mean sea level. Landslides are more prevalent in zones with moderate relative relief i.e., 37% of landslides have occurred in 37% of the area falling under the moderate relative relief category.

3.4. Effect of Anthropogenic Activities of Landslides

Anthropogenic factors are not included in the susceptibility model. Notwithstanding this, it is reasonable to construct a susceptibility map based on the geo-environmental factors and compare the susceptibility zones with the anthropogenic interferences. In fact, nearly 68% of the landslides have occurred in the zones heavily modified by anthropogenic activities. The gentle slopes are modified by various anthropogenic activities, mainly agricultural and construction activities, including infrastructure development and housing projects, because of their favorable topography. The gentle slope gradient does not allow rapid drainage of water during heavy rainfall periods increasing pore pressure and subsequent slope failure. The high density of settlements in favorable topography further increases the surcharge weight on the slopes, saturate soil due to improper drainage arrangements to carry stormwater, greywater, and sullage.

The principal land use categories are agriculture, forest (includes deciduous and evergreen forest), a forest plantation, land with scrub, built-up area, and water bodies (tanks and river). Tea plantation occupies nearly 64% of the region, forest and forest and forest plantation (24%), land with scrub (7%), built-up area (4%), and rest by water bodies. The vast area of land under agriculture points to intense anthropogenic interference. The slopes are continuously modified and irrigated for agricultural purposes, leading to saturated soil moisture conditions in the top 100–150 cms with shifted surface energy fluxes [60,61], causing significant concern in problems related to slope instability. Though the built-up area occupies only 4% of the study area, the built-up density is very high, making the region more prone to landslides. In a decade, an increase of nearly 28% in the built-up area i.e., the settlements, is witnessed, as seen from Figure 2g,h. The expansion of settlement zones indicates the land pressure caused due to urbanization and makes the built-up category more vulnerable to landslides and increases the risk associated with landslides. The losses in terms of life and property will be more when a landslide occurs in this area. The area under scrubland also does not protect the soil the slopes from sliding. Many of the landslides were also observed to have happened in these land use categories.

Linear infrastructure is a predominant factor causing slope instability in the region, particularly the railway lines. Around 90% of the landslides have occurred near the linear infrastructure facilities of which 48% have been reported along the rail route. Natural slopes are modified continuously to lay or widen the roads and regular maintenance activities for both the road and railway lines. These interferences have severe consequences on the slopes' stability as they usually steepen the natural slope reducing their shear resistance. Moreover, these modified slopes with made-up fills of borrow materials tend to have lower permeability, leading to pore pressure increase, which further decreases the slopes' shear resistance. The removal of forest cover for laying or widening of linear infrastructure further adds to slope instability problems as root cohesion can add to slope stability lost in these slopes. Major roads like the national and state highway with large traffic volumes appear to be most affected by these slope instability problems.

Average annual precipitation and land use are the two most dominant factors that cause landslides in the region. This study emphasizes that anthropogenic interference has played a major role in causing landslides in this environmental set-up. Particularly, linear infrastructure facilities like roads and railway lines have been very influential. These zones are more prone to the risk of landslides. Landslides have a significant social and economic effect in this region as the zones falling in the high susceptibility category are predominantly built-up area and intensely cultivated regions. A further study relating landslide occurrences to extreme climate events can add value to this study.

4. Conclusions

Coonoor is severely affected by landslides almost every year during periods of intense and prolonged rainfall, causing heavy social and economic losses to its residents. The economy of Coonoor is dependent mainly on tourism and tourism-related activities. Landslide susceptibility mapping can help identify zones that need immediate attention in terms of planning mitigation strategies and development activities. Logistic regression is a more reliable model to map landslide susceptibility compared to other heuristic models like the analytical hierarchy process model or statistical models like frequency ratio model and is hence used for this study.

The physical and environmental factors causing landslides were identified, and the logistic regression model was used to assess their impact on causing slope instability. The model shows that average annual rainfall, land use, slope morphometry, and soil type are important factors that contribute to slope instability. The landslide susceptibility map indicates the spatial distribution of areas' susceptibility to various degrees of landslide vulnerability. It is a crucial component to ascertain the temporal mapping of landslides. The spatial distribution of susceptibility classes in the region based on the logistic regression model shows that nearly 17.6% of the area is classified as highly unstable i.e., very high susceptible, and 48.6% of total landslides falls under this unstable category i.e., very highly susceptible. Anthropogenic interference is observed to be a very significant factor that has caused landslides in the region as most of the instable areas fall in the densely built-up zones, adjacent to major roads and railway line and in agriculture areas and where forests are disturbed by road infrastructure development like roads and forest plantations.

This study reinforces the need for providing landslide susceptibility maps in hill–town development and planning. It is an indispensable tool for planning land management and mitigation strategies. It will also aid the town planners in developing sustainable agriculture practices. It can help in locating regions for future growth in suitable areas of low susceptibility. It can help policymakers in hazard management and disaster planning and preparedness at the taluk level. It can also be further used at the block level with the availability of block boundaries. It can also help in drawing policies against land degradation and watershed deterioration.

Author Contributions: Conceptualization, E.R.S.; Data curation, E.R.S.; Formal analysis, E.R.S. and V.S.; Funding acquisition, E.R.S.; Investigation, E.R.S.; Methodology, E.R.S. and V.S.; Project administration, E.R.S.; Validation, E.R.S.; Writing—original draft, E.R.S.; Writing—review & editing, V.S. Both authors have read and agreed to the published version of the manuscript.

Funding: This research was funded by NRDMS, DST, grant number DST-NRDMS (155/18-2015) and the APC was funded by Hydrology, MDPI.

Data Availability Statement: All the data used in this study will be made available on request to the senior author.

Acknowledgments: This study was supported by DST-NRDMS (155/18-2015). The authors would like to acknowledge the financial support rendered by NRDMS, DST, for the research with thanks. We also would like to thank the Vice-Chancellor of SASTRA Deemed University, Thanjavur, India, for the support and the facilities to carry out this work. We acknowledge the partial support received by the corresponding author from the Virginia Agricultural Experiment Station (Blacksburg) and the Hatch Program of the National Institute of Food and Agriculture, U.S. Department of Agriculture (Washington, D.C.).

Conflicts of Interest: The authors declare no conflict of interest.

References

1. Varnes, D.J. *Commission on the Landslides of the IAEG, UNESCO. Landslide Hazard Zonation: A Review of Principles and Practice*; UN: New York, NY, USA, 1984; Volume 3, p. 61.
2. Froude, M.J.; Petley, D.N. Global fatal landslide occurrence from 2004 to 2016. *Nat. Hazards Earth Syst. Sci.* **2018**, *18*, 2161–2181. [CrossRef]
3. Lee, S.; Pradhan, B. Probabilistic landslide hazards and risk mapping on Penang Island, Malaysia. *J. Earth Syst. Sci.* **2006**, *115*, 661–672. [CrossRef]
4. Winter, M.G.; Barbara, S.; Derek, P.; David, P.; Clare, H.; Jonathan, S. The economic impact of landslides and floods on road networks. *Proc. Eng.* **2016**, *143*, 1425–1434. [CrossRef]
5. Schuster, R.; Highland, L.M. *Socioeconomic and Environmental Impacts of Landslides in Western Hemisphere*; U.S. Geological Survey: Denver, CO, USA, 2001.
6. Perera, E.N.C.; Jayawardana, D.T.; Jayasinghe, P.; Bandara, R.M.S.; Alahakoon, N. Direct impact of landslides on socioeconomic systems: A case study from Aranayake, Sri Lanka. *Geoenviron. Disas.* **2018**, *5*. [CrossRef]

7. Hervás, J.; Bobrowsky, P. Mapping: Inventories, Susceptibility, Hazard and Risk. In *Landslides—Disaster Risk Reduction*; Springer International Publishing: Berlin, Germany, 2008; pp. 321–349.

8. Jaiswal, P.; van Westen, C.J. Estimating temporal probability for landslide initiation along transportation routes based on rainfall thresholds. *Geomorphology* **2009**, *112*, 96–105. [CrossRef]

9. Sujatha, E.R.; Rajamanickam, G.V. Landslide Hazard and Risk Mapping Using the Weighted Linear Combination Model Applied to the Tevankarai Stream Watershed, Kodaikkanal, India. *Hum. Ecol. Risk Assess. Int. J.* **2014**, *21*, 1445–1461. [CrossRef]

10. Jaiswal, P.; Van Westen, C.J.; Jetten, V. Quantitative assessment of landslide hazard along transportation lines using historical records. *Landslides* **2011**, *8*, 279–291. [CrossRef]

11. Cloutier, C.; Locat, J.; Jakob, M.; Schornbus, M. Slope Safety Preparedness for Effects of Climate Change. In Proceedings of the Joint Technical Committee JTC-1TR3 Forum, Naples, Italy, 17–18 November 2015.

12. Gariano, S.L.; Guzzetti, F. Landslides in chaning climate. *Earth Sci. Rev.* **2016**, *162*, 227–252. [CrossRef]

13. Lee, S.; Hong, S.-M.; Jung, H.-S. A Support Vector Machine for Landslide Susceptibility Mapping in Gangwon Province, Korea. *Sustainability* **2017**, *9*, 48. [CrossRef]

14. Sangelantoni, L.; Gioia, E.; Marincioni, F. Impact of climate change on landslides frequency: The Esino river basin case study (Central Italy). *Nat. Hazards* **2018**, *93*, 849–884. [CrossRef]

15. Haque, U.; da Silva, A.P.F.; Devoli, G.; Pilz, J.; Zhao, B.; Khaloua, A.; Wilopo, W.; Andersen, P.; Lu, P.; Lee, J.; et al. The human cost of global warming: Deadly landslides and their triggers (1995–2014). *Sci. Total Environ.* **2019**, *682*, 673–684. [CrossRef]

16. Ayalew, L.; Yamagishi, H. The application of GIS-based logistic regression for landslide susceptibility mapping in the Kakuda-Yahiko Mountains, Central Japan. *Geomorphology* **2005**, *65*, 15–31. [CrossRef]

17. Hasekioğulları, G.D.; Ercanoglu, M. A new approach to use AHP in landslide susceptibility mapping: A case study at Yenice (Karabuk, NW Turkey). *Nat. Hazards* **2012**, *63*, 1157–1179. [CrossRef]

18. Zare, M.; Pourghasemi, H.R.; Vafakhah, M.; Pradhan, B. Landslide susceptibility mapping at Vaz Watershed (Iran) using an artificial neural network model: A comparison between multilayer perceptron (MLP) and radial basic function (RBF) algo-rithms. *Arab. J. Geosci.* **2013**, *6*, 2873–2888. [CrossRef]

19. Eker, R.; Aydın, A. Assessment of Forest Road Conditions in Terms of Landslide Susceptibility: A Case Study in Yığılca Forest Directorate (Turkey). *Turk. J. Agric. For.* **2014**, *38*, 281–290. [CrossRef]

20. Pourghasemi, H.R.; Moradi, H.R.; Aghda, S.M.F.; Gokceoglu, C.; Pradhan, B. GIS-based landslide susceptibility mapping with probabilistic likelihood ratio and spatial multi-criteria evaluation models (North of Tehran, Iran). *Arab. J. Geosci.* **2013**, *7*, 1857–1878. [CrossRef]

21. Trigila, A.; Iadanza, C.; Esposito, C.; Mugnozza, G.S. Comparison of Logistic Regression and Random Forests techniques for shallow landslide susceptibility assessment in Giampilieria (NE Sicily, Italy). *Geomorphology* **2015**, *249*, 119–136. [CrossRef]

22. Hong, H.; Naghibi, S.A.; Pourghasemi, H.R.; Pradhan, B. GIS-based landslide spatial modeling in Ganzhou City, China. *Arab. J. Geosci.* **2016**, *9*, 1–26. [CrossRef]

23. Pham, B.T.; Pradhan, B.; Bui, D.T.; Prakash, I.; Dholakia, M. A comparative study of different machine learning methods for landslide susceptibility assessment: A case study of Uttarakhand area (India). *Environ. Model. Softw.* **2016**, *84*, 240–250. [CrossRef]

24. Roodposhti MJShahabi, H.; Safarrad, T. Fuzzy Shannon entropy: Ahybrid GIS-based landslide susceptibility mapping method. *Entropy* **2016**, *18*, 343.

25. Sehgal, V.; Lakhanpal, A.; Maheswaran, R.; Khosa, R.; Sridhar, V. Application of multi-scale wavelet entropy and multi-resolution Volterra models for climatic downscaling. *J. Hydrol.* **2018**, *556*, 1078–1095. [CrossRef]

26. Bijukchhen, S.M.; Kayastha, P.; Dhital, M.R. A Comparative Evaluation of Heuristic and Bivariate Statistical Modeling for Landslide Susceptibility Mappings in Ghurmi-DhadKhola, East Nepal. *Arab. J. Geosci.* **2013**, *6*, 2727–2743. [CrossRef]

27. Kayastha, P.; Dhital, M.R.; Smedt, F.D. Evaluation and Comparison of GIS Based Landslide Susceptibility Mapping Procedures in Kulekhani Watershed, Nepal. *J. Geol. Soc. India* **2013**, *81*, 219–231. [CrossRef]

28. Nandi, A.; Shakoor, A. A GIS-based landslide susceptibility evaluation using bivariate and multivariate statistical analyses. *Eng. Geol.* **2010**, *110*, 11–20. [CrossRef]

29. Pradhan, B.; Mansor, S.; Pirasteh, S.; Buchroithner, M.F. Landslide hazard and risk analyses at a landslide prone catchment area using statistical based geospatial model. *Int. J. Remote Sens.* **2011**, *32*, 4075–4087. [CrossRef]

30. Pradhan, B.; Lee, S. Landslide susceptibility assessment and factor effect analysis: Back propagation artificial neural networks and their comparison with frequency ratio and bivariate logistic regression modelling. *Environ. Model. Softw.* **2010**, *25*, 747–759. [CrossRef]

31. Sujatha, E.R.; Kumaravel, P.; Rajamanickam, V. GIS based Landslide Susceptibility Mapping of Tevankarai Ar Sub-watershed, Kodaikkanal, India using Binary Logistic Regression Analysis. *J. Mt. Sci.* **2011**, *8*, 505–517.

32. Akgun, A. A comparison of landslide susceptibility maps produced by logistic regression, multi-criteria decision, and likelihood ratio methods: A case study at İzmir, Turkey. *Landslides* **2011**, *9*, 93–106. [CrossRef]

33. Kleinbaum, D.G.; Klein, M. *Logistic Regression: A Self-Learning Text*, 3rd ed.; Springer: New York, NY, USA, 2010; p. 701.

34. Talaei, R. Landslide susceptibility zonation mapping using logistic regression and its validation in Hashtchin Region, northwest of Iran. *J. Geol. Soc. India* **2014**, *84*, 68–86. [CrossRef]

35. Sujatha, E.R.; Suribabu, C.R. Rainfall Analyses of Coonoor Hill Station of Nilgiris District for Landslide Studies. *IOP Conf. Ser. Earth Environ. Sci.* **2017**, *80*, 012066. [CrossRef]

36. Seshagiri, D.N.; Badrinarayanan, S.; Upendran, R.; Lakshmikantham, C.B.; Srinivasan, V. *The Nilgiris Landslide—Miscellaneous Publication No. 57*; Geological Survey of India: Kolkata, India, 1982.

37. Chandrasekaran, S.S.; Sayed Owaise, R.; Ashwin, S.; Jain Rayansh, M.; Prasanth, S.; Venugopalan, R.B. Investigation on infrastructural damages by rainfall-induced landslides during November 2009 in Nilgiris India. *Nat. Hazards* **2013**, *65*, 1535–1557. [CrossRef]

38. Ganapathy, G.P.; Rajawat, A.S. Use of hazard and vulnerability maps for landslide planning scenarios: A case study of the Nilgiris, India. *Nat. Hazards* **2015**, *77*, 305–316. [CrossRef]

39. Pourghasemi, H.R.; Rahmati, O. Prediction of the landslide susceptibility: Which algorithm, which precision? *Catena* **2018**, *162*, 177–192. [CrossRef]

40. Sujatha, E.R.; Rajamanickam, V. Landslide susceptibility mapping of Tevankarai Ar sub-watershed, Kodaikkanal taluk, India, using weighted similar choice fuzzy model. *Nat. Hazards* **2011**, *59*, 401–425. [CrossRef]

41. Basu, T.; Pal, S. Identification of landslide susceptibility zones in Gish River basin, West Bengal, India. *Georisk Assess. Manag. Risk Eng. Syst. Geohazards* **2017**, *12*, 14–28. [CrossRef]

42. Youssef, A.M. Landslide susceptibility delineation in the Ar-Rayth area, Jizan, Kingdom of Saudi Arabia, using analytical hierarchy process, frequency ratio, and logistic regression models. *Environ. Earth Sci.* **2015**, *73*, 8499–8518. [CrossRef]

43. Van Den Eeckhaut, M.; Vanwalleghem, T.; Poesen, J.; Govers, G.; Verstraeten, G.; Vandekerckhove, L. Prediction of landslide susceptibility using rare events logistic regression: A case study in the Flemish Ardennes (Belgium). *Geomorphology* **2006**, *76*, 392–410. [CrossRef]

44. Bai, S.-B.; Wang, J.; Lü, G.-N.; Zhou, P.-G.; Hou, S.-S.; Xu, S.-N. GIS-based logistic regression for landslide susceptibility mapping of the Zhongxian segment in the Three Gorges area, China. *Geomorphology* **2010**, *115*, 23–31. [CrossRef]

45. Domínguez-Cuesta, M.J.; Jiménez-Sánchez, M.; Berrezueta, E. Landslides in the Central Coalfield (Cantabrian Mountains, NW Spain): Geomorphological features, conditioning factors and methodological implications in susceptibility assessment. *Geomorphology* **2007**, *89*, 358–369. [CrossRef]

46. Hosmer, D.W.; Lemeshow, S. *Applied Regression Analysis*; John Wiley and Sons: New York, NY, USA, 1989; ISBN 978-0-470-58247-3.

47. Allison, P.D. *Logistic Regression Using the SAS System: Theory and Application*; Wiley Interscience: New York, NY, USA, 2001; p. 288.

48. Cornell, R.G.; Clark, W.A.V.; Hosking, P.L.; Ebdon, D.; Shaw, G.; Wheeler, D.; Wilson, A.G.; Bennett, R.J. Statistical Methods for Geographers. *J. Am. Stat. Assoc.* **1988**, *83*, 575. [CrossRef]

49. Xiao, T.; Segoni, S.; Chen, L.; Yin, K.; Casagli, N. A step beyond landslide susceptibility maps: A simple method to investigate and explain the different outcomes obtained by different approaches. *Landslides* **2020**, *17*, 627–640. [CrossRef]

50. Bisht, D.S.; Chatterjee, C.; Raghuwanshi, N.S.; Sridhar, V. Spatio-temporal trends of rainfall across Indian river basins. *Theor. Appl. Clim.* **2018**, *132*, 419–436. [CrossRef]

51. Yalcin, A. GIS-based landslide susceptibility mapping using analytical hierarchy process and bivariate statistics in Ar-desen (Turkey): Comparisons of results and confirmations. *Catena* **2008**, *72*, 1–12. [CrossRef]

52. Parker, R.N.; Hales, T.C.; Mudd, S.M.; Grieve, S.W.D.; Constantine, J.A. Colluvium supply in humid regions limits the frequency of storm-triggered landslides. *Sci. Rep.* **2016**, *6*, 34438. [CrossRef]

53. Chen, W.; Han, H.; Huang, B.; Huang, Q.; Fu, X. A data-driven approach for landslide susceptibility mapping: A case study of Shennongjia Forestry District, China. *Geomat. Nat. Hazards Risk* **2018**, *9*, 720–736. [CrossRef]

54. He, Q.; Shahabi, H.; Shirzadi, A.; Li, S.; Chen, W.; Wang, N.; Chai, H.; Bian, H.; Ma, J.; Chen, Y.; et al. Landslide spatial modelling using novel bivariate statistical based Naïve Bayes, RBF Classifier, and RBF Network machine learning algorithms. *Sci. Total Environ.* **2019**, *663*, 1–15. [CrossRef] [PubMed]

55. Sørensen, R.; Zinko, U.; Seibert, J. On the calculation of the topographic wetness index: Evaluation of different methods based on field observations. *Hydrol. Earth Syst. Sci.* **2006**, *10*, 101–112. [CrossRef]

56. Sehgal, V.; Sridhar, V.; Tyagi, A. Stratified drought analysis using a stochastic ensemble of simulated and In-Situ soil moisture observations. *J. Hydrol.* **2017**, *545*, 226–250. [CrossRef]

57. Kang, H.; Sridhar, V. Assessment of Future Drought Conditions in the Chesapeake Bay Watershed. *JAWRA J. Am. Water Resour. Assoc.* **2018**, *54*, 160–183. [CrossRef]

58. Raduła, M.W.; Szymura, T.H.; Szymura, M. Topographic wetness index explains soil moisture better than bioindication with Ellenberg's indicator values. *Ecol. Indic.* **2018**, *85*, 172–179. [CrossRef]

59. Moeslund, J.E.; Arge, L.; Bøcher, P.K.; Dalgaard, T.; Ejrnaes, R.; Odgaard, M.V.; Svenning, J.-C. Topographically controlled soil moisture drives plant diversity patterns within grasslands. *Biodivers. Conserv.* **2013**, *22*, 2151–2166. [CrossRef]

60. Sridhar, V.; Wedin, D.A. Hydrological behaviour of grasslands of the Sandhills of Nebraska: Water and energy-balance assessment from measurements, treatments, and modelling. *Ecohydrology* **2009**, *2*, 195–212. [CrossRef]

61. Sridhar, V. Tracking the Influence of Irrigation on Land Surface Fluxes and Boundary Layer Climatology. *J. Contemp. Water Res. Educ.* **2013**, *152*, 79–93. [CrossRef]

Article

Common Pool Resource Management: Assessing Water Resources Planning for Hydrologically Connected Surface and Groundwater Systems

Francisco Muñoz-Arriola [1,2,3,*], Tarik Abdel-Monem [3] and Alessandro Amaranto [4]

[1] Department of Biological Systems Engineering, University of Nebraska-Lincoln, Lincoln, NE 68583, USA
[2] School of Natural Resources, University of Nebraska-Lincoln, Lincoln, NE 68583, USA
[3] Public Policy Center, University of Nebraska, Lincoln, NE 68588, USA; tabdelmonem2@unl.edu
[4] Department of Electronics, Information, and Bioengineering, Politecnico di Milano, Via Ponzio 34/5, 20133 Milano, Italy; alessandro.amaranto@polimi.it
* Correspondence: fmunoz@unl.edu

Abstract: Common pool resource (CPR) management has the potential to overcome the collective action dilemma, defined as the tendency for individual users to exploit natural resources and contribute to a tragedy of the commons. Design principles associated with effective CPR management help to ensure that arrangements work to the mutual benefit of water users. This study contributes to current research on CPR management by examining the process of implementing integrated management planning through the lens of CPR design principles. Integrated management plans facilitate the management of a complex common pool resource, ground and surface water resources having a hydrological connection. Water governance structures were evaluated through the use of participatory methods and observed records of interannual changes in rainfall, evapotranspiration, and ground water levels across the Northern High Plains. The findings, documented in statutes, field interviews and observed hydrologic variables, point to the potential for addressing large-scale collective action dilemmas, while building on the strengths of local control and participation. The feasibility of a "bottom up" system to foster groundwater resilience was evidenced by reductions in groundwater depths of 2 m in less than a decade.

Keywords: common pool resources; integrated water management; water governance; water resilience

Citation: Muñoz-Arriola, F.; Abdel-Monem, T.; Amaranto, A. Common Pool Resource Management: Assessing Water Resources Planning for Hydrologically Connected Surface and Groundwater Systems. *Hydrology* **2021**, *8*, 51. https://doi.org/10.3390/hydrology8010051

Received: 29 January 2021
Accepted: 17 March 2021
Published: 19 March 2021

Publisher's Note: MDPI stays neutral with regard to jurisdictional claims in published maps and institutional affiliations.

1. Introduction

Common pool resource (CPR) institutions have been the subject of extensive research for several decades. A CPR is defined as a consumable resource where it is difficult to exclude users and where one person's use depletes the pool for others [1]. Much of this commentary has focused on what the literature calls the collective action dilemma, defined as the tendency for actors to overexploit natural resources such as water, fisheries, and grazing forage in the absence of norms and rules developed by users to govern sustainable use [1,2] and her colleagues argued that while regulation by an external authority is necessary in some circumstances, empirical evidence shows that individual users can overcome self-interest and avert a "tragedy of the commons" through collective action [3]. Based on field research in settings such as small irrigation districts, [2] identified a framework of design principles which demonstrated that users in multiple, small-scale environments have successfully created and used CPR arrangements that work to their mutual benefit.

This study adds to that research by examining integrated surface and ground water management plans (IMP) in the Upper Platte River Basin, where Nebraska employs a statutorily enacted framework for state and local government cooperation in the integrated management of surface and ground water—the Ground Water Management and Protection Act (GWMPA) (Neb. Rev. Stat. §46-701 et seq.). In examining this framework, we relied on Ostrom's design principles for common pool resource institutions, because of its "bottom-up" perspective. Nebraska's unique system can represent an alternative to manage common pool of water resources worldwide. Water management in Nebraska includes a statewide agency, Nebraska Department of Natural Resources (NeDNR), with primary statewide authority over surface water, and 23 Natural Resource Districts (NRDs) public entities with taxing authority and primary responsibility for regulatory control over ground water (Figure 1). When state lawmakers established the NRD framework in 1972 there was a consensus that boundaries should follow surface watersheds and that local control was important to the citizens of Nebraska [4]. NeDNR and the NRDs are jointly responsible for facilitating the development of integrated water management plans.

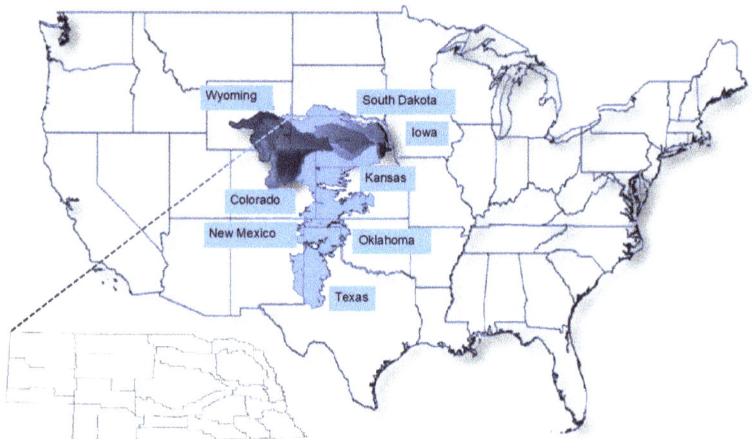

Figure 1. Platte River Basin (PRB) and Nebraska's natural resources. The deep blue tones evidence the main topographic features and PRB's sub basins. The light blue area is the High Plains Aquifer. At the bottom, it can be seen the state of Nebraska and its 23 Natural Resources Districts.

Examining Nebraska's approach is important for several reasons. Globally, water use for irrigation is the largest and key to develop sustainable water planning and management for food and energy production [5]. In the USA irrigation accounts for 62% of water withdrawals, being Nebraska the top state in irrigated acreage. Additionally, along with many other western states of the USA, Nebraska faces challenges in meeting competing demands for water by multiple in-state users and various interstate obligations. These challenges are exacerbated by episodes of severe drought and floods [6,7], the increasing likelihood of long-term changes in climate [8], the inherent risks to water supplies, and volatile crop markets driving resources' tradeoffs [9]. State policymakers are sensitive to the importance of managing water for its agricultural economy; however, its political culture values local control of natural resources, especially ground water, and the state has also experienced a history of conflict over water policy. These discrepancies in policies for CPR design and management water can also be evident in integrated water resources management frameworks and water governance across the globe [10–15]. CPR design principles based on principles of "bottom-up" governance are therefore a valuable lens through which to view the challenge of managing surface and ground water with a hydrological connection that can be exacerbated by a changing climate.

2. Building-Blocks for a Common Pool of Water Resources
Water Resources Management and Policy in Nebraska

The Ground Water Management Protection Act (GWMPA) was enacted in 2004 as a result of a growing recognition that Nebraska needed a strong proactive framework to manage integrated surface and ground water. The statute was passed with widespread support in the unicameral legislature, with 44 lawmakers voting in support of the bill and only two opposed [16]. The GWMPA was the result of a consensus recommendation of a gubernatorial task force representing a diverse range of water users across the state. The task force recognized that a major issue facing the state was harm to surface water appropriations from ground water irrigation [17]. The GWMPA requires development of IMPs in areas designated as fully or over appropriated through a joint process between the NeDNR and the applicable NRD. The NeDNR designated the Upper Platte Basin (UPB) as "over-appropriated," triggering a statutory requirement for NRDs in that basin to develop individual IMPs in their jurisdictions as well as a Basin Wide Plan across the five NRDs in the UPB (Figure 2).

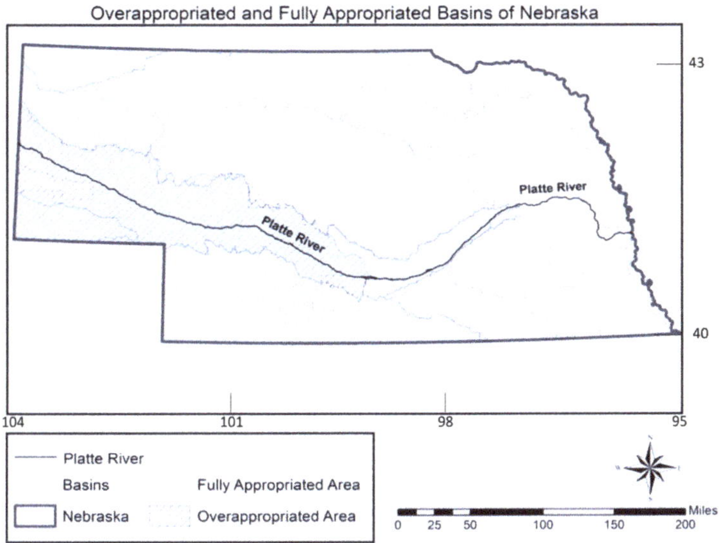

Figure 2. Nebraska's fully and over appropriated surface water boundaries in the Upper Platte Basin.

The IMP process creates a partnership between NRDs and NeDNR to maintain a sustainable balance between water supply and use, and to roll back over-appropriated usage to sustainable levels. Goals and objectives of the IMP are jointly determined by the NRD and NeDNR, including consultation and collaboration with stakeholders (Neb. Rev. Stat. §46-715 et seq.) Only NeDNR and NRDs have decision-making authority; however, the GWMPA requires them to consult and collaborate with public power and irrigation districts and other major stakeholders in development of an IMP. For example, the Central Nebraska Public Power and Irrigation District (CNPPID) uses surface water to generate electricity at a federally licensed hydropower dam in the UPB and delivers surface water for irrigation to over 400,000 hectares along the North Platte and Platte River sub-basins. Its service area cuts across several NRDs in the basin, and ground water use affects the delivery of surface water to irrigators served by CNPPID. The drafters of the GWMPA recognized that depletions to surface water appropriations from ground water use are a major challenge to integrated management, and that offsets to new depletions by NRDs are the primary solution to achieving a balance between water supply and use in areas with a hydrological connection [17].

NRDs are locally elected political entities whose boundaries follow the watersheds of the state's major river systems, and that develop their own priorities and programs for natural resources management based on local preferences and needs [4]. Nebraska's water governance system is unique among the fifty states. Some western states employ a highly centralized orientation, albeit with significant consultation from local entities [18–20]. Other states—like Texas—have historically taken a much more decentralized approach, with local entities driving water use and management [21]. In Nebraska, rules for managing ground water are formally nested within a system of state-wide facilitation by NeDNR and NRDs. The approach provides for local autonomy but situates local decision-making within a vertical structure of joint decision-making with the state's NeDNR that resembles a federal system—defined by [22] as jurisdictions that are nested across levels, e.g., counties within a state.

Following the adoption of an IMP, the Nebraska GWMPA requires the NeDNR to annually evaluate the expected long-term availability of water supplies. The ultimate test of the impact of the GWMPA will be sufficiency of the water supply in the long term for beneficial uses (Neb. Rev. Statute §46-713). This test is especially important in the UPB where a significant area is over-appropriated. While the GWMPA requires the five NRDs in the UPB to develop a basin-wide plan, there are distinct differences within each district. [23] also applied Ostrom's design principles to the Platte River Basin in their study of the perspectives of water users. In the present study, the Ostrom's eight principles listed and defined below represent an opportunity to identify the interdependency between water governance and distributed ground water-surface water interactions.

1. Clearly defined boundaries: This principle states that managers should clearly define the boundary of the CPR and who has rights to withdraw resources. In the absence of clearly defined boundaries there is little incentive to coordinate, because of the risk that "free riders" will benefit from, and eventually destroy, the resource.
2. Appropriation rules relevant to local conditions: Each CPR is unique in its conditions for water use. Incentives to cooperate depend on usage rules that are reasonable and reflect the situation. A "one-size-fits all" approach to managing water supply and use discourages cooperation at the local level.
3. Participation by users: The individuals who directly interact with the CPR and with one another on a local level are in the best position to modify operations over time, and therefore they are motivated to participate in decision-making.
4. Monitoring by users: Despite shared norms valuing compliance with cooperative arrangements, most cases of long-enduring common pool resources involve active investments in monitoring by the resource users themselves. Local users are bound by these arrangements to effectively monitor the common pool resource.
5. Graduated sanctions: Punishment for non-compliance by actors in robust self-governing settings occurs in graduated steps, because local monitors are familiar with the individuals and circumstances of the infraction.
6. Accessible conflict resolution: Conflicts are often resolved informally by local leaders in robust CPR settings.
7. Recognition of local rules: External government officials recognize the authority and legitimacy of rules that are developed by local actors.
8. Nested enterprises: Established rules for management of CPRs at the local level are nested within rules at higher-level governmental jurisdictions, creating a complete system of governance.

3. Methodology

This study is a qualitative analysis of both the text of the GWMPA, as well as local decision-maker and stakeholder perspectives, based on an in-depth case study of one NRD in the Upper Platte River Basin. The location of the NRD and the identities of the interviewees are held confidential. The questionaries (Table 1) were part of the proposal Cross-scale Common Pool Resources Linkages in Integrated Water Management Plan reviewed by the Institutional Review Board under the IRB#745-14-EX. The authors worked independently to code provisions of the GWMPA using Ostrom's design principles as an organizing framework, and ATLAS.ti as their analysis software. In addition, there were field interviews with nine decision-makers and stakeholders to ask how implementation was proceeding in the NRD. Interviewees were selected to represent the NeDNR, the NRD in question, and the stakeholders (i.e., users and societal sector representatives) who participated in the NRD's IMP development and implementation.

Table 1. Sections of the Ground Water Management Protection Act (GWMPA) reflecting common pool resource (CPR) design principles.

1. Let's begin with the development of the most recent IMP. What was your overall role in the process? Have you been involved in the development and implementation of the plan? What about the role, if any, of others in your organization?
2. Did you interact with other organizations and government agencies involved in the IMP process? Who was involved from other organizations and government agencies? How often did you meet during the development of the IMP?
3. The IMP process requires a map that delineates the geographic area. Who was involved and what were the considerations that went into the map? What issues or difficulties came up in delineating the area with a hydrological connection?
4. The IMP process also requires ground water and surface water controls. Who was involved and what were the considerations that went into deciding which controls to include in the plan? What issues or difficulties came up in deciding on those controls?
5. The IMP has been in place now for at least two years. Who has been involved in monitoring water supply and use? How would you say compliance with the plan is going? Do water users think that the plan spreads the costs and benefits fairly?
6. Have conflicts between surface and water users emerged during either the development or implementation of the IMP? Have any issues arisen because of requests for new water uses that may require offsets? How are those issues resolved?
7. Let's wrap up by asking you how effective you think the IMP process has been in managing water with a hydrological connection? What has worked especially well in your view? What improvements in the process are needed in your view?

Interview questions also followed the design principles framework, in order to prompt responses about the overall operational characteristics of the IMP, such as how IMP boundaries were delineated, how monitoring and compliance mechanisms worked "on the ground," and the nature and extent of interactions between decision-makers and stakeholders. The authors also probed for interviewees' perceptions of the IMP process overall, their criticisms, and suggestions for improvements. The interview questions asked about their experiences with the full range of CPR design principles, but responses varied depending on the roles played by the interviewees. Time limitations affected the extent to which the interviews captured experiences incorporating all of the CPR design principles; the field guide allowed the interviewers some discretion on allocating time to the various questions.

Finally, changes in ground water levels and recharge were estimated as evidence of the complexity of a coupled hydrological and human system. Such monitoring integrates the potential ground water recovery in response of addressing large-scale collective action dilemmas, while building on the strengths of local control and participation in a changing environment. Ground water level changes were obtained from [24] following [7] criteria for station selection. Measurements of precipitation were obtained from the Global Land Data Assimilation System (GLDAS [25]). The consumptive use of water was estimated from the MODerate resolution Imaging Spectroradiometer (MODIS [26]). Recharge was determined as the difference between precipitation and evapotranspiration assuming a constrained runoff generation at the location of the well. The recharge is normalized using statistics of dispersion (standard deviation) and central tendency (mean) obtained from data spanning between 2002 and 2010 (due to the availability of MODIS-ET).

4. Results and Discussion

The Legislative Framework is integrated in Table 2 indicating each of Ostrom's principles, and corresponding sections of the GWMPA, including requirements for IMPs. The most salient individual sections of the statute are identified for each principle. It should be noted that the last of Ostrom's principles—recognition of local rules and nested enterprises—

are not coded because their overall design purposes are integrated through the overall approach requiring state (NeDNR) and local (NRD) coordination and cooperation in IMP operationalization. This cross-jurisdictional approach recognizing both local and state responsibility is clearly reflected in the statute's legislative findings, which identify NRDs as the "preferred regulators" for groundwater (§46-702), and state that the objective is that "(a)ll involved natural resource districts, the department, and surface water project sponsors should cooperate and collaborate on the identification and implementation of management solutions" (§46-703 (6)).

Table 2. Sections of the GWMPA reflecting CPR design principles.

Boundaries	§46-715(1)(a), §46-715(1)(b), §46-715(2)(b), §46-718(2)	IMPs are mandated in over appropriated or fully appropriated areas as agreed-upon by NeDNR and impacted NRDs.
Appropriations	§46-715(2)(c), §46-715(2)(d), §46-715(4), §46-715(5)(c), §46-716(1)(b), §46-716(1)(c), §46 716(1)(d), §46-716(2), §46-718(2), §46-739	IMP must include one or more controls on both surface and ground water appropriation or use to sustain a balance between hydrologically connected water uses and supplies so that the economic viability, social and environmental health, safety, and welfare of the basin be achieved and maintained. Further, IMPs in over-appropriated basins must identify the amount of water necessary to offset the impact of stream flow depletions initiated after 1997 [1].
Participation	§46-715(3)(f), §46-715(5)(b), §46-715(5)(d)(ii), §46-717(2), §46-719(3), §46-719(4)	Stakeholder groups must be consulted with during development of the IMP. NeDNR and the NRDs may amend an IMP at annual review, for which there are no provisions for involving stakeholder groups.
Monitoring	§46-715(2)(e), §46-715(3)(d), §46-715(5)(d)(ii), §46-715(5)(d)(iii), §46-715(5)(d)(v), §46-715(6)	NeDNR and NRDs jointly progress toward meeting IMP goals and objectives. NeDNR forecasts the maximum water volume from stream flow for beneficial use in both the short and long term.
Sanctions	§46-707(1–3), §46-708(3), §46-745(1), §46-745(2)(a), §46-746 (1–2)	NRDs may require reporting, metering or decommission of wells, issue cease and desist orders, initiate lawsuits, and take other forms of action.
Conflict Resolution	§46-715(5)(b), §46-718(3), §46-719(2), §46-719(3), §46-719(4)	If the parties reach agreement on the plan, then the NeDNR and the NRD adopt it. NeDNR and NRDs develop and adopt the plan if participating parties disagree. If NeDNR and NRDs are in dispute, the matter may be taken to the Interrelated Water Review Board.

[1] The year 1997 refers to the signing date of the Cooperative Agreement that created the Platte River Recovery Implementation Program (PRRIP) beginning on 1 January 2007. The PRRIP covers the Basin of the Platte River within Colorado, Wyoming and Nebraska. Each state was responsible for developing a plan to mitigate effects of surface and ground water depletions initiated after 1997.

Six Ostrom design principles include references and field interviews relevant to sections of the GWMPA. The last two principles, Recognition of Local Rules and Nested Enterprises did not receive any mention from the interviewees.

4.1. Clearly Defined Boundaries

The GWMPA requires IMPs to include designation of the geographic area and inclusion of a map delineating its boundaries (Neb. Rev. Stat. §46715(1–2), §46-718(2)). The IMP includes a map of the geographic area covered and delineated over-appropriated and fully appropriated portions identified through modeling efforts, each of which is subject to different requirements. The boundary and associated regulations limit water use to those who have agreed to self-regulate ground water irrigation.

Almost all interviewees indicated that establishing a geographic basis for regulatory action was a key step to the IMP. Throughout plan development, participating decision makers and stakeholders were involved in modeling efforts to measure and identify areas under their jurisdiction that were hydrologically connected, and the extent to which those areas were fully or over appropriated. These modeling efforts are ongoing and have resulted in analysis of hydrological and geological conditions in the entire basin that are incorporating groundwater flow, soil-water balance, and surface water dynamics. These modeling efforts have been supported by multiple sponsors, including NRDs, state agencies, municipalities and power companies in the Platte River Basin, and have resulted in identified geographic boundaries of the IMP and extensive data on its hydrological characteristics that have driven decision making.

Mandatory IMP Interviewee #6. "The COHYST [Cooperative Hydrology Study] group, which stands for the conjunctive cooperative hydrology study group, which involved game and parks, DNR, all the NRDs, the two major irrigation districts, CNPPD and NPPD, kind of make up the COHYST study stuff. The Platte River program headwaters group is somewhat involved as well. We were developing the tools and DNR basically requested that we do the study, the COHYST group. So, we took the groundwater models to COHYST, and they ran all the models to generate the percent depletion by use".

Mandatory IMP Interviewee #4. "(T)he NRD didn't really have much control over the surface water. But then once they established the relationship in the COHYST between how groundwater pumping depletes the surface water. They became much more involved".

Mandatory IMP Interviewee #5. "Every 40-acre tract out here has a designated value that they have worked out through this COHYST model that shows the returns and the length of time that . . . obviously closer to the river water would get back there faster obviously than it would next to the canal...".

4.2. Appropriation Rules Relevant to Local Conditions

The GWMPA mandates that IMPs include one or more controls on surface and ground water appropriation or use to sustain a balance between hydrologically connected water uses and supplies, and to maintain the economic viability, social and environmental health, safety, and welfare of the basin. Further, IMPs in over-appropriated basins must identify the amount of water necessary to offset the impact of stream flow depletions initiated after 1997 (see §46-715(1–6)). The year 1997 refers to the signing date of the Cooperative Agreement creating the Platte River Recovery Implementation Program (PRRIP) beginning on 1 January 2007. The PRRIP covers the Platte Basin within Colorado, Wyoming and Nebraska. Each state is responsible for developing a plan to mitigate effects of surface and ground water depletions initiated after 1997. Thus, the PRRIP and IMPs in the Upper Platte Basin are interconnected documents.

The NRD's fully appropriated portion is under a moratorium on new well permits and expanded irrigation acres as per statutory requirements (§46-714(1–2)). The NRD is responsible for offsetting new or expanded ground water irrigation, as well as increases in consumptive municipal use from population growth and commercial/industrial consumptive use, up to limits of 25 million gallons per year. The NRD is also responsible for finding offsets to new or increased non-municipal industrial use up to 25 million gallons per year. The NeDNR has also placed a moratorium on new surface water appropriations. The over-appropriated portion is under the same moratorium; however, the NRD must also offset "new" depletions dating back to 1997. Appropriation rules allow for continued development through the use of offsets to new or expanded uses. One example of strategies to offset new depletions in the over-appropriated area is an agreement between the NRD and local irrigation districts. Surface water irrigators may switch to their (existing) wells, and the NRD applies to the NeDNR on their behalf for the right to divert excess river flows into canals for ground water recharge and retiming base flows to the river. NeDNR calculates the addition to the base flow and counts it as an offset to new depletions.

Participants had mixed but generally positive perceptions about appropriation rules and their relevance to local conditions. The IMP mandate to decrease over-appropriation drives restrictions and controls in the area, but also allows for collaborative mechanisms among decision makers and stakeholders to establish use arrangements that comply with IMP goals. This has led to the creation of some cooperative projects between the NRD and stakeholders that were perceived as win-win efforts to advance both the interests of water users in the basin, as well as overall IMP goals.

> **Mandatory IMP Interviewee #6.** "Basically, we have an agreement with each of the irrigation districts . . . . We have a lease agreement to put together the water rights, transfer the water rights. The irrigation district signs them, and we send them in. They total up the bills (for canal repairs) and we go half and half. They pay half and we pay half".

These agreements emerged based on trust after years of discussions: surveys of the land area; and calculations based on a hydrologic model of the interactive effects of surface and ground water in that area. Overcoming distrust between surface and ground water users took time, as did negotiations based on an equitable sharing of the investment costs associated with maintaining the canals for recharge purposes, and future benefits of the revenues from leasing unused surface irrigation water for other uses. While the agreements between the NRD and local irrigation districts require NeDNR approval to transfer surface water rights, and involve a lengthy approval process, the IMP facilitates implementation because it allows the DNR to treat transfers as a beneficial use. NeDNR's role is therefore one of facilitating the strategies developed at the local level by the NRD and irrigation districts. Thus, while the threat of regulatory controls on ground water irrigation may have been a prime motivator in bringing people together in the IMP process, local cooperation resulted in a proactive approach to controls on appropriations that were unique to local conditions and which mitigated conflict with some, though not all, users.

> **Mandatory IMP Interviewee #1.** "I think the nice thing about what they are doing is that they have become partners with the surface water folks, who at the beginning of this process, when we started IMP, they were still not partners. They were still thinking everyone was out to get them".

4.3. Participation by Users

The GWMPA mandates that stakeholder groups be consulted during development of the IMP (see, e.g., §46-715(3)(f), (5)(b) and §46-717(2)). The IMP reflects statutorily mandated decision-making by NeDNR and the NRD; requires meetings with stakeholders; and outlines the process for NeDNR and NRD to annually review the progress of the IMP and jointly agree upon any amendments. Although the NeDNR and the NRDs may amend an IMP at annual review, there are no explicit provisions for involving stakeholder groups in the amendment process (see §46-715(5)(d)(ii)). The goals and objectives for this IMP, as well as the major strategies for addressing depletions to the Platte River, evolved from ideas discussed among NRD staff, irrigation district board members, and municipal officials prior to the start of the planning process. During the planning process, the NRD held public meetings for stakeholders and members of the public. These meetings fulfilled the consultation requirements in the GWMPA.

Decision maker and stakeholder perspectives on participation varied widely. One interviewee reported that his engagement with the NRD and other stakeholders predated the IMP, and that a great deal of mutual exchange and education among surface and ground water users had already occurred. On the other hand, another interviewee reported that those who proposed increasing minimum accretions to stream flow were "laughed off the floor." Still another questioned whether the NeDNR and the NRD actually consulted and collaborated with stakeholders to a meaningful degree, as opposed to simply gathering input and then writing the plan on their own.

The most frequent comment was that IMP stakeholder meetings were infrequent compared to the other IMPs in the western part of the state, and that consultation was perfunctory. Several reported attending and listening, without offering any input. Some stakeholders had specific ideas to propose but had the sense that the NRD was controlling the agenda. These perspectives seem to reflect characteristics of the GWMPA that restrict decision making to select entities, or do not adequately define what appropriate collaboration is among stakeholders in the IMP:

Mandatory IMP Interviewee #8. "Collaboration in this sense was basically, "We will meet with you and take your input." We were told many times during the (name of NRD redacted) IMP process that the NRD board would make the decisions. We sent in comments. My recollection was that the NRD drafted the IMP and presented it to the stakeholders. In many cases the department responded the same as the stakeholders did. Everyone was feeling their way. There was no set process".

Mandatory IMP Interviewee #7. "The statutes say that they are to consult and collaborate with us. Those are two different words. They have two different meanings. And very often what we find is, they come and consult, and they say, "We are consulting and collaborating with you now." And we would often ask, "Where is the collaboration? Where is the part where you are asking us to be involved with and participate in finding solutions to this? Because it seems like really what you are doing is consulting only".

Other interviewees who participated in various IMP development meetings believed that the highly technical nature of discussions impeded participation. As one stakeholder commented:

Mandatory IMP Interviewee #9. "And I do know that the water professionals and irrigators came. I think the process would have benefited from a much more educational bent. Because not everyone was on the same level of education on how water works and how this whole thing gets put together. There was very little if I remember it right, very little effort to bring people up to speed with all the stakeholders in fact. And I think I came at it with a fairly decent knowledge, but there was a lot of jargon and acronyms and things like that that probably limited how well people could participate".

4.4. Monitoring by Users

The GWMPA mandates that NeDNR and NRDs jointly progress toward meeting IMP goals and objectives (§46-715(3)). NeDNR forecasts the maximum water volume from stream flow for beneficial use in both the short and long term. In the IMP, the NRD tracks yearly certification of ground water use, water well construction, and consumptive uses by municipal and non-municipal industrial water systems within its jurisdiction. It also tracks the number and location of retired irrigated acres and offsets for new uses, including depletions dating to 1997 in the over-appropriated part of the basin within its jurisdiction (§46-715(2)(e)). NeDNR tracks changes in permits for surface water (§46-716). The NRD board is elected by local ground water users.

Mandatory IMP Interviewee #1. "So, there is a reporting and monitoring section in the plan. So basically, the NRD and my department come together and say, "OK here are all the activities that have taken place in the last year" just in a checklist fashion, have we caused more depletions? Are there more accretions? Where are we in the permitting process? And that is telling us on an annual basis are we getting where we want to be".

The NeDNR relies on NRD records for tracking certified acres, including transfers from a water rights holder associated with retired acres and/or transferred ground water use from one tract of land to another. The NRD also uses aerial photography to insure that irrigators are staying within their certified number of acres. At the time of this study, decision-makers were finalizing plans to run an updated ground water (hydrologic) model in order to verify the number of acre feet per year that will be needed to offset depletions dating back to 1997. However, some interviewees expressed skepticism of benchmarks and incremental approaches used for monitoring and assessment of accretions or depletions under the IMP, believing that the GWMPA should require NRDs to offset depletions dating back prior to 1997, because there were prior (surface water) appropriations predating the introduction of widespread use of central pivot irrigation that were impacted by those ground water wells. The GWMPA, however, requires only voluntary efforts to offset depletions prior to 1997 as part of an incremental approach. (§46-715(5)(d)(i)).

> **Mandatory IMP Interviewee #2.** "What is fully appropriated? Is it where your development is affecting streamflow? These are measures of degree. In our mind, what is that difference? The fact that it wasn't (fully defined) when all these plans were done was disappointing, and of major concern to us. The difference between fully and over. We still don't agree with the way the department is proposing to do that. Basically, we are not really allowed to participate any more".

Those concerns were connected to perceptions that the structure of the IMP did not allow for full participation among all stakeholders, especially surface water providers, and that there were few avenues to air such grievances. Thus, perceptions of the efficacy of monitoring activities varied depending on the interests of those involved and whether those interests were represented in the development and implementation of the IMP.

4.5. Graduated Sanctions

The GWMPA identifies a variety of sanctions on individual water users for non-compliance with its mandates (see, e.g., Neb. Rev. Statute §46-746). The NRD was in the early stages of implementing its IMP during this study, including monitoring for compliance with ground water controls. Nevertheless, the NRD adopts the IMP, including the moratorium on ground water use, in consultation with local users and stakeholders. Thus, local monitors are familiar with the individuals and circumstances of the infraction In fact, several interviewees pointed to the IMP and the importance of enforcement as a hedge against further restrictions on ground water use.

Interviewees had little more to say about sanctions, partially because the ten-year timeframe to review progress towards plan goals had yet to occur. As the first increment of the current IMP was due at the end in 2019, it is possible that developments in regard to non-compliance and sanctions may emerge after the formal technical analysis of plan progress is completed. As discussed previously, there were concerns that the statutory framework of the GWMPA failed to address the effects of ground water depletions on prior (surface water) appropriations that predated the 1997 benchmark in the law.

> **Mandatory IMP Interviewee #1.** "The way the law is set up, this first increment, which is 10 years, is that we will get back to 97. So the triggers you see are built to get back to 1997. But there are still shortages in the system just because we are still . . . well you have drought anyway, and you have wells that have existed before 1997 that are impacting the system as well, and those are not at this point being addressed. The interests in the part of the surface water parties is that those should all be addressed right away. But the plan isn't set up that way, its set up to do it in incremental fashion, so there is conflict and tension going on there. So it's not that there isn't a shortage, it's that we don't have to address all the shortage right now".

4.6. Accessible Conflict Resolution

The GWMPA provides for conflict resolution as a two-step process. If the parties reach agreement on the plan, then the NeDNR and the NRD adopt it (§46-715 (1)(a)). If NeDNR and NRDs are in dispute, the matter may be taken to the Interrelated Water Review Board (§46-719(2)). Interviewees expressed a range of opinions about the accessibility of conflict resolution mechanisms through the current plan structure. Not surprisingly, perceptions of conflict resolution mechanisms varied depending on an interviewee's overall perceptions of how well the plan advanced either individual interests, or the overall goals of the IMP to reduce over-appropriation. For example, there was general agreement that the conjunctive management approach of the IMP was beneficial, and that individual actions taken under the plan were successful.

> **Mandatory IMP Interviewee #1.** "I think it's going very well. It's nice to see everyone being very conscientious about what the plan says, and how to be in compliance with that plan. (Name of NRD redacted) has been making great strides to get all those conjunctive management pieces in place. When they started the process, they purchased a lot of easements and buying out groundwater and surface water rights and retiring them. So, they have been a leader in Platte NRDs in implementing various types of practices in getting us to where we need to be".

> **Mandatory IMP Interviewee #7.** "Often, we are in disagreement in terms of whether they have actually set something that will actually meet their objective. But it sounds like their objective, their intent, in areas that are not yet fully appropriated … try and identify where they will occur, try and head them off. That's good. But that's not really our area. Our concern is they are not really directly trying to resolve the conflict that was already created. We think that there is an obligation to try to do that".

However, there were distinct criticisms about the scope of the plan's mandate as well as a perceived absence of a mechanism to resolve issues before the ten-year increment ends. Interviewees who represented surface water users indicated that although current conjunctive management practices under the IMP were generally positive, the GWMPA does not adequately address perceived inequities between ground and surface water purveyors that existed prior to the enactment of the 2004 GWMPA amendments, because the law calls only for voluntary efforts, subject to the availability of funds, to offset depletions to streamflow dating back prior to 1997 (§46-715(5)(d)(i). Another criticism was that the plan's ten-year incremental structure does not provide an adequate means to resolve conflict that would happen in the interim period before the first ten-year phase would end.

> **Mandatory IMP Interviewee #2.** "What is the best way to achieve results then? Is it to get things out there, or just wait 10 years until they get new plans done and then hope we potentially see some change? Then you see lower lake levels. Do you just have to say, 'I will keep quiet and wait my ten years because that is the only option that is out there?' We are disappointed in those options".

Both stakeholders and decision makers voiced concern that conflict management under the current GWMPA statute—and IMPs derived from its requirements—were not sufficient. For example, the Interrelated Water Review Board has never been convened, nor does it review disputes that a stakeholder may have in regard to the Plan. On the contrary, the statute and Plan allow for the NeDNR and NRD to move forward with the plan regardless of whether all stakeholders impacted by it are supportive:

> **Mandatory IMP Interviewee #1.** "Well there is the Interrelated Water Review Board. Yeah, that is not if the stakeholders can't agree, but if the department and the NRD can't agree what the plan should be. As we go through the stakeholder process even with the consultation and collaboration the statutes clearly say that DNR and NRD can go back and say, 'OK you guys couldn't agree, so we are going to see if we are going to agree,' and so that is where we ended up. We could agree, so we could move forward with that plan even though not all the stakeholders were on board with it".

4.7. Recognition of Local Rules

The GWMPA establishes a joint decision-making process between state and local officials, and the NeDNR facilitates the planning process and approves local IMP plans.

4.8. Nested Enterprises

Rules for developing and implementing IMPs are nested within provisions of the GWMPA. [2] saw the volatility of climate as a common characteristic across multiple CPRs. However, [27] reflect on the absence of environmental accounting within Ostrom's CPR design principles. In our case, the changes in ground water levels and recharge in three NRDs encompass the complexities of interdependent hydroclimate, water management, and soil physical properties [6,7,28] used crops' evapotranspirative demands, rainfall, and streamflow as variables that integrate irrigation management, rainfall variability, and soil infiltration capacities. These observed data were inputs to a data-driven model developed that successfully reproduces the changes in groundwater well-levels. Thus, the difference among Figure 3A–C can represent the differences in water governance across UPB's NRDs. For example, consistent depletion of groundwater levels in NRD-1, -2, and -3 (2, 4 and 5 m, respectively) started in 2002 indicate how consistent streamflow withdrawals affect aquifer recharge. On the other hand, inflections in the negative trends that occurred in 2005 and 2007 in the NRD-1 and the NRD-2, respectively, illustrate how changes in diverted excess of river flows increase aquifer recharge. In comparison, the rise in ground water well levels in the NRD-1 (2 m) responds to the agreement among the NRDs, irrigation districts, and NeDNR. In locations like the NRD-2 and NRD-3, the less conspicuous rise (<1 m) may be attributed to intraseasonal increments in rainfall. Depletion of ground water depth after 2012 can be attributed to droughts. The 2012 flash drought reported by [29] was evident in NRD-1, -2, and -3.

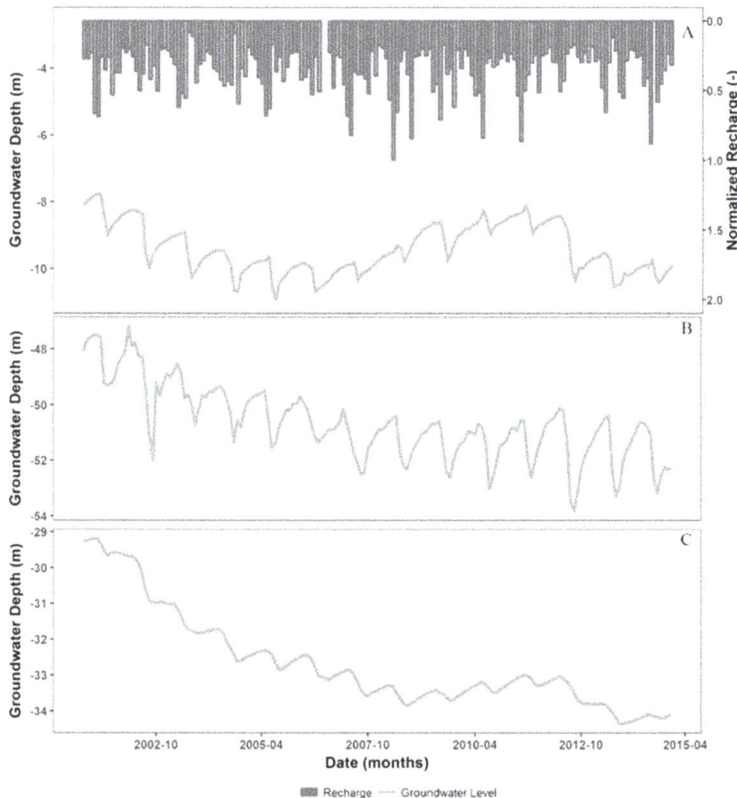

Figure 3. Interannual changes of integrated (surface and sub-surface) hydrological responses in three Natural Resources Districts. (**A**) illustrates NRD 1; (**B**) illustrate temporal changes in groundwater well levels in NRD 2; and (**C**) illustrate temporal changes in groundwater well levels in NRD 3. The location of the NRD is not disclosed due to security constrains.

The GWMPA, its IMP process, and data on variations in hydrological connectivity across the UPB provided the background for this study. Nebraska's decentralized framework in which each NRD develops an IMP based on its unique conditions, is consistent with Ostrom's "bottom-up" approach to the management of common pool resource institutions, although this study makes no claim that following Ostrom's design principles is a predictor of successful outcomes. As [30] point out, they are relevant for simple common pool resources, but additional research is needed in more complex social-ecological systems. Conditions in the UPB are more complex than the small-scale irrigation districts that were the focus of Ostrom's original work. Nevertheless, these design principles can work as a heuristic device—helping to focus on key elements in common pool resource management like surface and ground water having a hydrological connection.

Common pool resource principles help to explain why it is in the collective interest of actors to decrease the likelihood of exploiting and exhausting resources, thereby obtaining long-term benefits for all [2]. The emphasis in Ostrom's original work was on processes of local self-governance in small-scale situations. As [31] point out, however, Ostrom recognized that when local common pool resources are part of larger systems, the organizations that govern them are more successful when linked in a nested fashion, that is, when actors at different scales share rules or strategies through formal means. Following [31], we argue that Ostrom's design principles can be used to examine the IMP process in Nebraska.

5. Conclusions

Water governance structures like those in Nebraska indicate that, even in a large-scale, complex common pool resource such as hydrologically connected surface and ground water, a "bottom up" system is feasible. Evidence from the field interviews suggests that local ground water users have accepted the moratorium imposed by the NRD, because it reduces uncertainties about the future, including the possibility of more devastating restrictions if previous patterns of consumptive use had continued unabated as those observed in Figure 3B,C. The exception to these findings, however, is that the GWMPA framework limits participation by surface water providers in the IMP process. This limitation becomes relevant in cases where surface water contributes to the recovery of ground water levels (Figure 3A).

Interviews revealed concern among some stakeholders with this arrangement, because they perceived it as resulting in a less-than-equitable process and outcome in terms of water appropriation. This tension between surface and ground water providers and users stems from the bifurcated system of water laws in Nebraska. Laws governing surface water use according to the doctrine of prior appropriation with its associated principle of seniority evolved independently of doctrines of reasonable use and correlative rights governing access to ground water in the state. An over-appropriated designation requires offsets to depletions of surface water flows from ground water use dating back only to 1997, even though there are older surface water appropriations impacted by those earlier depletions. Conflicts stemming from this bifurcated system, especially in over-appropriated areas, are beyond the scope of the GWMPA and the IMP process [32].

The tension that results from this bifurcated system has complicated the implementation of the IMP process. In fact, the significance of the hydrological connection between surface and ground water wasn't fully appreciated by decision-makers during passage of the GWMPA and the framework splitting jurisdiction between NeDNR and the NRDs [4]. As a result, there is a gap in the alignment of the legal framework with CPR design principles, in particular the principle that individuals who directly interact with the common pool resource and with one another are in the best position to modify operations over time, and that they are therefore motivated to participate in decision-making [2]. As the experiences of some interviewees have suggested, their participation in the development and implementation of the IMP is limited in scope, especially decisions about the extent of controls on ground water use that impact surface water supplies. These limitations, in turn, affect perceptions of an inequitable system for imposing sanctions and resolving conflicts. Ultimately, these limitations could impact the sustainable management of hydrologically connected surface and ground water supplies in the UPB.

The effectiveness of integrated management strategies may depend on the extent of connectivity and the inherent complexity of the drivers of ground water-level changes Nonetheless such analysis is beyond the scope of this study it is evident that approaches such as those proposed by [6,7,28] could be predict changes in ground water levels in response to water policies and climate variability, and consequently the CPR design principles.

Ostrom's principles emerged from years of empirical research demonstrating their effectiveness as an alternative to hierarchical government or private market allocation of common pool resources. This study used Ostrom's framework to study the implementation of the IMP process in the field, and it identified major areas of alignment suggesting that there is potential for Nebraska's decentralized approach to achieve sustainable levels of surface and ground water. Nevertheless, the future effects of severe droughts and long-term climate change are largely unknown at this time, and more comprehensive reforms may be necessary to involve surface water providers and users in the integrated management of hydrologically connected common pool water resources. These efforts should lead to the design and creation of a more climate-resilient water infrastructure based on a better understanding of the socio-ecological functionalities of the surface water and ground water resources.

Author Contributions: Conceptualization, F.M.-A. and T.A.-M.; methodology, F.M.-A., T.A.-M. and A.A.; software, F.M.-A., T.A.-M. and A.A.; validation, F.M.-A., T.A.-M. and A.A.; formal analysis, F.M.-A., T.A.-M. and A.A.; investigation, F.M.-A., T.A.-M. and A.A.; resources, F.M.-A., T.A.-M. and A.A.; data curation, F.M.-A., T.A.-M. and A.A.; writing—F.M.-A.; writing—review and editing, F.M.-A. and T.A.-M.; visualization, A.A. and F.M.-A.; supervision, F.M.-A. and T.A.-M.; project administration, F.M.-A. and T.A.-M.; funding acquisition, F.M.-A. and T.A.-M. All authors have read and agreed to the published version of the manuscript.

Funding: The authors wish to thank the Nebraska Department of Natural Resources for funding this research. Some research ideas and components were also developed within the framework of the USDA National Institute of Food and Agriculture, Hatch project NEB-21-166 Accession No.1009760, the Robert B. Daugherty Water for Food Global Institute at the University of Nebraska and the University of Nebraska-Lincoln Institute of Agriculture and Natural Resources-Agricultural Research Division.

Institutional Review Board Statement: The study was conducted according to the guidelines of the Declaration of Helsinki, and approved by the Institutional Review Board (or Ethics Committee) of University of Nebraska-Omaha Institutional Review Board (IRB#745-14-EX).

Informed Consent Statement: Informed consent was obtained from all subjects involved in the study.

Data Availability Statement: Restrictions apply to the availability of interview data. As reflected in our Institutional Review Board protocol, the interview data was obtained from individuals on a confidential basis, and further disclosure of this data could directly or indirectly reveal their identities.

Acknowledgments: The authors also wish to thank Jesse Bradley and Jennifer Schellpeper from NeDNR for fact-checking the original report.

Conflicts of Interest: The authors declare no conflict of interest.

References

1. Leuenberger, D.; Reed, C. Social capital, collective action and collaboration. In *Advancing Collaboration Theory: Models, Typologies and Evidence*; Morris, J., Miller-Stevens, K., Eds.; Routledge Press: New York, NY, USA, 2015; p. 238.
2. Ostrom, E. *Governing the Commons: The Evolution of Institutions for Collective Action*; Cambridge University Press: Cambridge, UK, 1990.
3. Hardin, G. The Tragedy of the Commons. *Science* **1968**, *162*, 1243–1248. [CrossRef]
4. Bleed, A.; Babbitt, C. *Nebraska's Natural Resources Districts: An Assessment of a Large-Scale Locally Controlled Water Governance Framework*; Robert, B., Ed.; Daugherty Water for Food Institute: Lincoln, NE, USA, 2015.
5. Scanlon, B.R.; Ruddell, B.L.; Reed, P.M.; Hook, R.I.; Zheng, C.; Tidwell, V.C.; Siebert, S. The food-energy-water nexus: Transforming science for society. *Water Resour. Res.* **2017**, *53*, 3550–3556. [CrossRef]

6.	Amaranto, A.; Munoz-Arriola, F.; Solomatine, D.P.; Corzo, G. A Spatially Enhanced Data-Driven Multimodel to Improve Semiseasonal Groundwater Forecasts in the High Plains Aquifer, USA. *Water Resour. Res.* **2019**, *55*, 5941–5961. [CrossRef]
7.	Amaranto, A.; Pianosi, F.; Solomatine, D.; Corzo, G.; Muñoz-Arriola, F. Sensitivity analysis of data-driven groundwater forecasts to hydroclimatic controls in irrigated croplands. *J. Hydrol.* **2020**, *587*, 124957. [CrossRef]
8.	Ou, G.; Munoz-Arriola, F.; Uden, D.R.; Martin, D.; Allen, C.R.; Shank, N. Climate change implications for irrigation and groundwater in the Republican River Basin, USA. *Clim. Chang.* **2018**, *151*, 303–316. [CrossRef]
9.	Troy, T.J.; Pavaozuckerman, M.A.; Evans, T.P. Debates-Perspectives on socio-hydrology: Socio-hydrologic modeling: Tradeoffs, hypothesis testing, and validation. *Water Resour. Res.* **2015**, *51*, 4806–4814. [CrossRef]
10.	Cselenyi, M.N. Policy and Management of Water as a Common-Pool Resource in Spain: The Case of the Aquifer' Carbonatado de la Loma de Úbeda'. Master's Thesis, Stockholm University, Stockholm, Sweden, 2014; 70p.
11.	Skurray, J.H. The scope for collective action in a large groundwater basin: An institutional analysis of aquifer governance in Western Australia. *Ecol. Econ.* **2015**, *114*, 128–140. [CrossRef]
12.	Dickin, S.; Bisung, E.; Savadogo, K. Sanitation and the commons: The role of collective action in sanitation use. *Geoforum* **2017**, *86*, 118–126. [CrossRef]
13.	Yu, H.H.; Edmunds, M.; Lora-Wainwright, A.; Thomas, D. Governance of the irrigation commons under integrated water resources management—A comparative study in contemporary rural China. *Environ. Sci. Policy* **2016**, *55*, 65–74. [CrossRef]
14.	Bernacchi, L.A.; Fernandez-Bou, A.S.; Viers, J.H.; Valero-Fandino, J.; Medellín-Azuara, J. A glass half empty: Limited voices, limited groundwater security for California. *Sci. Total. Environ.* **2020**, *738*, 139529. [CrossRef]
15.	Whaley, L.; Cleaver, F.; Mwathunga, E. Flesh and bones: Working with the grain to improve community management of water. *World Dev.* **2021**, *138*, 105286. [CrossRef]
16.	Nebraska Legislature. *Second Regular Session Journal for the 98th Legislature*; Nebraska Legislature: Lincoln, NE, USA, 2004.
17.	Nebraska Water Policy Task Force. *Report of the Nebraska Water Policy Task Force to the 2003 Nebraska Legislature*; Nebraska Water Policy Task Force: Lincoln, NE, USA, 2003.
18.	Bell, C.; Taylor, J. *Water Laws and Policies for a Sustainable Future: A Western States' Perspective*; Western Governors' Association: Murray, UT, USA, 2008.
19.	Johnson, N.K.; DuMars, C.T. A survey of the evolution of western water law in response to changing economic and public interest demands. *Nat. Resour. J.* **1989**, *29*, 347–387.
20.	Sabatier, P.; Weible, C.; Ficker, J. Eras of water management in the United States: Implications for collaborative water-shed approaches. In *Swimming Up-Stream: Collaborative Approaches to Watershed Management*; Sabatier, P., Focht, W., Lubell, M., Trachtenberg, Z., Vedlitz, A., Matlock, M., Eds.; MIT Press: Cambridge, MA, USA, 2005; pp. 23–52.
21.	Terrell, B.L.; Johnson, P.N.; Segarra, E. Ogallala aquifer depletion: Economic impact on the Texas high plains. *Hydrol. Res.* **2002**, *4*, 33–46. [CrossRef]
22.	Koontz, T.; Gupta, D.; Mudliar, P.; Ranjan, P. Adaptive institutions in socio-ecological systems governance: A synthesis framework. *Environ. Sci. Policy* **2015**, *53*, 139–151. [CrossRef]
23.	Babbitt, C.H.; Burbach, M.; Pennisi, L. A mixed-methods approach to assessing success in transitioning water manage-ment institutions: A case study of the Platte River Basin, Nebraska. *Ecol. Soc.* **2015**, *20*, 54. [CrossRef]
24.	USGS. National Water Information System: USGS Groundwater Data for the Nation. 2015. Available online: https://waterdata.usgs.gov/nwis/gw (accessed on 4 December 2020).
25.	Rodell, M.; Houser, P.R.; Jambor, U.E.A.; Gottschalck, J.; Mitchell, K.; Meng, C.-J.; Arsenault, K.; Cosgrove, B.; Radakovich, J.; Bosilovich, M. The global land data assimilation system. *Bull. Am. Meteorol. Soc.* **2004**, *85*, 381–394. [CrossRef]
26.	Running, S.; Mu, Q.; Zhao, M. *MOD16A2 MODIS/Terra Net Evapotranspiration 8-Day L4 Global 500m SIN Grid V006*; NASA EOSDIS Land Processes DAAC: Sioux Falls, SD, USA, 2017.
27.	Steins, N.A.; Röling, N.; Edwards, V.M. Re-'Designing' the Principles: An Interactive Perspective to CPR Theory. Presented at Constituting the Commons: Crafting Sustainable Commons in the New Millenium, the Eighth Conference of the International Association for the Study of Common Property, Bloomington, IN, USA, 31 May–4 June 2000; pp. 1–4.
28.	Amaranto, A.; Munoz-Arriola, F.; Corzo, G.; Solomatine, D.P.; Meyer, G. Semi-Seasonal Groundwater Forecast Using Multiple Data-Driven Models in an Irrigated Cropland. *J. Hydroinformatics* **2018**, *20*, 1227–1246. [CrossRef]
29.	Hoerling, M.P.; Eischeid, J.K.; Kumar, A.J.; Leung, R.; Mariotti, A.W.A.; Mo, K.; Schubert, S.D.; Seager, R. Causes and Predictability of the 2012 Great Plains Drought. *Bull. Am. Meteorol. Soc.* **2014**, *95*, 269–282. [CrossRef]
30.	Anderies, J.M.; Janssen, M.A.; Ostrom, E. A Framework to Analyze the Robustness of Social-ecological Systems from an Institutional Perspective. *Ecol. Soc.* **2004**, *9*, 18. [CrossRef]
31.	Heikkila, T.; Schlager, E.; Davis, M. The role of cross-scale linkages in common pool resource management: Assessing inter-state river compacts. *Policy Stud. J.* **2011**, *39*, 121–145. [CrossRef]
32.	Reed, C.; Abdel-Monem, T. Integrated Surface and Groundwater Management in Nebraska: Reconciling Systems of Prior Appropriation and Correlative Rights. *Am. Bar Assoc. Sect. Environ. Energy Resour. Water Resour. Comm. Newsl.* **2015**, *18*, 10–12.

Article

Assessment of the Readiness and Resilience of Czech Society against Water-Related Crises

Štěpán Kavan [1,*], Šárka Kročová [2] and Jiří Pokorný [2]

[1] Faculty of Health and Social Studies, University of South Bohemia in Ceske Budejovice, 370 11 Ceske Budejovice, Czech Republic
[2] Faculty of Safety Engineering, VSB—Technical University of Ostrava, 700 30 Ostrava, Czech Republic; sarka.krocova@vsb.cz (S.K.); jiri.pokorny@vsb.cz (J.P.)
* Correspondence: stepan.kavan@email.cz

Abstract: This assessment of societal readiness and resilience to water-related situations in the Czech Republic focuses on an interdisciplinary approach in the Czech Republic for solving this problem. The goal of the article is to evaluate and characterize the preparedness for handling water-related crises. The analysis is carried out via a SWOT analysis, which is a universal analytical method used to understand and interpret strengths and weaknesses and to identify opportunities and threats. For the calculation of the weight factor of the SWOT analysis, an assessment was determined based on the multicriteria analysis. The pair comparison method was used to determine the relative importance of the parameters of the strengths and weaknesses, opportunities and threats. The Fuller Triangle method was chosen for the system used to make the comparisons of the individual criteria. The uniqueness of the study consists of the issue of water management, which is thus reflected from a non-traditional perspective, being a contemporary model—the paradigm of the view on the preparedness of the planning documentation as one of the characteristics of societal resilience for water-related crises. The result of the research is the fact that a positive approach prevails in the researched area from the perspective of preparedness for water-related crises. For the creation of the conditions, the factors arising from the internal environment currently prevail slightly over those arising from the external environment.

Keywords: society; risk analysis; water-related crises; resilience; security; floods; drinking water; crisis planning

Citation: Kavan, Š.; Kročová, Š.; Pokorný, J. Assessment of the Readiness and Resilience of Czech Society against Water-Related Crises. *Hydrology* **2021**, *8*, 14. https://doi.org/10.3390/hydrology8010014

Academic Editors: Tamim Younos, Tammy Parece, Juneseok Lee, Jason Giovannettone and Alaina J. Armel

Received: 9 December 2020
Accepted: 21 January 2021
Published: 22 January 2021

Publisher's Note: MDPI stays neutral with regard to jurisdictional claims in published maps and institutional affiliations.

1. Introduction

Water has many facets. In general, water is often characterized as one of the basic components required for life. A second viewpoint is that water is a natural element which threatens human society via floods. Human activities have significantly interfered with the environment and in doing so have altered the behavior of surface as well as underground water in the landscape. Water is thus becoming an increasingly precious liquid and simultaneously a feared element. In order to protect the values it has built up, society strives to adopt measures with the aim of preventing or minimizing risks. Water management in the area of drinking water and protection against its negative aspects are some of the important activities carried out by a dynamically developing society on the level of communities and countries, and also on an international level.

The information obtained provides an assessment of water-related crises for which a type plan has been prepared based on a risk analysis within crisis management. The water management area is thus reflected on from a non-traditional perspective and a contemporary model—the paradigm of considering societal resilience for water-related crises. The assessment was carried out within the scope of a case study focusing on the Czech Republic. In this aspect, the presented study is unique, innovative and original by incorporating and combining a reflection on crisis management and on water-related crises.

International literature in the area usually offers reflections of specific separate components which can be classified into texts focusing on securing drinking water, handling floods and crisis management. The topic of the emergency supply of drinking water is handled, for instance, from the viewpoint of a strategy for the supply of drinking water, the area of water management systems and the associated operating risks [1], the possible use of underground water for emergency supply and an assessment of the sources of backup drinking water sources [2,3], Preventing Secondary Disasters through Providing Emergency Water Supply [4]; yet another approach is the analysis of risks and sensitivity of potential sources of the emergency water supply [5], emergency supply of drinking water with the use of available technology [6]. One important document with an international impact in this area is "Water Safety Plans", authored by the World Health Organisation; the document focuses on Managing drinking-water quality from catchment to consumer [7].

The second area that has been the focus for researchers is that of floods. There is an ample number of documents and papers; however, these are very diverse in nature, covering topics such as various means of protecting against floods [8–10], the framework for flood risk communication [11,12], strategies for reducing risks during floods [13–16], and the implementation of the flood risk management directive in selected European countries [17,18]. Some documents also focus on selected aspects of floods such as the use of mobile Mose barriers to protect Venice [19] or the household insurance system for households located in areas at risk from floods [20].

One specific area which is also covered by the presented study and is often reflected on separately in the literature is the area of crisis management. The management of people and crises represents an important component of the study. Crisis management must be based on a risk analysis and react to potential threats [21]. International studies usually focus on a single topic, such as the government's position on crisis management [22], team-based handling of catastrophes [23], coordination and collaboration when handling floods [24] and the sustainability of critical infrastructure [25]. A link between crisis management and water-related catastrophes in the scope covered by this article cannot be found in international databases. This specific angle thus provides a new perspective on the given area and contributes to the development of crisis management theory. It will certainly be interesting to carry out a similar study in other countries and to compare the results obtained for this specific area of safety within the scope of further systemic research.

In the Czech Republic, as a country situated in Central Europe with a predominantly continental climate and without access to a sea, the experience with the solution of water-related crises is primarily connected with the extensive floods in 1997 and 2002. These floods first hit the eastern part of the country (1997) and subsequently the center and west of the country (2002). From the perspective of the supply of drinking water, no wide-reaching problems with a lack of water have occurred. In recent years, however, with regard to lower rainfall and a decrease in the level of groundwater, there were local problems with the supply of drinking water and also partially with a restriction on the use of surface water for irrigation. The given facts were reflected in the area of crisis management when preparing risk analyses and were also exhibited during the processing of type plans as part of the crisis plans.

Society strongly emphasizes the prevention of extraordinary events [26–28]. Understanding the risks and threats has a direct impact on the stance of the public administration and consequently also the general populace to planned or adopted measures and the associated economic costs. The selection of topics for this study was based on the risks that were identified and evaluated in the "Risk Analysis for the Czech Republic" and approved by the Czech government's ruling of 27 April 2016 after discussion in the Country's Security Council [29].

A total of 72 types of hazards were identified in the Risk Analysis for the Czech Republic, divided into three basic categories [29]. These are:

- Acceptable risks, no extraordinary measures are expected, these are usually situations that can be handled in the scope of standard activities of integrated emergency services and the appropriate administrative bodies;
- Conditionally acceptable risks, which require the adoption of measures aimed at their elimination; these occur in the areas of preparing for extraordinary events and include notably emergency planning and the preparation of the integrated emergency system and its components;
- Unacceptable risks, where the measures aimed at their elimination are part of the preparation for the handling of crises and notably include crisis planning.

The risk analysis identified 22 types of hazards/risks that were evaluated as unacceptable and where, if they occur, these are expected to lead to a state of crisis and need to be prioritized. Risks that are unacceptable for society include (in the category of antropogeneous and natural risks, the subcategory of abiotic and technogeneous risks): floods, flash floods, catastrophic rainfall, disruption of dams of prominent waterworks (special floods), and large-scale disruptions of the delivery of drinking water. When preparing the study, the authors decided to adopt a certain level of generality and focus the article on events whose theme is associated with water, with the aim of increasing the awareness of and subsequently improving the quality of prevention within these risk categories (see also Czech government directive 369/2016).

In connection with the risk analysis for the purpose of system access, it is important to formulate the perception of safety and to specify deviations from safety and disruptions of safety, i.e., defining various types of events. Safety and security are not stand-alone constructs without relations to other notions—they are always considered with respect to a referential object [30]. The referential object may be a country, company, community, person, building, secret information, etc. It is then possible to speak about the safety and security of a country, information security, etc. The safety and security environment consists of a range of elements of the referential object, which are then linked to other elements [31]. This work focuses on country-level safety and security in relation to critical situations related to water.

In terms of safety, the actual safety environment consists of real and usually well-known referential objects, such as a country, its organizational elements, or international organizations. Safety then represents a state of calmness, where the referential object is not experiencing interference and is not under threat, and hence it is not necessary to make proactive steps to ensure the required state. A joint foundation for all areas dealing with safety is the adoption of a set of measures aimed at ensuring safety. These measures are aimed at reaching a state where damage is minimized and the referential object can continue performing its intended function [32]. The measures aimed at handling crises that have been prepared for the Czech Republic are handled within the crisis management and planning system.

Planning is one of the basic functions of management. Crisis planning in the Czech Republic is a crisis management tool that represents a summary of planning activities, procedures and links implemented by crisis management bodies and their designated government or public institutions, legal bodies or entrepreneurs in order to implement the goals and tasks associated with ensuring the safety of the country and its populace during emergencies [33].

Humans live in a society that consists of various people who all have their own interests, capabilities and preferences. That is why it is necessary to regulate our mutual cohabitation, for instance via legal regulations, and during crises this system also incorporates measures that arise from the crisis plan. In the case of water crisis management, the crisis plan focuses on risks which were assessed as unacceptable risks. These groups include catastrophes caused by floods, flash floods, excessive rainfall, breaches of dams of significant waterworks (special floods) and significant disruptions of drinking water supply.

The aim of the article is to assess and characterize crisis situations related to water, for a type plan that has been prepared within crisis planning based on a risk analysis. The assessment of this topic is carried out primarily from the viewpoint of prepared measures targeting society and individuals. The assessment is implemented as a case study under the conditions of the Czech Republic. From a scientific viewpoint, this represents a research limitation and means that we focus on the conditions and approach to the given topic implemented by the specific considered country. In view of the aims of the study, an assessment of the technical quality of specific equipment of emergency services and of preventive technical-structural measures is not part of the study's scope. A certain limitation on the study comes from the composition of the expert group for performing the multicriteria analysis. Their own experience from the resolution of crisis situations (especially water-related) could influence the objectivity and a certain degree of detachment. For this reason, the expert group was comprised of experts from different areas. The essential characteristics were included as part of the performed SWOT analysis; in light of the scope of the study, it was not possible to resolve the individual type plans individually, but a summary evaluation of the plans that relate to water-related catastrophes was performed.

The theoretical part of the article delimits and characterizes selected type plans which are included in crisis plans and which are related to water. The practical part of the article focuses on a concretization of selected measures within type plans and possible impacts on life in the Czech Republic. The article assesses crisis preparedness within the scope of a case study focusing on the national level.

The research question was formulated as the process of creating organized knowledge based on systematic research. Specifically, the following research questions were formulated: how has the preparedness for crisis situations associated with water been incorporated in the planning documentation on the national level within the Czech Republic? How are individual type plans handling water-related catastrophes designed? An analysis of the topic was carried out in order to reflect these scientific problems.

2. Materials and Methods

2.1. Methods

The input information used to study the topic was obtained primarily via literature research, from provided internal sources of selected crisis management entities, and the activities of one of the authors as the group leader for preparing the Crisis Plan of the South-Bohemian Region. The practical part uses multicriteria analysis methods. Synthesis was used to connect the individual partial findings. The actual assessment of the preparedness (readiness) for crisis situations related to water listed in the planning documentation on the national level of the Czech Republic and the assessment of individual type plans handling catastrophes associated with water was carried out via a SWOT analysis.

The SWOT analysis (of Strengths, Weaknesses, Opportunities and Threats) was developed in the second half of the 20th century in the United States. It is a useful and highly universal analytical technique that provides us with an understanding and interpretation of strengths and weaknesses as well as of opportunities and threats. It is most frequently used in the area of business, as a strategic tool that can be used to help a company's development [34,35].

A SWOT analysis is based on an evaluation and analysis of the current state of the evaluated entity/topic, its internal environment and the current situation surrounding the assessed entity and outside environment. At its core, it identifies the strengths and weaknesses in the internal environment, meaning what the given entity is good at and where its weaknesses lie, and also analyzes the threats and opportunities located in the outside environment—i.e., factors which the entity itself cannot influence [36,37]. It begins by analyzing strengths, since these are considered an internal force. Above-standard skills, knowledge, and potentials are identified along with resources that may be used in the future to benefit the entity. Weaknesses represent an opposite of the strengths and notably include internal weaknesses of the organization/considered area, areas where it would be possible to achieve better results. Opportunities are used to identify potential options that may lead to improvement if they are correctly used. External circumstances were identified that may lead to future success. Threats then represent external influences that may make it more difficult to achieve the intended goals. Threats are used to identify aspects that bring negatives, factors which need to be taken into account and systematically prevented.

In order to determine the weight factor, it was necessary to identify the items of assessment. The assessment was carried out based on multicriteria decision-making. Decision-making here refers to choosing or classifying a value of one variant based on selected criteria. Decision-making leads to conflicts of interest, where in socially-economic systems it is difficult to identify priority values. Different groups of people give preference to different consequences of a decision, and hence the assessment of an optimal decision may be carried out via different criteria. The following criteria were selected to assess the individual properties of the SWOT analysis: societal impact, impact on health and lives of the population, economic regression, environmental impact and traffic limitations. A multicriteria evaluation was carried out by a group of experts from the Regional Office of the South-Bohemian Region—Crisis Management Department, the Czech Firefighting Services—Civil Protection Department, the Czech Police, Emergency Medical Services of the South Bohemian Region, Czech Army, Czech Hydrometeorological Institute, Povodí Vltavy and VSB—Technical University of Ostrava.

The pair comparison method was used to determine the relative importance of the parameters of the strengths and weaknesses, opportunities and threats. The individual criteria were compared with each other when calculating the weight coefficient. The more important criterion for the given problem was chosen from each pair. The Fuller Triangle method was chosen for the system used to make the comparisons. The research design scheme can be summarized in the following steps:

- specification of research goals and their limits;
- determination of water-related crises for which a type plan has been prepared in crisis planning on the basis of a risk analysis;
- SWOT analysis—specification of concrete criteria for individual areas;
- performance of multicriteria analysis—determination of concrete criteria values (score);
- Specification of relative importance of the parameters of individual characteristics and SWOT areas using the Fuller Triangle method (weight factor);
- calculation of criterion strength—product of score and weight factor;
- comparison of criterion strength for the individual SWOT areas (specification of partial sums);
- summary comparison and assessment of positive or negative approach;
- discussion and interpretation of results.

2.2. Materials—Characteristics of Selected Type Plans

2.2.1. Type Plan—Large-Scale Disruptions of the Supply of Drinking Water

A large-scale disruption of the supply of drinking water is a critical situation which is unlikely to occur unexpectedly, without a preceding event. If the cause of the disruption is simply a standard defect of the water supply network, its delivery can be secured by the appropriate operator of the water mains via a backup supply. However, if there is a critical situation that has a large-scale impact on society related to a large-scale interruption of the delivery of drinking water (where it will not be possible to secure a sufficient reserve supply), the supply of drinking water will be handled within the scope of the system for the emergency drinking water supply. The emergency drinking water supply here refers to securing drinking water for the population in the amount necessary to ensure survival, for the period necessary to restore the standard drinking water supply systems within the affected area.

Disruptions of the drinking water supply could be caused by contamination of the water source (intentional or unintentional), damage to the water treatment system used to create drinking water, extensive disruptions of the water network, or long periods of drought leading to a lack of underground as well as above-ground water. The possible impacts on society include: threats to the life and health of citizens due to drinking contaminated drinking water or water from other, unverified sources; lack of hygiene; the occurrence of epidemics of mass illnesses; intentional damage to water containers or other drinking water dispensing machinery in the case of potential panic or unrest; looting at drinking water retailers and others.

2.2.2. Type Plan—Flood

Floods mean a temporary significant increase in the level of water in rivers or other surface water bodies which causes the water to flood an area outside of the riverbed and may cause damage. Due to its significant amount of broken terrain, the Czech Republic has a very dense hydrographic network spanning a total of about 85 thousand kilometers. It is located in the temperate climate zone with a regular seasonal temperature and precipitation cycle. Aside from these long-term swings, short-term temperature changes are caused by frequent transitions of atmospheric fronts, which separate colder and warmer air and are often accompanied with rainfall.

In general, the long-term amount of rainfall increases at higher altitudes, but the predominant wind direction in mountainous areas is a significant factor as well. The average annual volume of outflow from the territory of the Czech Republic is about 15.1 billion m^3, corresponding to a nominal outflow of 6.1 $L{\cdot}s^{-1}{\cdot}km^{-2}$. The outflow ratios are, however, not equally distributed. The ratio between the average and maximum flow during a 100-year flood is 1:20 to 1:50 on larger rivers, may reach nearly 1:100 on smaller rivers, and small mountainous rivers may have even larger ratios. In the vast majority of cases, floods in the Czech Republic are decisively influenced by hydrological causes and phenomena that also occur in the Czech Republic. Floods coming from abroad may only be considered on the river of Ohra (flow into the Skalka reservoir), Lužnice (flow into the Třeboň pond system), Dyje (flow into the Vranov reservoir) and Stěnava (flow from Poland).

The floods caused by long-lasting regional rain (1997, 2002) caused the rise of floods on a large scape on a regional level. They usually occur on all watercourses in the afflicted region and propagate on medium-sized and larger watercourses (e.g., the Berounka, Vltava, Elbe, Morava, Thaya river basins).

Extreme winter and spring floods are caused by the melting of the snow cover, especially in combination with heavy rains. It can affect both mountain regions, where they happen most often at the end of the winter season, when intensive rains fall on the remains of snow cover, then primarily in foothills and middle elevations, but also in lowlands (Znojmo region 2006). In the second aforementioned case, the negative factors for the rise of significant flooding are, in particular: a large amount of snow cover in lower and middle elevations that can thaw over the course of a few short days; rainfall; the temperature of the air exceeding 8 °C on average and even persisting in the night on a level of about 4 °C and more; freezing soil under the snow cover, which limits the permeation and speeds up the runoff; and the strong convection increasing rainfall on the windward side and the increasing melting of snow.

Melting significant enough for a flood to arise can occur practically from the end of November until April. In a year with an abundance of snow, approximately 5 billion m^3 of water accumulates in snow throughout the entire territory. The height of the snow cover reaches an average of 10–20 cm in the lowlands, 40–60 cm in middle elevations and more than 100 cm in the mountains. The period of the melting of the snow cover is not regular, over the long-term, the maximum average snow cover in lowlands is in the middle of February, in mountains in the second half of March. The risk of flooding from spring thaws has decreased when compared to the period of the first half of the 19th century as a result of warming, the regularity of spring flooding has decreased, though the threat of significant spring flooding remains.

Flooding caused by ice phenomena occur even at relatively lower flow rates in sections of the flow susceptible to the creation of ice sheets and ice jams in sites with lower flow rates or in a narrow section of the riverbed (dangerous ice phenomena can arise on various flows and tend to occur very often on, for example, the rivers Berounka, Otava, Bečva, Moravská Sázava, Sázava, Divoká Orlice, etc.).

A critical situation may be announced if the ongoing critical water or flow levels and their impacts threaten the operation of critical infrastructure and where the handling of such situations is beyond the capabilities of individual regions in view of unfavorable weather forecasts. The impacts of large floods may include significant material damage and may have a long-term impact on the affected territory, and the impacts may require the use of all the country's available resources including the possible use of international humanitarian aid even beyond the duration of the flood itself.

2.2.3. Type Plan—Flash Flood

Flash floods are most frequently caused by sudden surface flow caused by significant and concentrated rainfall, which quickly concentrates in the river network due to broken terrain.

Flash floods are characterized by high rainfall over a shorter duration, i.e., a high intensity and limited areal extent. In general, potentially dangerous precipitation is usually connected to the transition of frontal disorders or the influx of warm and humid air, which supports the development of convective clouds. The probability of the occurrence of heavy precipitation increases in mountain ranges, which create conditions for the occurrence of extreme rainfall on windward sides thanks to the increased flow of air. The resulting flood flows are, in addition to the intensity and duration of the causal rain and size of the afflicted area, also dependent on the physical/geographical characteristics of the afflicted territory. The main factors here represent the size and shape of the river basin, the inclination ratios of the terrain, soil permeability and saturation of the river basin prior to the rainfall. In some cases of flood situations of a regional nature, which are caused by long-lasting, less heavy precipitation, there is a considerable worsening of the course of the flood in the afflicted territory by locally limited heavy rainfall. Flash floods can be expected more in rural areas.

Damage caused by excessive rainfall in rural areas including the erosion of agricultural areas and subsequent transport of sediments into built-up areas are, for the purposes of preparing type plans, considered to be a part of flash floods. The course of a flood is characterized by the type of the flood itself, the value of the culmination flow, the shape and volume of the flood wave and the season of its occurrence.

Summer flash floods caused by short but intense rains usually affect a small area. They may affect any smaller river, but have catastrophic consequences especially when affecting sloping fan-like rivers. They exhibit a very rapid rise in water levels, followed by a very quick decrease. In addition to high precipitation intensity, an important role is also played by the ability of the soil to hold/soak rainfall water, which is affected by vegetation and anti-erosion measures as well as the current saturation of the soil due to previous precipitation. The most dynamic onset and largest death tolls are usually caused by rivers that are approximately 1–20 km^2 large, with a concentration period of less than 1 h.

The risk factors for flash floods are:

- intensity of rain;
- speed of storm movement—the slower the movement, the greater the risk;
- chain effect—the transition of several storms in rapid succession over one river basin; these storms do not need to be extremely strong;
- synergy of the movement of storms in the direction of the water flow in the river basin; if the storm moves in the direction of the water runoff from the river basin, the risk of flooding is higher than movement in the opposite direction;
- occurrence of impermeable and poorly permeable surfaces or the great previous saturation of the area supporting rapid runoff;

- configuration of the terrain with steep slopes, narrow river valleys.

The possibility of predicting flash floods is very limited due to the sharp dynamic of the development of convective clouds from which heavy precipitation comes. Even if meteorological conditions for the occurrence of heavy precipitation can be forecast relatively successfully, the precise localization of the occurrence, the duration and intensity of the heavy precipitation, and thus the specific threat to a locality can essentially not be predicted.

The societal impacts may include:

- significant limitations of social services and securities of citizens evacuated from flooded areas and other citizens in the immediate vicinity of flooded areas;
- damage or destruction of houses and apartments, which will remain uninhabitable for a long period of time, long-term emergency accommodation of evacuated citizens;
- impact on the function of waste water treatment plants;
- damage to cultural landmarks;
- traffic restrictions caused by flooded roads, direct damage to roads and associated buildings (notably bridges and floodgates), landslides or power outages. In case of rail transport, damage to, e.g., rails, sidings, tractional rail or obstacles is expected, and this may also lead to traffic disruptions.

2.2.4. Type Plan—Excessive Precipitation (Rainfall)

For the purposes of crisis planning, excessive precipitation and its consequences refer to: the occurrence of intense precipitation in rural areas, leading to exceeding the capacity of the outflow network, flooding of lower floors of buildings and technical infrastructure by outflowing water. Damage caused by sudden excessive precipitation in rural areas including the erosion of agricultural areas and the subsequent transport of sediments into built-up areas is considered to be part of the flash flood or flood.

In unfavorable conditions, rainfall may lead to a quick build-up of flowing water, especially on reinforced, less permeable or saturated surfaces, and to the flooding of lower-placed areas, buildings or underground areas, or to a rise of the water level in rivers and subsequent flooding. Excessive precipitation connected with storms is a fairly common phenomenon in the summer, but in most cases lasts only a short time (up to 30 min). However, in some cases storm cells may remain exceptionally active and emit an extreme amount of precipitation over a very short period of time. Moreover, storm cells may agglomerate into a group of cells which repeatedly pass over the affected area. In addition to flash floods, storms are also usually accompanied with sudden gusts of wind, electric discharges, and possibly hailstorms.

The occurrence of excessive precipitation is very random, and so it is very difficult to forecast the specific affected areas. They may also cause other unfavorable phenomena, notably soil erosion and landslides, which may then also damage traffic infrastructure, clog sewage systems, reduce flow capacities of river beds and the retention area of water recipients.

2.2.5. Type Plan—Disruptions of Dams of Significant Waterworks Associated with a Special Flood

This category covers floods caused by artificial factors, which are situations that may occur on certain kinds of water works. The owners (users) or managers of water works are obliged to ensure that these are supervised by specialized technical and safety officers with the aim of continuously tracking the technical condition of the water works in view of its stability, safety and possible disruptions along with the proposal of suitable corrective measures.

A special flood needs to be distinguished from natural floods, notably due to the different characteristics of its occurrence, course and the different flood measures that are to be used before and during a special flood. The handling of a special flood requires the careful preparation of bodies involved in the management of floods and crises and the preparation of a basic planning document (plan for protecting the landscape below a selected water facility against a special flood) aimed at handling arising critical situations. This critical situation can only be forecasted in the case of an unmanageable waterworks defect, gradually increasing overflow or water inflow into the waterworks following after long-lasting and/or sudden precipitation.

Disruption of a water work (dam) and adjacent buildings due to accumulated surface water associated with the occurrence of a flood wave via a special flood will, due to its extensive destructive effects, be a reason for the announcement of a critical situation and subsequent declaration of an emergency/crisis. The area where the emergency will be announced directly depends not only on the amount of held water, the landscape configuration below the waterworks and the characteristics of the flood wave, but also on the size of the cities and towns (in terms of residential zones and production facilities) that may be affected by the special flood. There will be immense loss of human life, health, property and infrastructure, but the precise scope of these will, to a significant extent, depend on how well prepared the local bodies and citizens are for this critical situation.

Table 1 contains an overview of the plan characteristics, the application scale and the societal impacts

Table 1. Overview of the plan characteristics.

Type Plan	Application Scale (Urban or Rural)	Societal Impacts
Large-scale disruptions of the supply of drinking water	urban or rural	threats to the life and health of citizens due to drinking contaminated drinking water or water from other unverified sources; lack of hygiene; occurrence of epidemics of mass illnesses; intentional damage to water containers or other drinking water dispensing machinery in case of a potential panic or unrest; looting at drinking water retailers
Flood	urban or rural	significant material damage and may have a long-term impact on the affected territory, and the impacts may require the use of all the country's available resources, including the possible use of international humanitarian aid even beyond the duration of the flood itself.
Flash flood	rural	limitations of social services and securities of citizens evacuated from flooded areas; damage or destruction of houses and apartments; impact on the function of waste water treatment plants, damage to cultural landmarks; traffic restrictions caused by flooded roads, direct damage to roads and associated buildings (notably bridges and floodgates), landslides or power outages
Excessive precipitation (rainfall)	urban	may also cause other unfavorable phenomena, notably soil erosion and landslides, which may then also damage traffic infrastructure, clog sewage systems, reduce flow capacities of river beds and the retention area of water recipients.
Disruptions of dams of significant waterworks associated with a special flood	urban or rural	immense loss of human life, health, property and infrastructure, but the precise scope of these will, to a significant extent, depend on how well prepared the local bodies and citizens are for this critical situation.

Source: own research.

3. Results

3.1. Factors of SWOT Analysis

SWOT analysis was used to perform the actual assessment. In order to make the organization of the factors into individual segments of the analysis easier to read, we provide Table 2 of preparedness for water-related critical situations listed in crisis documentation and focusing on type plans. The table is divided into four quadrants: the left side contains factors that have a positive impact on the topic while the right side contains negative effects and undesirable factors. The upper part maps internal factors, while the lower part presents external influences. Individual factors of the analysis were processed based on multi-criteria analysis by a group of experts in the area of crisis management, planning and risk analysis.

Table 2. SWOT analysis of preparedness for water-related crises listed in the planning documentation with a focus on type plans.

<table>
<tr><th></th><th>Strengths</th><th>Weaknesses</th></tr>
<tr><td rowspan="5">Internal environment</td><td>Flexibility and alternative implementations of the water-related crisis plan</td><td>Ambiguous competences, especially on the strategic level of regional management (regional president—crisis measures from the national level)</td></tr>
<tr><td>Preparation of individual type plans up to the level of concrete water-related crises</td><td>Low engagement of management with respect to the process of preparing crisis plans (low knowledge of the links and connections between individual parts of the planning documentation focusing on water)</td></tr>
<tr><td>Setup of a systemic hierarchical approach on the town/regional/government levels for handling crises</td><td>Direct link between crisis management, consisting of elected officials, to political parties and special interests in society</td></tr>
<tr><td>An ethical approach as a foundation for correct behavior is always part of the plans and the decision processes of management</td><td>High demands on solutions in areas with high concentrations of inhabitants—municipal agglomerations—and the need of ensuring a timely and effective reaction in stationary medical and social facilities</td></tr>
<tr><td>Risk analysis carried out by a team of experts, generally recognized and accepted</td><td>Lack of information transfer and processing of planning documentation on the municipal level</td></tr>
<tr><th></th><th>Opportunities</th><th>Threats</th></tr>
<tr><td rowspan="4">External environment</td><td>Use of experience from previous crises in the Czech Republic—notably floods</td><td>Incorrect generalization of approaches within crisis management and difficulty of forecasting the occurrence of a large-scale water-related crisis sufficiently in advance</td></tr>
<tr><td>The specified resources intended for handling catastrophes in crisis plans are gradually purchased under the auspices of the State Material Reserve Management</td><td>Lack of experience of crisis management and the population in general with extensive interruptions of drinking water supply</td></tr>
<tr><td>Implementation of training with a focus on the correct handling of and communication during water-related crises on the government, regional and municipal levels</td><td>Difficulty of coordinating a large number of different entities participating in the crisis and its management</td></tr>
<tr><td>Timely preparation and implementation of preventive anti-flood measures and measures for emergency supply of drinking water</td><td>Water-related crises may lead to secondary societal impacts (interruption of power supply, traffic restrictions, disruptions of internal security, disruption of water treatment plants, epidemiological complications, etc.)</td></tr>
<tr><td>Increase the international exchange of experience in the area of catastrophe management for water-related crises, opportunities for cross-border and internal collaboration</td><td>Lack of focus on water-related crisis management at the expense of other catastrophes (COVID-19, migration)</td></tr>
</table>

Source: own research.

3.2. Calculation of SWOT Analysis

After listing the strengths, weaknesses, opportunities and threats, each item was assessed individually. Based on the carried-out interpretation, a corresponding significance and value was assigned to the investigated area, leading to the creation of a comprehensive image. For strengths and opportunities, we used the positive scale of 1 to 5, where 5 represented highest satisfaction and 1 lowest satisfaction. For weaknesses and threats, a negative scale of -1 (representing the smallest dissatisfaction) to -5 (representing the greatest dissatisfaction) was used instead. The assessment was performed on the basis of a multicriteria decision that is based on the selection or classification of values of one variation of the specified criteria. For the assessment of the individual characteristics of the SWOT analysis, the criteria of the social impact, impact on the health and lives of the population, economic regression, impact on the environment and restriction of transportation were specified. The multicriteria assessment was performed by a group of experts from the Regional Council of the South Bohemian Region—Crisis Management Department, Fire Rescue Service of the Czech Republic—Civil Protection Department, Police of the Czech Republic, Emergency Medical Services of the South Bohemian Region, Army of the Czech Republic, Czech Hydrometeorological Institute, Povodí Vltavy and VSB—Technical University of Ostrava.

The pair comparison method was used to determine the relative importance of the parameters of the strengths and weaknesses, opportunities and threats. A weight factor was calculated for each item, and this captured the importance of individual items in the given category. The greater the weight, the greater the importance of the item within its category, and vice-versa. The sum of the weights in each category was always 1. The weight coefficients were obtained by comparing the individual criteria to each other. From each pair, the more important criterion with respect to the given problem was selected. The Fuller Triangle method was used to make the comparisons. The scores and weights were then multiplied with each other. The results in each category are summed up, after which the internal components (strengths and weaknesses) are summed up, while the external components (opportunities and threats) are summed up separately. These two results are then deducted from each other, resulting in the final value of the SWOT analysis [35,36]. The aforementioned calculation for the SWOT analysis of preparedness for water-related crises in the planning documentation focusing on type plans is presented in Tables 3 and 4.

Table 3. Calculation for the SWOT analysis of preparedness for water-related critical situations in planning documentation with a focus on type plans.

	Strengths	Weight Factor	Score	Criterion Strength
	S.1 Flexibility and alternative implementations of the water-related crisis plan	0.27	4	1.07
	S.2 Preparation of individual type plans up to the level of concrete water-related crises	0.33	5	1.67
	S.3 Setup of a systemic hierarchical approach on the town/regional/government levels for handling crises	0.13	3	0.40
	S.4 An ethical approach as a foundation for correct behavior is always part of the plans and the decision processes of management	0.07	2	0.13
	S.5 Risk analysis carried out by a team of experts, generally recognized and accepted	0.20	3	0.60
	Sum for this component			3.87
	Weaknesses			
	W.1 Ambiguous competences, especially on the strategic level of regional management (regional president—crisis measures from the national level)	0.13	-2	-0.27
	W.2 Low engagement of management with respect to the process of preparing crisis plans (low knowledge of the links and connections between individual parts of the planning documentation focusing on water)	0.27	-3	-0.80
	W.3 Direct link between crisis management, consisting of elected officials, to political parties and special interests in society	0.07	-1	-0.07
	W.4 High demands on solutions in areas with high concentrations of inhabitants—municipal agglomerations—and the need of ensuring a timely and effective reaction in stationary medical and social facilities	0.33	-4	-1.33
	W.5 Lack of information transfer and processing of planning documentation on the municipal level	0.20	-3	-0.60
	Sum for this component			-3.07

(The leftmost spanning label for all rows reads: Internal environment)

Table 3. *Cont.*

	Strengths	Weight Factor	Score	Criterion Strength
	Opportunities			
	O.1 Use of experience from previous crises in the Czech Republic—notably floods	0.27	4	1.07
	O.2 The specified resources intended for handling catastrophes in crisis plans are gradually purchased under the auspices of the State Material Reserve Management	0.07	1	0.07
	O.3 Implementation of training with a focus on the correct handling of and communication during water-related crises on the government, regional and municipal levels	0.20	3	0.60
	O.4 Timely preparation and implementation of preventive anti-flood measures and measures for emergency supply of drinking water	0.33	5	1.67
	O.5 Increase the international exchange of experience in the area of catastrophe management for water-related crises, opportunities for cross-border and internal collaboration	0.13	3	0.40
	Sum for this component			3.80
	Threats			
	T.1 Incorrect generalization of approaches within crisis management and difficulty of forecasting the occurrence of a large-scale water-related crisis sufficiently in advance	0.13	−3	−0.40
	T.2 Lack of experience of crisis management and the population in general with extensive interruptions of drinking water supply	0.27	−4	−1.07
	T.3 Difficulty of coordinating a large number of different entities participating in the crisis and its management	0.07	−1	−0.07
	T.4 Water-related crises may lead to secondary societal impacts (interruption of power supply, traffic restrictions, disruptions of internal security, disruption of water treatment plants, epidemiological complications etc.)	0.33	−5	−1.67
	T.5 Lack of focus on water-related crisis management at the expense of other catastrophes (COVID-19, migration)	0.20	−2	−0.40
	Sum for this component			−3.60

(Row label spanning Opportunities and Threats sections: External environment)

Source: own research.

Table 4. Resulting values of the SWOT analysis.

Internal environment (S + W)	0.80
External environment (O + T)	0.20
Total (difference)	+0.60
Positive environment (S + O)	7.67
Negative environment (W + T)	−6.67
Total (difference)	+1.00

Source: own research.

3.3. Assessment of SWOT Analysis

The first step in the assessment of the SWOT analysis was to perform an evaluation based on multi-criteria decision-making carried out by a specialized group. This was followed by a calculation of the value of the weight criterion using Fuller's triangle, where the expert group compared the significance/priority of the individual elements of the SWOT analysis. A subsequent product of the identified values was then used to determine the strength of each criterion. All the data are provided in Table 3. An overview is provided by the sums for individual components of the SWOT areas, which are compared against each other in Table 4.

The resulting values of the SWOT analysis show that water-related crises for which a type plan has been prepared within crisis management based on a risk analysis received a slightly positive score—a value of +0.60. The factors arising from the internal environment are stronger than those arising from the external environment; that being said, the resulting score difference of 0.60 can also be considered very small. It is clear that it remains necessary to support and expand on the strengths, such as the preparation of detailed type plans covering specific water-related crises. Furthermore, the analysis also suggests that it would be helpful to take a closer look at problematic areas, notably the preparedness of all societal tools and solutions used for locations with high concentrations of inhabitants, municipal agglomerations and the need to implement timely and correct reactions in stationary medical and social facilities.

From another perspective, a comparison between the positive (S + O) and negative (W + T) environments was carried out. Here it is clear that the more dominant environment is the positive one, with a value of 7.67 compared to −6.67. The resulting difference is +1.00, indicating that the positive environment is dominant.

The final values also indicate that the system for handling water-related crises in the Czech Republic for which a type plan has been prepared based on risk analysis within crisis management is mildly positive. Further improvements are possible by reinforcing the system's strengths and minimizing its weaknesses in a way which makes use of available opportunities.

4. Discussion

The SWOT analysis methodology allows us to create a matrix summarizing the strengths, weaknesses, opportunities and threats and to evaluate these aspects. One advantage of this procedure is that it is possible to search for links between the identified items, and these links can then be used in the management strategies for the risks arising from the weaknesses and threats. The achieved results of the SWOT analysis indicate a small advantage of the positive approach and environment.

On the basis of the performed investigation, it can be said that planning documentation for managing water-related crises has been prepared in the Czech Republic. The individual plans contain selected measures that react to possible crisis scenarios. The possible impacts on society in the Czech Republic are reflected in the plans. The necessity of supporting the strengths and opportunities is evident from the performed SWOT analysis. The foundation is the analysis of the risks that ensue from this study and that are also one of the assessed characteristics of the strengths. When preparing the multicriteria analysis, which formed the basis of the creation of the score, attention was drawn to the imprecise resolution of the individual crisis situations and their considerable overlap. This applied to the examined crises of:

- floods,
- flash floods,
- excessive precipitation.

From the perspective of Opportunities, an emphasis was placed on the use of experience from the resolution of previous crises in the Czech Republic. The performed assessment was the source material for the update and innovation of the approach of the crisis management bodies on the level of the state administration and local governments. Another characteristic that was assessed is the implementation of exercises with a focus on solution possibilities and proper communication during water-related crises. The crisis plans were updated following the extreme floods in the Czech Republic in 1997 and 2002. Subsequently, however, a social position began to predominate that there is sufficient experience with floods and preventive flood measures were adopted and it is not necessary to spend much time and effort on this area. There is an exception, with regard to changes in climate conditions, to the approach to the preparation of the possible crisis situation— large-scale disruptions of the supply of drinking water. In recent years, attention has been focused during exercises on the issue of a nuclear accident and the long-term interruption of the power supply.

When evaluating the Weaknesses, there is evidently a higher assessment of the characteristic of the higher demands on the solution:

- in areas with a high concentration of living persons (city agglomerations),
- the need for a timely and substantive reaction in health and social facilities of a stationary type.

The large-scale organization of crisis measures in sites with a high concentration of persons and health and social facilities is highly demanding and complicated for all types of events. Thus, it is important for the authorities and population on a communal level to have sufficient information on the planned measures and to also contribute to the acceptance of the preventative measures.

The results of the SWOT analysis for the area of Threats show the highest score for the characteristic of the threat of secondary impacts for crisis situations, e.g., the interruption of the power supply, traffic restrictions, disruptions of internal security, disruption of water treatment plants, epidemiological complications, etc. The minimization of the possibility of domino effects is a basic goal of crisis management and must also be resolved during water-related crises.

Safety is one of the basic characteristics of quality of life, for individuals as well as society as a whole. It is clear that society has a collective as well as individual dimension. The article focuses predominantly on the collective sphere of safety and the assessment of societal preparedness and resilience to water-related crises in the Czech Republic. However, the individual dimension of safety must also be taken into account. It is crucial to allow citizens to access information about possible risks and threats [38,39], and it is also important to ensure that the appropriate government bodies actively disseminate this information.

The society's preparedness for crises associated with a lack of drinking water or floods is based on two basic dimensions. On one hand, these are measures implemented by crisis management bodies and the resources and know-how available to the emergency system and its components. The second dimension consists of the education of the populace, whereas people know the principles on how to act in the event of such emergencies and can react in a timely and appropriate way. Collective and individual preparedness are prerequisites for the successful handling of crises and for minimizing death, property damage and damage to the environment.

The society's resistance to water-related crises is part of the wider concept of the society's safety and quality of life. Within this concept, it is necessary to distinguish between problems that are specifically related to safety—the protection of lives and health—or whether they are rather related to ensuring satisfactory living conditions, within the wider scope of human security. In some cases—notably, in well-developed countries—the impact may be primarily attributed to the latter, i.e., to maintaining a certain quality of life.

Safety can never be guaranteed completely and perfectly, and this naturally also applies to water-related crises. Aiming for complete safety and imagining that one can completely eliminate or prevent serious threats is dangerous and incorrect. The notion of societal vulnerability may be used to describe the level of safety of a society; this maps, categorizes and measures the amount of vulnerability.

The results of the presented study provide a new perspective on a case study covering the preparedness and resilience of society to water-related crises in the Czech Republic.

With the use of the SWOT analysis, it was shown that water-related crises for which a type plan has been prepared as part of the crisis planning on the basis of a risk analysis point to a slightly positive score—a value of +0.60. The identified positives and negatives, and also the identified vulnerabilities, are the best ways to identify safety policy priorities. It is clear that the handling of water-related crises needs to be carried out via the implementation of preventive as well as reactive tools. Safety problems in general, and hence also water-related crises, cannot be handled exclusively via safety measures. Measures represent only concrete reactions to specific threats, but do not usually tackle the core (source) of the threat. Aside from strengthening safety capabilities, it is necessary to create a safe space and in general a safe environment by, e.g., proper zoning, landscape protection, and the protection of river basins.

5. Conclusions

This assessment of societal readiness and resilience to water-related situations in the Czech Republic focuses on an interdisciplinary approach in the Czech Republic for solving this problem. The goal of the article is to evaluate and characterize the preparedness for handling water-related crises, which are covered by a type plan based on a risk analysis within the scope of crisis planning. The assessment is carried out via a case study in the Czech Republic and focuses notably on the viewpoint of prepared societal measures. The theoretical part of the article delimits and characterizes selected type plans that are part of crisis plans and are related to water. The practical part of the article focuses on an assessment of selected measures of type plans with a focus on water.

The analysis is carried out via a SWOT analysis, which is a universal analytical method used to understand and interpret strengths and weaknesses and to identify opportunities and threats. For the calculation of the weight factor of the SWOT analysis, an assessment was determined based on the multicriteria analysis that a group of selected experts performed on crisis management, planning and risk analysis. The pair comparison method was used to determine the relative importance of the parameters of the strengths and weaknesses, opportunities and threats. The Fuller Triangle method was chosen for the system used to make the comparisons of the individual criteria. The obtained information led to the results of the assessment of water-related crises which are covered by a type plan based on a risk analysis within the scope of crisis planning.

On the basis of the performed investigation, it can be said that planning documentation for managing water-related crises has been prepared in the Czech Republic. The individual plans contain selected measures that react to possible crisis scenarios. The possible impacts on society in the Czech Republic are reflected in the plans.

From the perspective of Opportunities, an emphasis was placed on the use of experience from the resolution of previous crises in the Czech Republic. Another characteristic that was assessed is the implementation of exercises with a focus on solution possibilities and proper communication during water-related crises. When evaluating the Weaknesses, there is evidently a higher assessment of the characteristic of the higher demands on the solution in areas with a high concentration of living persons (city agglomerations) and the need for a timely and substantive reaction in health and social facilities of a stationary type.

With the use of the SWOT analysis, it was shown that water-related crises for which a type plan has been prepared as part of the crisis planning on the basis of a risk analysis point to a slightly positive score—a value of +0.60. The identified positives and negatives, and also the identified vulnerabilities, are the best ways to identify safety policy priorities. It is clear that the handling of water-related crises needs to be carried out via the implementation of preventive as well as reactive tools. Safety problems in general, and hence also water-related crises, cannot be handled exclusively via safety measures.

The uniqueness of the study consists of the issue of water management, which is thus reflected from a non-traditional perspective, being a contemporary model—the paradigm of the view on the preparedness of the planning documentation as one of the characteristics of societal resilience for water-related crises.

Author Contributions: Data curation, Š.K. (Štěpán Kavan), J.P.; Formal analysis, Š.K. (Štěpán Kavan), Š.K. (Šárka Kročová); Investigation, Š.K. (Štěpán Kavan), Š.K. (Šárka Kročová), J.P., Methodology, Š.K. (Štěpán Kavan); project administration, J.P.; Editing, Š.K. (Šárka Kročová). All authors have read and agreed to the published version of the manuscript.

Funding: This research was funded by the Ministry of the Interior of the Czech Republic Project No. VH20182020042 Population Protec-tion in a Spatial Planning and within Setting Technical Conditions for Building Engineering.

Institutional Review Board Statement: To exclude this statement - the study did not involve humans or animals. Not applicable.

Informed Consent Statement: Not applicable.

Data Availability Statement: Not applicable.

Conflicts of Interest: The authors declare no conflict of interest.

References

1. Kročová, Š. *Strategie Dodávek Pitné Vody*; Sdružení požárního a bezpečnostního inženýrství: Ostrava-Poruba, Czech Republic, 2009; 158p, SPBI Spektrum 63, ISBN 978-80-7385-072-2.
2. Bozek, F.; Bozek, A.; Bumbova, A.; Bakos, E.; Dvorak, J. Classification of Ground Water Resources for Emergency Supply. World Academy of Science, Engineering and Technology. *Int. J. Geol. Environ. Eng.* **2012**, *6*, 728–731. Available online: http://waset.org/publications/9966 (accessed on 21 November 2020).
3. Bozek, F.; Bumbova, A.; Bakos, E.; Bozek, A.; Dvorak, J. Semi-quantitative risk assessment of groundwater resources for emergency water supply. *J. Risk Res.* **2014**, *18*, 505–520. [CrossRef]
4. Bross, L.; Krause, S. Preventing Secondary Disasters through Providing Emergency Water Supply. In *World Environmental and Water Resources Congress 2017*; American Society of Civil Engineers: Reston, VA, USA, 2017; pp. 431–439. [CrossRef]
5. Bumbová, A.; Čáslavský, M.; Božek, F.; Dvořák, J. The Analysis of Hazard and Sensitivity of Potential Resource of Emergency Water Supply. *World Acad. Sci. Eng. Technol. Int. J. Environ. Ecol. Eng.* **2013**, *7*, 634–639. Available online: https://waset.org/publications/16845/the-analysis-of-hazard-and-sensitivity-of-potential-resource-of-emergency-water-supply (accessed on 12 November 2020).
6. Loo, S.L.; Fane, A.G.; Krantz, W.B.; Lim, T.T. Emergency water supply: A review of potential technologies and selection criteria. *Water Res.* **2012**, *46*, 3125–3151. [CrossRef]
7. Word Health Organization. *Water Safety Plans. Managing Drinking-Water Quality from Catchment to Consumer*; WHO/SDE/WSH/05.06; Word Health Organization: Geneva, Switzerland, 2005.
8. Sayers, P.; Yuanyuan, L.; Galloway, G.; Penning-Rowsell, E.; Fuxin, S.; Kang, W.; Yiwei, C.; Le Quesne, T. *Flood Risk Management: A Strategic Approach*; Asian Development Bank: Paris, France, 2013; ISBN 978-92-3-001159-8.
9. Drotárová, J.; Vacková, M.; Mesároš, M. Voluntery Municipal Fire Brigades as a part of Crisis Management in the context of Flood Protection. In Proceedings of the SGEM 2019 Conference Proceedings (5.1. Ecology, Economics, Education and Legislation): Ecology and Environmental Protection, Sofia, Bulgaria, 28 June–7 July 2019; pp. 809–816. [CrossRef]
10. Esposito, G.; Parodi, A.; Lagasio, M.; Masi, R.; Nanni, G.; Russo, F.; Alfano, S.; Giannatiempo, G. Characterizing Consecutive Flooding Events after the 2017 Mt. Salto Wildfires (Southern Italy): Hazard and Emergency Management Implications. *Water* **2019**, *11*, 2663. [CrossRef]
11. Intrieri, E.; Dotta, G.; Fontanelli, K.; Bianchini, C.; Bardi, F.; Campatelli, F.; Casagli, N. Operational framework for flood risk communication. *Int. J. Disaster Risk Reduct.* **2020**, *46*, 101510. [CrossRef]
12. Pan, T.-Y.; Lin, H.-T.; Liao, H.-Y. A Data-Driven Probabilistic Rainfall-Inundation Model for Flash-Flood Warnings. *Water* **2019**, *11*, 2534. [CrossRef]
13. Yang, T.; Liu, W. A General Overview of the Risk-Reduction Strategies for Floods and Droughts. *Sustainability* **2020**, *12*, 2687. [CrossRef]
14. Akpoti, K.; Antwi, E.O.; Kabo-Bah, A.T. Impacts of Rainfall Variability, Land Use and Land Cover Change on Stream Flow of the Black Volta Basin, West Africa. *Hydrology* **2016**, *3*, 26. [CrossRef]
15. Tuladhar, D.; Dewan, A.; Kuhn, M.; J Corner, R. The Influence of Rainfall and Land Use/Land Cover Changes on River Discharge Variability in the Mountainous Catchment of the Bagmati River. *Water* **2019**, *11*, 2444. [CrossRef]
16. Morán-Tejeda, E.; Fassnacht, S.R.; Lorenzo-Lacruz, J.; López-Moreno, J.I.; García, C.; Alonso-González, E.; Collados-Lara, A.-J. Hydro-Meteorological Characterization of Major Floods in Spanish Mountain Rivers. *Water* **2019**, *11*, 2641. [CrossRef]
17. Müller, U. Implementation of the flood risk management directive in selected European countries. *Int. J. Disaster Risk Sci.* **2013**, *4*, 115–125. [CrossRef]
18. Molenveld, A.; van Buuren, A. Flood Risk and Resilience in the Netherlands: In Search of an Adaptive Governance Approach. *Water* **2019**, *11*, 2563. [CrossRef]
19. Mose Venezia. *Per la Difesa di Venezia e Della Laguna Dalle Acque Alte*; Consorzio Venezia Nuova: Venice, Italy, 2020. Available online: https://www.mosevenezia.eu/ (accessed on 2 November 2020).
20. FloodRe. *Flood Re is Helping Insurers to Help Householders at Risk of Flooding*; Flood Re Limited: London, UK, 2020. Available online: https://www.floodre.co.uk/ (accessed on 9 November 2020).
21. Maléřová, L. Assessment of Safety Level Risk. In *Crisis Management A Leadership Perspective*; Nova Science Publishers. Inc.: New York, NY, USA, 2016; pp. 127–139. ISBN 978-1-63483-395-0.
22. Anttiroiko, A.-V. Successful government responses to the pandemic: Contextualizing national and urban responses to the COVID-19 outbreak in east and west. *Int. J. E-Plan. Res.* **2021**, *10*, 1–17. [CrossRef]
23. Moon, J.; Sasangohar, F.; Son, C.; Peres, S.C. Cognition in crisis management teams: An integrative analysis of definitions. *Ergonomics* **2020**, *63*, 1240–1256. [CrossRef]
24. Mcmaster, R.; Baber, C. Multi-agency operations: Cooperation during flooding. *Appl. Ergon.* **2012**, *43*, 38–47. [CrossRef]
25. Garschagen, M.; Sandholz, S. The role of minimum supply and social vulnerability assessment for governing critical infrastructure failure: Current gaps and future agenda. *Nat. Hazards Earth Syst. Sci.* **2018**, *18*, 1233–1246. [CrossRef]
26. Dušek, J. International Cooperation of Regional Authorities of the Czech Republic: History, Presence and Future. In Proceedings of the 18th International Colloquium on Regional Sciences, Masaryk University, Hustopece, Czech Republic, 17–19 June 2015; pp. 300–305. [CrossRef]

27. Kováčová, L.; Vacková, M. Achieving of environmental safety through education of modern oriented society. In Proceedings of the 14th SGEM GeoConference on Ecology, Economics, Education and Legislation, Albena, Bulgaria, 17–26 June 2014; Volume 2, pp. 3–8. [CrossRef]
28. KAVAN, Š. Výsluhové nároky příslušníků bezpečnostních sborů v České republice. In Proceedings of the International Conference Safe and Secure Society 2020, College of European and Regional Studies, Ceske Budejovice, Czech Republic, 14–15 October 2020; pp. 52–60. [CrossRef]
29. Usnesení Vlády ČR č. 369 ze dne 27. Dubna 2016 k Analýze Hrozeb pro Českou Republiku. 2020. Available online: https://apps.odok.cz/djv-agenda?date=2016-04-27 (accessed on 2 November 2020).
30. Ivančík, R. Quo Vadis európska obrana a bezpečnosť. *Politické Vedy* **2019**, *22*, 47–67. [CrossRef]
31. Pokorný, J.; Tomášková, M.; Balažiková, M. Study of changes for selected fire parameters at activation of devices for smoke and heat removal and at activation of fixed extinguishing device. *MM (Mod. Mach.) Sci. J.* **2015**, *2015*, 764–767. [CrossRef]
32. Lukáš, L. Vývoj Českého bezpečnostního pojmosloví. In *Conference Proceedings of VII. Medzinárodnej Vedeckej Konferencie Bezpečné Slovensko a Eúropská únia*; University of Security Management in Košice: Košice, Slovac Republic, 2013; ISBN 978-80-89282-88-3.
33. Svoboda, J.; Václavík, V.; Dvorský, T.; Klus, L. The design of flood protection in Kobeřice municipality. *IOP Conf. Ser. Earth Environ. Sci.* **2017**, *92*, 012062. [CrossRef]
34. Newton, P.; Bristoll, H.B.R. *SWOT Analysis: Free eBook*; Free Management Books: USA, 2013. Available online: http://www.free-management-ebooks.com/dldebk/dlst-swot.htm (accessed on 2 November 2020).
35. Fine, L.G. *The SWOT Analysis: Using Your Strength to Overcome Weaknesses, Using Opportunities to Overcome Threats*; Createspace Independent Publishing Platform: North Charleston, SC, USA, 2009; ISBN 1449546757.
36. Shewan, D. How to Do a SWOT Analysis for Your Small Business (with Examples). *WordStream: Online Advertising Made Easy*. 2018. Available online: https://www.wordstream.com/blog/ws/2017/12/20/swot-analysis (accessed on 8 November 2020).
37. Sarsby, A. *SWOT Analysis: A Guide to Swot for Business Studies Students*; Spectaris Ltd.: Stowmarket, UK, 2016; ISBN 0993250424.
38. Kročová, Š.; Kavan, Š. Cooperation in the Czech Republic border area on water management sustainability. *Land Use Policy* **2019**, *86*, 351–356. [CrossRef]
39. KAVAN, S. Selected social impacts and measures resulting from the Covid-19 epidemic in the Czech Republic on the specific example of the South Bohemian Region. *Health Social Care Community* **2021**. [CrossRef] [PubMed]

hydrology

MDPI

Article

How Perceptions of Trust, Risk, Tap Water Quality, and Salience Characterize Drinking Water Choices

Madeline A. Grupper [1,*]**, Madeline E. Schreiber** [2] **and Michael G. Sorice** [1]

[1] Department of Forest Resources & Environmental Conservation, Virginia Tech, Blacksburg, VA 24061, USA; msorice@vt.edu
[2] Department of Geosciences, Virginia Tech, Blacksburg, VA 24061, USA; mschreib@vt.edu
* Correspondence: maddyag@vt.edu

Abstract: Provision of safe drinking water by water utilities is challenged by disturbances to water quality that have become increasingly frequent due to global changes and anthropogenic impacts. Many water utilities are turning to adaptable and flexible strategies to allow for resilient management of drinking water supplies. The success of resilience-based management depends on, and is enabled by, positive relationships with the public. To understand how relationships between managers and communities spill over to in-home drinking water behavior, we examined the role of trust, risk perceptions, salience of drinking water, and water quality evaluations in the choice of in-home drinking water sources for a population in Roanoke Virginia. Using survey data, our study characterized patterns of in-home drinking water behavior and explored related perceptions to determine if residents' perceptions of their water and the municipal water utility could be intuited from this behavior. We characterized drinking water behavior using a hierarchical cluster analysis and highlighted the importance of studying a range of drinking water patterns. Through analyses of variance, we found that people who drink more tap water have higher trust in their water managers, evaluate water quality more favorably, have lower risk perceptions, and pay less attention to changes in their tap water. Utility managers may gauge information about aspects of their relationships with communities by examining drinking water behavior, which can be used to inform their future interactions with the public, with the goal of increasing resilience and adaptability to external water supply threats.

Keywords: drinking water; behavior; trust; risk; tap water; salience

Citation: Grupper, M.A.; Schreiber, M.E.; Sorice, M.G. How Perceptions of Trust, Risk, Tap Water Quality, and Salience Characterize Drinking Water Choices. *Hydrology* **2021**, *8*, 49. https://doi.org/10.3390/hydrology8010049

Received: 26 February 2021
Accepted: 13 March 2021
Published: 18 March 2021

Publisher's Note: MDPI stays neutral with regard to jurisdictional claims in published maps and institutional affiliations.

1. Introduction

Challenges to the safety of public drinking water are increasing as external drivers, including climate and land-use change, adversely impact the stability of drinking water sources [1–3]. Increasingly frequent disturbances, such as algae blooms, hypoxic conditions, or rising metal concentrations, jeopardize water quality in the lakes and reservoirs sourced for public drinking water, requiring drinking water managers to respond with new technologies and strategies [4,5]. One strategy with the potential to help utilities respond to changing environmental conditions is resilience-based management, a style that focuses on adaption and flexibility amidst change [6–8].

Disturbances to raw water sources not only present challenges to treating those water sources but can also threaten the social capital that resilient systems rely on. Concern about the quality of lakes and reservoirs and household drinking water has risen in recent years among Americans [9,10]. As these concerns rise, there is potential for them to spill over to impact the relationship between communities and the water utility managers, who supply their drinking water [11].

The ability of water utility managers to maintain resilient systems and quickly adapt to water quality issues often depends on their relationship with the community they serve [12]. Managers with high levels of community trust have the flexibility to adapt and respond

77

more quickly [12,13]. Trust can encourage citizen cooperation with managers through support for management plans, increased speed and effectiveness of the enaction of those plans, and reduced backlash to management shifts [14–17].

One behavioral indicator of community trust is the acceptance of tap water as a drinking source [18,19], although trust is not the only factor linked to water drinking behavior. Risk perceptions and personal experience with water quality (i.e., taste, smell, and appearance) are also related to the decision to drink tap water, utilize additional filters to treat it, or avoid it altogether e.g., [20,21]. For instance, residents in West Virginia who perceived higher risk associated with their tap water were more likely to drink bottled water compared to tap water, but not home-filtered water compared to tap [22]. Additionally, changes in taste, smell, and appearance of tap water are related to an individual's decision to drink from both bottled and home-filtered sources. Less favorable evaluations of tap water are commonly linked to the decision to drink from alternative water source choices [18]. Taste, in particular, has been linked to decisions to drink from bottled water sources as opposed to tap [23,24].

In areas where issues with water quality are slight or infrequent, the strength of associations and ease with which a person brings to mind water quality issues becomes a relevant factor. This issue of salience is related to the degree to which an individual may drink directly from the tap [25,26]. For instance, Grupper [27] found that residents' trust in a water utility to deliver safe drinking water varies based on two indicators of salience: (1) their familiarity with the water utility providing the water, and (2) the amount of attention they pay to variations in their water quality. Finally, other factors, including industry efforts to market sink and appliance filters as well as bottled water as "pure" sources of water, also contribute to the preference of bottled water and home-filtered water over tap water [28,29].

We focused this study on the roles of risk, trust, and salience in the choice of drinking tap water. Our goal was to explore in-home drinking water behavior in a community in southwest Virginia as a way to understand public trust in water utilities to provide safe and clean drinking water.

1.1. Conceptual Framework

Our study's conceptual framework integrates conceptualizations of risk from Saylor (2011) [21] and Debbler et al. (2018) [30], the salience framework of Stewart (2009) [31], and the concept of trust based on Mayer et al. (1995) [32] (Figure 1).

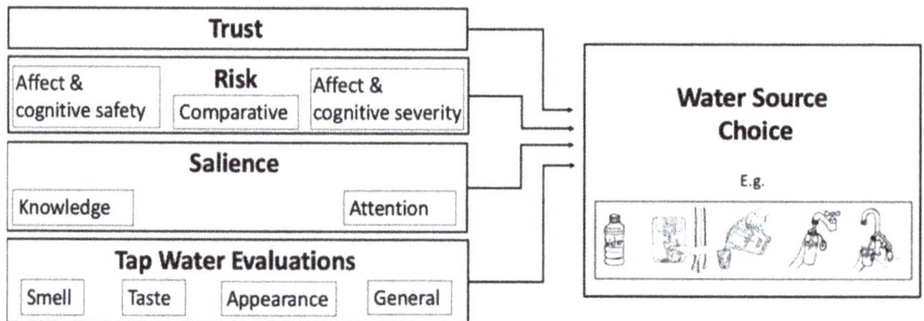

Figure 1. The water source, or sources, that individuals choose to drink from is a function of their trust in the utility to deliver safe drinking water to them, salience of their home water quality, risk judgments of water safety (alone and relative to bottled water), emotional concern, severity of consequence should their water become unsafe, and quality judgments of their tap water.

1.1.1. Trust

The intention to drink tap water directly, as opposed to drinking from bottled or personally filtered sources, has been tied to an individual's trust in their utility to deliver safe drinking water to them. Trust is a psychological state when one actor, the trustor, accepts vulnerability based on the expectation that another entity, the trustee, will perform a certain action or behavior [32]. In a review of literature and surveys that examined individuals' decision to drink tap or bottled water, Doria (2006) found that when trust in tap water providers or bottled water companies is eroded, they are less likely to drink from those sources [18]. Saylor (2011) found that university students who trusted their local water utility to deliver safe drinking water were more likely to drink from tap water sources, while those who lacked trust in their government and university were more likely to drink bottled water [21]. These results were supported in a survey of residents in West Virginia that found residents' decisions to drink from tap water sources were related to higher levels of trust in the water utility [33].

1.1.2. Risk Perceptions

Alongside trust, risk perceptions have also been established as a driving factor in the acceptance of tap water [17,18]. Risk is conceptualized as a function of the perceived likelihood of an unwanted event occurring and the severity of consequence should that unwanted event occur [34]. Risk perceptions are a frequent object of study in natural resource management, where institutions are charged with protecting the public from hazards such as poor water quality [35]. Saylor (2011) found that barriers to drinking tap water include both low trust and perceived health risks associated with the tap [21]. Similarly, Hu et al.'s (2011) national survey found that people who have safety concerns with their tap water are more likely to reject it as a drinking source [36]. This was also true when household filtered water was included as a drinking choice [37]. Triplett et al. (2019) found that the more residents believe their tap water is unsafe, contaminated, or likely to cause sickness, the more they consume bottled water compared to tap and household filtered water [37].

Risk has been conceptualized in drinking water studies most frequently as a cognitive indicator in surveys, asking how safe people believe their water is to drink [38,39]. Debbler et al. (2018) added to this by including a measure of affect-based risk [30]. While cognitive risk measures assess probability beliefs, affect risk measures assess worry, concern, or other emotional components associated with the occurrence of unwanted events. By examining both cognitive and affect measures, researchers can get a more complete picture of risk perceptions. Additionally, Saylor (2011) found that when people's perceived risk of bottled water is lower than their perceived risk of tap, they are less likely to accept tap water, introducing the concept of comparative risk to the water risk framework [21]. We conceptualize risk as a function of cognitive and affective perceptions of water safety and impressions of severity should one's drinking water be compromised. We included comparative risk of bottled and tap water as an additional measure of overall risk.

1.1.3. Salience

Both risk and trust perceptions are moderated by salience, or the degree to which drinking water topics are readily brought to mind or have strong associations for an individual [25,26]. For instance, residents with low salience about drinking water quality issues had a relatively high baseline trust in the water utility [27]. In that case, trust was positively related to familiarity with one's utility and negatively related to higher attention to changes in water quality. Anadu and Harding (2000) also found that, in cases where a water safety violation, such as coliform contamination or filtration issues, had occurred, increased awareness of that problem over longer durations was related to increased risk perceptions [40]. Because salience plays an important role in risk and trust evaluations, we expected that it may also be a significant contextual factor for an individual's behavior. For this study, we adapted aspects of Stewart's (2009) weather forecasting salience framework, which includes indicators such as event noticeability, event frequency, and an individual's knowledge, to form our drinking water salience conceptualization [31].

1.1.4. Tap Water Quality Evaluations

The taste, smell, and appearance of drinking water, referred to as organoleptic characteristics, are commonly cited factors for choosing filters or bottled water. Doria (2009) identified these organoleptic factors as the primary determinants of an individual's judgment of the quality of their tap water, showing a relationship between organoleptic factors and individuals' risk perceptions, trust, and decision to drink tap water [39]. Huerta-Saenz et al (2011) demonstrated that organoleptic ratings drive preferences for bottled water consumption over tap water [41]. Triplett et al. (2019) also found that organoleptic perceptions are important in distinguishing between tap and household filtered water drinkers, even when respondents have similar risk perceptions about the safety, contamination, and health risks of unfiltered tap water [37]. Although organoleptic evaluations are clearly important in impacting drinking water behaviors, it remains unclear whether or when they are a reflection of objective preference, familiarity with a water source, or, in part, an artifact of other traits and perceptions [18,41,42].

1.2. Hypotheses

Overall, the majority of the literature on drinking water choices focuses on the specific public preference for bottled water over tap, or vice versa. Since bottled water has a substantially higher carbon cost to produce and dispose of [43–45], and a higher economic cost for consumers [46], it is natural that this dichotomy of tap or bottled water has been the focus of previous studies. However, focusing on the dichotomy does not acknowledge the complex patterns of use that characterize how people consume water in their homes. While some studies have included home-filtered water [22,41], and some assessed the difference between exclusive and preferred tap or bottled water use e.g., [47], water use is still often determined by asking about primary water source choice, rather than assessing patterns in use frequency.

To better understand this complexity and the patterns that characterize how people consume water in their homes, we examined the frequency and patterns of use of four water source choices in a population in Roanoke, Virginia, USA. We looked at the factors of risk, salience, trust, and water quality evaluations to get a more comprehensive view of which factors are related to patterns of water consumption in the home. We hypothesized that:

H1:   Those with higher trust in their water utility to provide safe drinking water to them will be more likely to drink tap water.

H2:   Those with greater risk perceptions associated with their tap water will be less likely to drink tap water.

H3:   Salience is related to in-home drinking water patterns. Specifically:

    H3a: Those who are more knowledgeable about their tap water quality will be more likely to drink tap water.

    H3b: Those who notice changes to water quality more frequently will be less likely to drink from tap water sources.

H4:   Those with more positive organoleptic water quality evaluations will be more likely to drink from tap water sources.

Previous research has highlighted the importance of these factors to water source choice; however, to the best of our knowledge, this is the first study to relate all four to patterns of in-home water drinking behavior. We explored these factors' contributions to explaining patterns of drinking water sources in one's home with the ultimate goal of better evaluating potential indicators of community trust.

2. Materials and Methods

2.1. Study Area

The greater Roanoke area in western Virginia contains the city of Roanoke and the neighboring Roanoke and Botetourt counties. The urban and suburban area is home to roughly 227,000 people, the majority of whom are white (78%), lower-middle-class (median income from \$41,483 in Roanoke City to \$64,733 in Botetourt County) citizens [48,49]. Approximately 50% of households in the area are customers of the local water utility. Since its formation in 2004, the water utility that provides public water for the greater Roanoke area has had a record of safe and timely drinking water delivery that meets both the utilities' and government's standards.

Because our study focused on the decision to drink municipal tap water in the context of threats to surface water sources, we limited our population to customers of the local water utility whose water was sourced from surface water reservoirs. We formed our sampling frame using publicly available addresses, cross-listing our data with the utility to exclude residents who sourced their home water from wells. After randomly selecting a sample of 800 residents from this list and removing eight due to invalid addresses, a sample of 792 residents remained. This sample required a 40% response rate (n = 385) to attain 95% confidence intervals and $\pm$5% sampling error [50].

2.2. Distribution

We used a four-stage drop-off pick-up method adapted from Trentelman et al. (2016) to distribute the questionnaire [51]. Compared to other distribution methods, drop-off pick-up methods have an increased emphasis on social exchange and personal rapport created from in-person interactions, allowing one to garner a higher response rate than mail or phone survey methods [50–52]. We expected that the lack of water quality issues in the Roanoke area would result in low salience of the topic among residents and reduce interest in survey participation. We adopted the drop-off pick-up method to offset this effect and heighten the survey response. Both the questionnaire and distribution methods were approved for use by Virginia Tech's Institutional Review Board.

We sent introduction letters to each resident between September and November 2019, informing them of our study and intended visit. One week after the letters were sent, our research team visited residents' homes to provide them with information about the study and invite them to participate. We mailed introductions to groups of about 100 houses per week to allow us time to visit all residents' homes no more than two weeks after they received the letter. To maximize the likelihood that residents would be home from work, we scheduled visits between 4 p.m. and 8 p.m. on weekdays and between 10 a.m. and 4 p.m. on weekends. To ensure uniform interactions, team members adhered to a script adapted from the guidelines that Dillman et al. (2014) and Trentelman et al. (2016) presented [50,51].

If they agreed to participate, residents were given a survey to complete at their convenience and a doorknob bag to deposit their survey and leave outside for us to pick up two days later. We made up to two follow-up visits, as needed to contact residents and pick up completed surveys, waiting two days between visits to allow ample time for them to complete the survey. If a resident was not home when researchers arrived at their door and a survey was not left outside for pick up, researchers would leave a note on the door saying, "Sorry we missed you, we will be back on [insert date]," and attempt that visit again two days later. The note established the research team's reliability because the participants had been told what date to expect the team's return. We made up to three attempts per visit to contact a resident. After a third failed visit, we left a packet by the resident's door with a note explaining our contact attempts, a cover letter, the survey, and a postage-paid envelope for the survey's return. If the third failed attempt was on an introductory visit, we also included a letter introducing the study and requesting participation.

We first visited homes closest to our base location west of Roanoke and progressed east. We started data collection in September 2019 but halted operations in November when daylight saving time restricted sunlit working hours. The COVID-19 pandemic prevented us from collecting data again in spring 2020. This reduced our sample from 792 to 611. The remaining 181 addresses were located to the northeast of Roanoke. Potential differences between neighborhoods sampled and unsampled may reduce the generalizability of our results to all residents in the greater Roanoke area. To examine the potential for this, we conducted a test of proportions to compare the respondents in our reduced sample with the 2017 data on race from the U.S. Bureau of Census. Race is a demographic variable commonly associated with water quality issues [19,53], and our results indicated no differences see [27].

2.3. Measurement

The questionnaire assessed resident's patterns of drinking water source choice, trust in the water utility, their salience of water topics, perceived risk, evaluations of their tap water quality, and demographic factors. We pre-tested and refined the questionnaire ($n = 60$) before conducting the full survey.

2.3.1. Tap, Filtered, or Bottled Water

We assume in this study that people may prefer a particular source of drinking water, but their utilization of that source may vary in degree. Thus, preference for filtered water does not necessarily equate to an exclusive reliance on filtered water. To determine the degree to which residents utilize one or more water sources, we asked them to indicate how often in the past six months they drank bottled water, tap water, filtered water from an appliance, such as water filters in pitchers or refrigerators, or filtered water from a sink attachment in their home. We differentiated appliance filters from sink filters because appliance filters are often standard features built into modern refrigerators or are more affordable to purchase, as is the case with water filter pitchers, thus requiring low effort in comparison with sink filters to obtain. Because the focus of this study was on how water source choice could reflect trust in municipal utility, we focused on drinking behavior in the respondents' homes, ensuring that the respondents would associate tap water with the water their local utility provided. We measured responses on a five-point scale from 1 = *Never or almost never* to 5 = *All of the time or almost all of the time*. The residents responded to each of the four potential water source choices.

2.3.2. Trust

Two items gauged resident trust based on the conceptualization of trust as a willingness to accept vulnerability [27,54]. The residents first marked the extent to which they "trusted [their] local water utility to provide drinking water to [their] home that is safe to drink." We then asked their degree of comfort "with [their] local water utility controlling the quality of water delivered to [their] home." This second item measured vulnerability as a dimension of trust [32].

To ensure the trust variable would capture variability in the sample, we pre-tested the versions of the two indicators on Virginia residents using Amazon Turk (N = 111) [55]. We then conducted an in-person pilot study ($n = 20$) in our study area. Based on this pretesting we selected a 9-point scale for both trust items from 1 = *Do not trust at all / Not comfortable at all* to 9 = *Completely trust / Completely comfortable*.

2.3.3. Risk

We measured the residents' perceived risk of tap water using five indicators assessing cognitive safety, affective safety, cognitive severity, affective severity, and comparative risk. Cognitive safety assessed the residents' beliefs about the likelihood of tap water safety issues by asking the respondents to rate how safe they believed their water was from 1 = *Completely unsafe to drink* to 5 = *Completely safe to drink*. The affective safety item asked how concerned the residents were about their tap water safety (1 = *No concern* to 5 = *Extremely concerned*). To identify how severely the residents believed a public water disturbance could impact their lives, the respondents indicated how they would react cognitively (inconvenience and daily routine change) and affectively (worry and anger) and if they found themselves unable to access tap water in their homes. We measured these items on a 5-point Likert-type scale from 1 = *Not at all* to 5 = *A great deal*. Lastly, to measure comparative risk, we chose to focus on the risk of drinking bottled water compared to tap water. The respondents ranked tap water on a 5-point Likert-type scale from 1 = *More safe* to 5 = *Less safe*.

2.3.4. Salience

Drawing from Stewart's (2009) salience framework, we measured attention by asking how often respondents noticed unacceptable changes in their tap water in terms of taste, smell, and appearance (1 = *Never* to 5 = *Extremely often*) and unacceptable changes in their tap water in general (1 = *Never* to 4 = *Often*) [31]. We measured knowledge of water topics by asking the residents the amount of information they could provide to a friend or family about their neighborhood water quality (1 = *No information* to 5 = *A great deal of information*).

2.3.5. Tap Water Quality Evaluations

Tap water quality evaluations were measured on a 5-point Likert-type scale of acceptability. Respondents rated how acceptable their water was overall and in specific terms of taste, smell, appearance, and safety (1 = *Not acceptable* to 5 = *Completely acceptable*).

2.4. Analysis

We conducted a hierarchical cluster analysis to partition the sample into mutually exclusive groups based on broad patterns of in-home drinking water consumption. We used Ward's minimum variance method, which groups items into clusters based on the similarities and differences between each data point [56]. To determine an optimum cluster solution to use in subsequent analysis, we used both the Calinski/Harabasz test and the Duda/Hart test to assist in identifying an optimal cluster solution [57].

We explored the internal consistency and dimensionality of multiple indicators using Cronbach's alpha and factor analysis. We conducted one-way analyses of variance (ANOVAs) to look for differences across in-home drinking water patterns based on trust, salience, risk, and water quality evaluation perceptions, in addition to a resident's drinking water behavior. We report comparisons of cluster mean differences (ANOVA contrast) and standardized mean differences (Cohen's d is a quantitative measure of the magnitude of the effect). A general guideline for interpreting effect size for Cohen's d is that a standardized difference <0.20 represents a small effect, a difference of 0.50 represents a medium effect, and a difference >0.80 represents a large effect.

3. Results

3.1. Response

Our household contact rate was 75%, communicating with 538 residents out of 611 attempts between September and November 2019. Of those 538 residents, 114 residents refused to participate, 7 were determined ineligible to participate, and 57 agreed to participate but failed to return a survey. We received 352 surveys for a 59% response rate [58].

3.2. Descriptive Statistics

Bottled water had the highest frequency of use among residents and drinking from a sink filter the lowest (Table 1). A substantial portion of residents chose to drink from a mixture of two or more sources rather than from one source alone, as indicated by the low to moderate means and medians for each water source.

Table 1. Descriptive statistics for water source choice item. Statistics for each water source include number of observations (*n*), mean, median, standard deviation (SD), minimum scale value (minimum), and maximum scale value (maximum).

Water Source	*n*	Mean	Median	SD	Minimum: 1 = *Never or Almost Never*	Maximum: 5 = *All the Time or Almost All the Time*
Bottled	339	2.84	3	1.55	1	5
Sink filter	312	1.38	1	1.03	1	5
Appliance filter	331	2.44	2	1.65	1	5
Tap	337	2.32	2	1.50	1	5

Most respondents (61%) reported that they *mostly* or *completely* trusted their utility to deliver safe drinking water to them (Table 2). The majority of residents believed the safety of their water was a low risk both cognitively and affectively and believed they would experience moderate to severe consequences if their access was compromised. The majority of residents considered bottled water equally as safe as tap water (45%). Drinking water was a low salience topic for the majority of respondents, with most (63%) reporting that they *never* noticed changes in their tap water and had *little* to *no* information about their neighborhood water quality (61%). Finally, the majority of residents (53%) had generally favorable impressions of their tap water quality, ranking their tap's smell, odor, taste, and general characteristics as *very* or *extremely* acceptable.

Table 2. Descriptive statistics for predictor variables. Statistics for each survey item include number of observations (*n*), mean, median, standard deviation (SD), minimum scale value (minimum) and maximum scale value (maximum).

Item	*n*	Mean	Median	SD	Minimum	Maximum
Trust	345	6.34	7	1.97	1 = No trust	9 = Complete trust
Risk						
Cognitive safety	340	1.71	1	0.89	1 = Completely safe	4 = Completely unsafe
Affective safety	344	1.87	1	1.09	1 = Not concerned	5 = Extreme concern
Cognitive severity	339	3.25	5	1.40	1 = None	5 = A great deal
Affective severity	326	3.23	4	1.28	1 = None	5 = A great deal
Comparative	339	3.52	3	0.98	1 = Bottled muchless safe than tap	5 = Bottled muchsafer than tap
Salience						
Attention	348	1.68	1	0.74	1 = Never	4 = Often
Knowledge	343	2.35	2	0.98	1 = None	5 = A great deal
Tap water quality evaluations	348	3.70	4	0.93	1 = Not acceptable	5 = Extremely acceptable

3.3. In-Home Drinking Water Behavioral Choice Patterns

We selected a six-cluster solution for water source choice behavior groups, which was the solution with the best fit (Figure 2; Table S1).

We labeled the clusters based on the mean frequency patterns of bottled, sink filter, tap, and appliance filter water use (Figure 3). Almost two-thirds of the respondents (64%) drank primarily from a single water source. The largest cluster reflected people who drink bottled water exclusively (24% of the respondents; Figure 3A). The next largest cluster (22% of the sample) drank exclusively from an appliance filter (Figure 3B), followed by a cluster exclusively drinking tap water (18% of the participants; Figure 3C). Combined patterns included a cluster that equally drinks bottled and tap water (16%; Figure 3D), followed by mixes of all sources except for a sink filter (12% of the participants; Figure 3E) and a cluster that mixes sources but predominantly uses a sink filter (8%; Figure 3F).

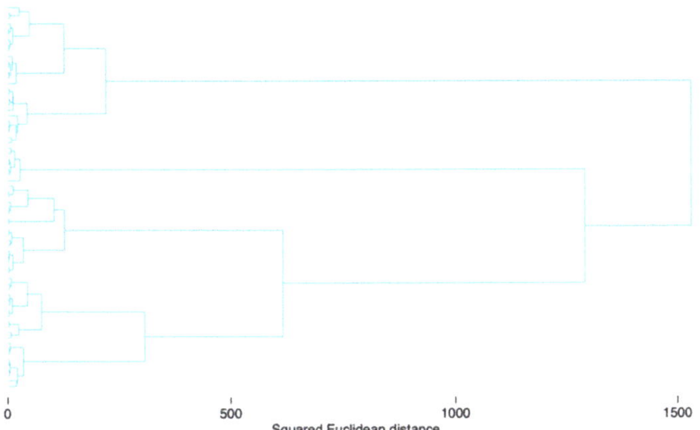

Figure 2. Dendrogram for the hierarchical cluster analysis.

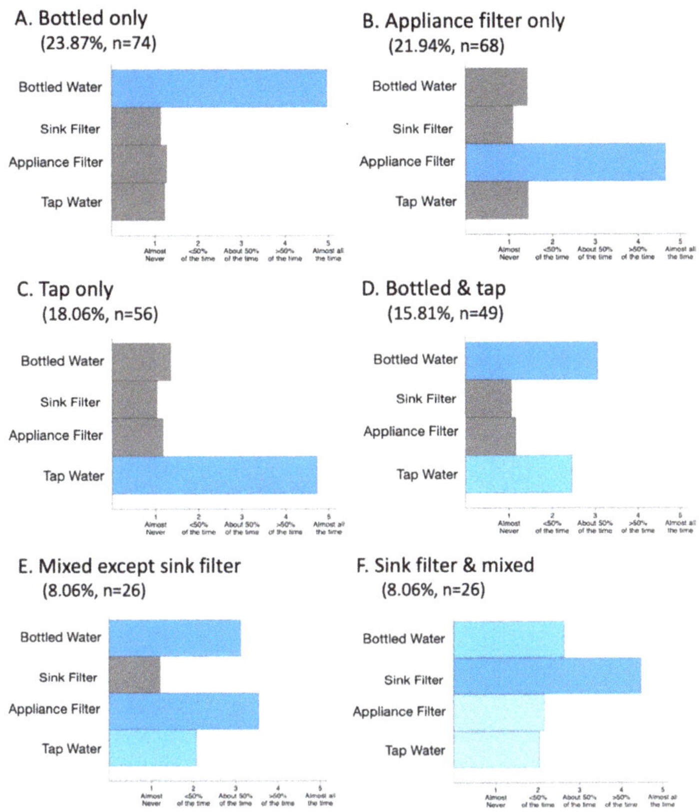

Figure 3. Means of the frequency of water source use for each cluster. Dark blue bars represent water source usage at or above *50% of the time*. Medium blue bars represent water usage between *<50% of the time* and *50% of the time*. Light blue bars represent water usage near *<50% of the time*. Grey bars represent water usage at or close to *almost never*.

3.4. Comparing Groups

The six clusters—*bottled only, tap only, appliance filter only, bottled & tap, sink filter & mixed*, and *mixed except sink*—differed not only in their water source choice patterns but also in terms of trust, salience, risk, and tap water evaluation variables (see Figure 4 for a comparison of differences significant at a $p < 0.05$ level; Table S2 contains full ANOVA results; Table S3 contains full pairwise comparison results).

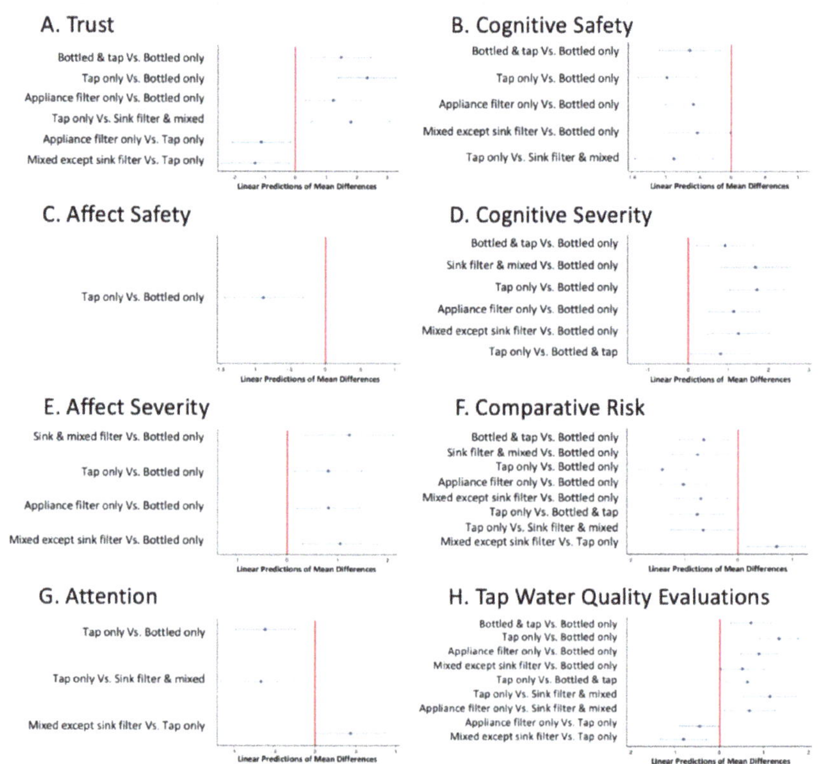

Figure 4. Post hoc pairwise comparisons of adjusted linear predictions of mean differences (first cluster–second cluster) between clusters for each predictor variable. Predictions were calculated using a Bonferroni test with 95% confidence intervals. Only significant differences at a $p < 0.05$ level are displayed.

Clusters were most similar in terms of affect risk perceptions (Figure 4C: affect safety and Figure 4E: affect severity) and salience variables (Figure 4G: attention and information known), evidenced by the lower number of mean differences between pairs in those variables. In contrast, comparative risk perceptions (Figure 4F) and organoleptic evaluations (Figure 4H) were most effective at distinguishing between groups.

3.4.1. H1: Those with Higher Trust in Their Water Utility to Provide Safe Drinking Water to Them Will Be More Likely to Drink Their Tap Water in Their Homes as Opposed to Drinking from Other Sources

The results of the pairwise comparisons mostly support our hypothesis that higher trust is related to the increased choice of tap water. Residents in the *tap only* cluster trusted their utility to deliver safe drinking water to them more than those in all other clusters except *bottled & tap* (Figure 4A; Table S3). Exclusive *tap water* drinkers' trust was much higher (mean (m) = 7.60) than that of exclusive *bottled water* drinkers (m = 5.26) (Cohen's d = 1.32) or *sink filter & mixed* water source drinkers (m = 5.75; Cohen's d = 1.23), moderately higher than that of *mixed except for sink filter* drinkers (m = 6.25; Cohen's d = 0.91), and slightly higher than that of *appliance filter* drinkers (m = 6.45; Cohen's d = 0.70).

3.4.2. H2: Those with Greater Risk Perceptions Associated with Their Tap Water Will Be Less Likely to Drink Their Tap Water in Their Home as Opposed to Drinking from Other Sources

We found partial support for this hypothesis. Clusters with tap water central to their drinking water choices had lower risk perceptions than some other behavioral clusters, but not all of them. Residents who drank *tap water* exclusively believed more strongly that their tap water was safe to drink (m = 1.21) than exclusive *bottled water* drinkers (m = 2.18; Cohen's d = −1.19) or *sink filter & mixed* water drinkers (m = 2.08; Cohen's d = −1.31) did (Figure 4B; Table S3). Exclusive tap water drinkers were on average *not* concerned about the safety of their tap water (m = 1.43). While they were less concerned than *bottled only* drinkers (m = 2.30; Cohen's d = −0.79), they did not differ in concern level from other behavioral groups (Figure 4C).

Residents who drank *tap* water exclusively reported that they would be more worried or angry if they lost access to drinking water (m = 3.38) than exclusive *bottled* water drinkers (m = 2.58; Cohen's d = 0.66; Figure 4E). Residents who drank *tap* water exclusively were also much more likely to think that not having access to their home tap water would directly impact their lives (m = 3.94) than people who drank *bottled water only* (m = 2.22; Cohen's d = 1.40) and moderately more likely than those who drank both *bottled & tap* water (m = 3.13; Cohen's d = 0.66; Figure 4D).

The comparative risk variable, asking about the relative safety of tap water compared to bottled water, differentiated tap water drinkers from almost every other cluster (Figure 4F). Residents who drank *tap water* exclusively believed that tap water is no different from bottled water in terms of safety (m = 2.88). Those who drank *bottled water* exclusively believed much more strongly that bottled water is safer than tap (m = 4.28; Cohen's d = −1.78). Most other groups also believed that bottled water is safer than tap, including those who drank *bottled & tap* water (m = 3.64; Cohen's d = −0.92), *sink filter & mixed* water sources (m = 3.52; Cohen's d = −0.71), and from *mixed water sources excepting a sink filter* (m = 3.58; Cohen's d = −0.80). There was no difference between the *tap water only* cluster and the *appliance filter only* cluster in terms of their beliefs about bottled versus tap water safety.

Overall, cognitive and comparative risk perceptions (Figure 4F) differed more than affective risk perceptions (Figure 4C,E) between clusters, as demonstrated by the larger effect sizes between means. Exclusive *tap water* drinkers had lower risk perceptions than several non-tap water clusters, but it was not universal. Exclusive *bottled water* drinkers and exclusive *tap water* drinkers differed across all variables, unlike other pairs of clusters, and had stronger effect size differences than with other groups. Exclusive tap water and appliance filter clusters did not show different means across any of the risk indicators.

3.4.3. H3: Issue Salience Is Related to In-Home Drinking Water Choice. Specifically

H3a: Those who are more knowledgeable about their tap water quality will be more likely to drink tap water in their homes.

H3b: Those who notice changes to water quality more frequently will be less likely to drink from tap water sources in their homes.

The pairwise comparison results did not support our first salience hypothesis (H3a), as information known about water quality was similar across the six clusters (F (5, 298) = 1.85, p = 0.103). The results partially support the hypothesis that attention paid to tap water quality is related to lower use of tap water (H3b; Figure 4G). Exclusive *tap water* drinkers noticed changes in their tap water (m = 1.31) less frequently than exclusive *bottled water* drinkers (m = 1.92; Cohen's d = −0.95), *sink filter & mixed* water source drinkers (m = 1.98; Cohen's d = −1.05), and *mixed water source excepting sink filter* drinkers (m = 1.75;

Cohen's d = −0.75). The *appliance filter only* cluster and *bottled & tap* cluster did not differ from the *tap only* cluster.

3.4.4. H4: Those with more Positive Organoleptic Water Quality Evaluations Will Be More Likely to Drink from Tap Water Sources in Their Home

Exclusive *tap water* drinkers had more favorable perceptions of their tap water quality (m = 4.35) than every other behavioral cluster (Figure 4H), supporting our fourth hypothesis. This difference was particularly pronounced with exclusive *bottled water* drinkers (m = 3.01; Cohen's d = 1.77) and *sink & mixed water source* drinkers (m = 3.28; Cohen's d = 1.48), and least pronounced with *appliance filter* drinkers, although still with a moderate effect difference (m = 3.91; Cohen's d = 0.62).

Overall, the ANOVA results (Figure 4 for selected comparison; Table S2 for full results) generally supported the hypotheses that tap water drinkers would have lower risk perceptions, higher trust, and higher water quality perceptions than other clusters. Exclusive tap water drinkers had the lowest number of differences with exclusive appliance water drinkers, suggesting a similarity between those clusters. Exclusive tap water drinkers showed the strongest differences when compared to exclusive bottled water drinkers and sink and mixed water source drinkers. This was evidenced by the high mean differences and effect sizes between those clusters compared to other pairs and the consistency with which those groups differed between variables.

4. Discussion

Our study examined patterns of in-home drinking water behavior and the degree to which it relates to perceptions of water and trust in water managers. We found that water behavior is best described by examining a range of drinking source choice patterns. Higher propensities to drink tap water were related to more positive perceptions of water quality and water utilities, including increased levels of trust in the water utility, more favorable evaluations of water quality, lower perceptions of risk, and lack of attention to salience to perceived changes in tap water.

4.1. Water Use Patterns: Characterizing Behavior

Our characterization of residents' water drinking behavior into clusters reliant on one water source exclusively and clusters who mix their water sources helped capture patterns of behavior more thoroughly than a dichotomy (e.g., tap water vs. bottled water) would have. Less than half of our sample (42%) drank either bottled or tap water exclusively, although many respondents did indicate that they employed an exclusive water source. Much of the previous literature examining drinking water source choice reduced water use to a comparison between bottled water use and tap water use e.g., [21,39]. Our results showed that exclusive tap water drinkers have the largest differences in perceptions with exclusive bottled water drinkers across variables (Figure 4). Despite these differences, we found that a more detailed picture could be painted by including mixed clusters, which accounted for over one-third (36%) of the respondents' behavior. Looking at patterns of behavior recognizes that people may rely on multiple sources of drinking water at home.

Distinguishing between filter types was also useful as differences in perceptions of risk, trust, salience, and water quality depended on the type of household filter individuals employed. While some previous research has added exclusive household filter water drinkers to their study scope e.g., [41], we are unaware of other studies that differentiate between household filter types. In one example of this characterization of filter types, Leveque and Burns (2017) found that individuals who report higher risk perceptions of tap water are equally likely to drink household filtered water compared to tap [22]. In contrast, we found the appliance filter cluster to be one of the most similar clusters to tap water, while the sink filter and mixed cluster had some of the strongest differences of all clusters, topped only by the bottled water cluster. It may be important to distinguish between the type of household filter individuals employ as it can indicate vastly different perceptions with regard to drinking water and water municipalities.

4.2. Perception Differences

4.2.1. Trust

Our results demonstrated that residents who primarily utilize tap water as their drinking source display higher levels of trust in their water utility than those who utilize mixed sources or bottled water only. Groups that relied partially on tap water, such as *bottled & tap* and *appliance filter*, had higher trust than the *bottled water* cluster. Yet those who drank tap water exclusively had higher trust levels than every cluster except for the *bottled & tap* cluster. Previous studies comparing tap water and bottled water groups have also found that tap water drinkers have higher trust levels [18,21,33]. Our results add to these previous findings by demonstrating how trust can increase with even partial reliance on tap water.

The difference in trust between those who drink bottled water exclusively and those who drink it half the time could be explained by examining why people are choosing to drink bottled water. Ward et al. (2009) found that, while those who drink bottled water have a range of beliefs about the health benefits of bottled water, these beliefs are not necessarily key drivers of bottled water purchases. In that case, convenience is the most motivating factor [59]. Similarly, Saylor et al. (2011) found that lack of convenience is a barrier to drinking tap water [21]. Consequently, a number of additional factors, including convenience and marketing effort, may impact drinking water choices. Future research could explore how these additional factors interact with trust, risk beliefs, and salience to explain drinking water behavior.

4.2.2. Risk

We found partial support for our hypothesis that tap water drinkers would have decreased risk perceptions, in terms of cognitive safety, affect safety, cognitive severity, affect severity, and comparative risk, compared to other clusters. Cognitive safety and severity were more useful at distinguishing between other clusters than affect safety and severity, and comparative risk most useful. Similar to previous studies, all risk items reliably and substantively differentiated between tap and bottled water e.g., [36,37].

Given the stronger risk perceptions of bottled water users, we were not surprised that comparative risk demonstrated strong differences between clusters but were surprised that those differences did not reflect patterns in the cognitive safety variable. The comparative risk variable introduced by Saylor (2011) asks respondents to contrast the cognitive safety of tap water relative to bottled water [21]. As a result, we expected perceptions to be similar across groups with these two variables. However, the strongest difference in comparative risk was found between tap water and bottled water groups, while the strongest difference in cognitive safety was between tap water and sink filter and mixed groups. Additionally, the comparative risk variable distinguished between tap water and all other clusters except for appliance filter drinkers, while cognitive safety only differentiated tap water from bottled water and sink filter and mixed groups. Future research might benefit from expanding avenues of relative risk to compare other water sources besides just bottled and tap or comparing water sources using other measures besides cognitive safety.

4.2.3. Salience

Out of the four variables we examined, salience contributed the least to understanding behavioral group differences. The respondents' salience perceptions did not support the hypothesis that increased knowledge about water quality would impact tap water usage and partially supported the hypothesis that increased attention to tap water changes would be related to decreased tap water usage. Grupper [27] surveyed respondents to determine the role that salience plays in trust formation and found that salience acts as an important contextual variable for trust determinants. While salience may not have a strong influence on drinking water behavior directly, it is indirectly relevant through the formation of trust, which is more proximately related to behavioral choice.

4.2.4. Water Quality Evaluations

Our findings supported the hypothesis that higher tap water quality evaluations in terms of taste, smell, and appearance would be associated with an individual's choice to drink from tap water sources instead of alternates. Of the perceptions we measured, water quality evaluations had as strong an effect as trust or comparative risk and was the only factor that differed between those who exclusively drank tap water and all other clusters. These results are similar to those of previous literature that cites organoleptic properties as a crucial determinant of tap water drinking [18,37,41]. While trust in water safety and risk perceptions both focus on water security, organoleptic evaluations relate to enjoyment of water. March et al. (2020) found that perception of taste is a stronger motivational factor in bottled water drinking than safety concerns [60]. Our results support the observation that these factors of enjoyment are strongly tied to each type of drinking water behavior.

5. Conclusions

Effective management of drinking water resources benefits when water utilities can adapt and be flexible amid changing environmental conditions. Such resilient management can be bolstered by positive relationships between water managers and the communities they serve [12]. In this study, we examined how community perceptions that can strengthen these relationships relate to drinking water behavior. We specifically examined the role of trust, risk perceptions, salience of drinking water, and water quality evaluations in their choice of in-home drinking water sources. We were particularly interested in the role of trust, a concept often studied in natural resource management because of its positive relationship with resilient management support, effective communication, and technology acceptance [13,14,16]. We allowed for the fact that people may vary in their in-home drinking water behaviors, enabling us to examine further differences that characterize behavioral patterns.

While other studies have demonstrated relationships between demographic variables such as income, educational level, and gender with water source consumption [41,61,62], it was not part of our research question. However, when we examine our data, we do not find a relationship between income, education level, or gender, and cluster membership.

Our results show that those who drink tap water often have more favorable tap water quality evaluations, have lower perceptions of its risk, pay more attention to tap water changes, and have higher trust in their water managers, especially compared to those who drink bottled water exclusively or from a sink filter. Managers may be able to use the link between high trust and tap water choice as an indicator of public trust. This can help them determine their need to promote this trust through efforts to increase communication between utilities and the public. As utilities shift to employing more resilience-based management strategies, it is increasingly important to understand the ties between the public perceptions that define community relationships with their utilities and the drinking water behavior those communities engage in.

Supplementary Materials: The following are available online at https://www.mdpi.com/2306-5338/8/1/49/s1: Table S1: Calinski/Harabasz test and Duda/Hart test for K means cluster analysis on water source choice; Table S2: One-way ANOVA results comparing drinking water behavior cluster means for each independent variable; Table S3: Post hoc pairwise comparisons of mean differences between pairs of clusters for each independent variable.

Author Contributions: Conceptualization, M.A.G. and M.G.S.; methodology, M.A.G., M.G.S., and M.E.S.; validation, M.A.G. and M.G.S.; formal analysis, M.A.G. and M.G.S.; investigation, M.A.G.; resources, M.A.G., M.G.S., and M.E.S.; data curation, M.A.G. and M.G.S.; writing—original draft preparation, M.A.G.; writing—review and editing, M.A.G., M.G.S., and M.E.S.; visualization, M.A.G., and M.G.S.; supervision, M.A.G., and M.G.S., M.E.S.; project administration, M.A.G., M.G.S.; funding acquisition, M.G.S., and M.E.S. All authors have read and agreed to the published version of the manuscript.

Funding: This work was financially supported by the National Science Foundation grant CNS-1737424.

Data Availability Statement: The data that support the findings of this study are not publicly available because of confidentiality restrictions with human subjects research. Ethical approval for this research study was granted by the Virginia Tech Institutional Review Board (IRB #19-023).

Acknowledgments: We would like to thank the Western Virginia Water Authority for their cooperation and help with our efforts. We also thank the students who distributed our survey and assisted with data entry: E. Bowlin, B. Wilson, A. Alls, A. Donnelly, A. Bosley, and Q. Hair.

Conflicts of Interest: The authors declare no conflict of interest. The funders had no role in the design of the study; in the collection, analyses, or interpretation of data; in the writing of the manuscript; or in the decision to publish the results.

References

1. Dunn, S.M.; Brown, I.; Sample, J.; Post, H. Relationships between Climate, Water Resources, Land Use and Diffuse Pollution and the Significance of Uncertainty in Climate Change. *J. Hydrol.* **2012**, *434–435*, 19–35. [CrossRef]
2. Garrote, L. Managing Water Resources to Adapt to Climate Change: Facing Uncertainty and Scarcity in a Changing Context. *Water Resour. Manag.* **2017**, *31*, 2951–2963. [CrossRef]
3. Pouget, L.; Escaler, I.; Guiu, R.; Mc Ennis, S.; Versini, P.A. Global Change adaptation in water resources management: The Water Change project. *Sci. Total Environ.* **2012**, *440*, 186–193. [CrossRef] [PubMed]
4. Ho, J.C.; Michalak, A.M. Challenges in tracking harmful algal blooms: A synthesis of evidence from Lake Erie Jeff. *J. Great Lakes Res.* **2015**, *41*, 317–325. [CrossRef]
5. Olsen, C.S.; Shindler, B.A. Trust, acceptance, and citizen agency interactions after large fires: Influences on planning processes. *Int. J. Wildland Fire* **2010**, *19*, 137–147. [CrossRef]
6. Carpenter, S.; Walker, B.; Anderies, J.M.; Abel, N. From Metaphor to Measurement: Resilience of What to What? *Ecosystem* **2001**, *4*, 765–781. [CrossRef]
7. Folke, C. Resilience (Republished). *Ecol. Soc.* **2016**, *21*, 44. [CrossRef]
8. Holling, C.S. Resilience and Stability of Ecological Systems. *Annu. Rev. Ecol. Syst.* **1973**, *4*, 1–23. [CrossRef]
9. Swanson, E. *AP-GfK Poll: About Half of Americans Confident in Tap Water*. 2016. Available online: https://apnews.com/article/eedf886daa334d7c871e531d804620a1 (accessed on 11 March 2021).
10. Water Quality Association. Summary & Highlights National Study of Consumers' Opinions & Perceptions Regarding Water Quality. 2019. Available online: https://wqa.org/Portals/0/Publications/ConsumerStudy2019_Public.pdf (accessed on 11 March 2021).
11. Slovic, P. Perceived risk, trust, and democracy. *Risk Anal.* **1993**, *13*, 675–682. [CrossRef]
12. Bondelind, M.; Markwat, N.; Toljander, J.; Simonsson, M.; Säve-Söderbergh, M.; Morrison, G.M. Building trust: The importance of democratic legitimacy in the formation of consumer attitudes toward drinking water. *Water Policy* **2019**, *21*, 1–18. [CrossRef]
13. Stern, M.J.; Baird, T.D. Trust ecology and the resilience of natural resource management institutions. *Ecol. Soc.* **2015**, *20*, 14. [CrossRef]
14. Davenport, M.A.; Leahy, J.E.; Dorothy, A.E.; Ae, H.A.; Jakes, P.J. Building Trust in Natural Resource Management Within Local Communities: A Case Study of the Midewin National Tallgrass Prairie. *Environ. Manag.* **2007**, *39*, 353–368. [CrossRef] [PubMed]
15. Lachapelle, P.R.; McCool, S.F. The Role of Trust in Community Wildland Fire Protection Planning. *Soc. Nat. Resour.* **2012**, *25*, 321–335. [CrossRef]
16. Song, A.M.; Temby, O.; Kim, D.; Saavedra Cisneros, A.; Hickey, G.M. Measuring, mapping and quantifying the effects of trust and informal communication on transboundary collaboration in the Great Lakes fisheries policy network. *Glob. Environ. Chang.* **2019**, *54*, 6–18. [CrossRef]
17. Ross, V.L.; Fielding, K.S.; Louis, W.R. Social trust, risk perceptions and public acceptance of recycled water: Testing a social-psychological model. *J. Environ. Manag.* **2014**, *137*, 61–68. [CrossRef]
18. Doria, M.d.F. Bottled water versus tap water: Understanding consumers' preferences. *J. Water Health* **2006**, *4*, 271–276. [CrossRef] [PubMed]
19. Doria, M.d.F. Factors influencing public perception of drinking water quality. *Water Policy* **2010**, *12*, 1–19. [CrossRef]
20. Jakus, P.M.; Shaw, W.D.; Nguyen, T.N.; Walker, M. Risk perceptions of arsenic in tap water and consumption of bottled water. *Water Resour. Res.* **2009**, *45*. [CrossRef]
21. Saylor, A.; Prokopy, L.S.; Amberg, S. What's Wrong with the Tap? Examining Perceptions of Tap Water and Bottled Water at Purdue University. *Environ. Manag.* **2011**, *48*, 588–601. [CrossRef]
22. Levêque, J.G.; Burns, R.C. A Structural Equation Modeling approach to water quality perceptions. *J. Environ. Manag.* **2017**, *197*, 440–447. [CrossRef] [PubMed]
23. De Doria, M.F.; Pidgeon, N.; Hunter, P. Perception of tap water risks and quality: A structural equation model approach. *Water Sci. Technol.* **2005**, *52*, 143–149. [CrossRef]
24. Levallois, P.; Grondin, J.; Gingras, S. Evaluation of consumer attitudes on taste and tap water alternatives in Quebec. *Water Sci. Technol.* **1999**, *40*, 135–139. [CrossRef]
25. Gray, S.; Shwom, R.; Jordan, R. Understanding factors that influence stakeholder trust of natural resource science and institutions. *Environ. Manag.* **2012**, *49*, 663–674. [CrossRef]

26. Higgins, E.T.; Kruglanski, A.W. Knowledge Applicability, Activation: Accessibility, and Salience. In *Social Psychology: Handbook of Basic Principles*; Higgins, T., Kruglanski, A.W., Eds.; The Guilford Press: New York, NY, USA, 1996; pp. 133–168.

27. Grupper, M.A. Exploring the Role of Trust in Drinking Water Systems in Western Virginia. Master's Thesis, Virginia Tech, VTechWorks, Blacksburg, VA, USA, 2020. Available online: http://hdl.handle.net/10919/99860 (accessed on 1 March 2021).

28. Ferrier, C. Bottled Water: Understanding a Social Phenomenon. *Ambio J. Hum. Environ.* **2001**, *30*, 118–119. [CrossRef]

29. Jaffee, D.; Newman, S.A. More Perfect Commodity: Bottled Water, Global Accumulation, and Local Contestation. *Rural Sociol.* **2012**, *78*, 1–28. [CrossRef]

30. Debbeler, L.J.; Gamp, M.; Blumenschein, M.; Keim, D.; Renner, B. Polarized but illusory beliefs about tap and bottled water: A product- and consumer-oriented survey and blind tasting experiment. *Sci. Total Environ.* **2018**, *643*, 1400–1410. [CrossRef]

31. Stewart, A.E. Minding the weather: The measurement of weather salience. *Bull. Am. Meteorol. Soc.* **2009**, *90*, 1833–1841. [CrossRef]

32. Mayer, R.C.; Davis, J.H.; Schoorman, F.D. An Integrative Model of Organizational Trust. *Acad. Manag. Rev.* **1995**, *20*, 709–734. [CrossRef]

33. McSpirit, S.; Reid, C. Residents' perceptions of tap water and decisions to purchase bottled water: A survey analysis from the Appalachian, Big Sandy coal mining region of West Virginia. *Soc. Nat. Resour.* **2011**, *24*, 511–520. [CrossRef]

34. Anthony Cox, L. What's wrong with risk matrices? *Risk Anal.* **2008**, *28*, 497–512. [CrossRef]

35. Poortinga, W.; Pidgeon, N.F. Trust in Risk Regulation: Cause or Consequence of the Acceptability of GM Food? *Risk Anal.* **2005**, *25*, 199–209. [CrossRef] [PubMed]

36. Hu, Z.; Morton, L.W.; Mahler, R.L. Bottled water: United States consumers and their perceptions of water quality. *Int. J. Environ. Res. Public Health* **2011**, *8*, 565–578. [CrossRef]

37. Triplett, R.; Chatterjee, C.; Johnson, C.K.; Ahmed, P. Perceptions of Quality and Household Water Usage: A Representative Study in Jacksonville, FL. *Int. Adv. Econ. Res.* **2019**, *25*, 195–208. [CrossRef]

38. Bratanova, B.; Morrison, G.; Fife-Schaw, C.; Chenoweth, J.; Mangold, M. Restoring drinking water acceptance following a waterborne disease outbreak: The role of trust, risk perception, and communication. *J. Appl. Soc. Psychol.* **2013**, *43*, 1761–1770. [CrossRef]

39. De Doria, M.F.; Pidgeon, N.; Hunter, P.R. Perceptions of drinking water quality and risk and its effect on behavior: A cross-national study. *Sci. Total Environ.* **2009**, *407*, 5455–5464. [CrossRef]

40. Anadu, E.C.; Harding, A.K. Risk perception and bottled water use. *J. Am. Water Work. Assoc.* **2000**, *92*, 82–92. [CrossRef]

41. Huerta-Saenz, L.; Irigoyen, M.; Benavides, J.; Mendoza, M. Tap or bottled water: Drinking preferences among urban minority children and adolescents. *J. Community Health* **2012**, *37*, 54–58. [CrossRef]

42. Harmon, D.; Gauvain, M.; Reisz, A.; Arthur, I.; Story, S.D. Preference for tap, bottled, and recycled water: Relations to PTC taste sensitivity and personality. *Appetite* **2018**, *121*, 119–128. [CrossRef] [PubMed]

43. Cole, M.; Lindeque, P.; Halsband, C.; Galloway, T.S. Microplastics as contaminants in the marine environment: A review. *Mar. Pollut. Bull.* **2011**, *62*, 2588–2597. [CrossRef] [PubMed]

44. U.S. Government Accountability Office. *Bottled Water: FDA Safety and Consumer Protections Are Often Less Stringent Than Comparable EPA Protections for Tap Water*; Report No. GAO-09-610; United States Government Accountability Office Testimony; 2009. Available online: https://www.gao.gov/assets/gao-09-610.pdf (accessed on 11 March 2021).

45. Gleick, P.H.; Cooley, H.S. Energy implications of bottled water. *Environ. Res. Lett.* **2009**, *4*, 1–6. [CrossRef]

46. Güngör-Demirci, G.; Lee, J.; Mirzaei, M.; Younos, T. How do people make a decision on bottled or tap water? Preference elicitation with nonparametric bootstrap simulations. *Water Environ. J.* **2016**, *30*, 243–252. [CrossRef]

47. Delpla, I.; Legay, C.; Proulx, F.; Rodriguez, M.J. Perception of tap water quality: Assessment of the factors modifying the links between satisfaction and water consumption behavior. *Sci. Total Environ.* **2020**, *722*, 137786. [CrossRef]

48. U.S. Census Bureau. ACS Questionnaire 2017. 2017. Available online: http://www.census.gov/acs (accessed on 13 July 2020).

49. U.S. Census Bureau. American Community Survey (ACS). 2017. Available online: https://www.census.gov/quickfacts/fact/table/roanokecountyvirginia,roanokecityvirginia,botetourtcountyvirginia/PST045219 (accessed on 11 March 2021).

50. Dillman, D.A.; Smyth, J.D.; Christian, L.M. *Internet, Phone, Mail, and Mixed Mode Surveys: The Tailored Design Method*, 4th ed.; John Wiley & Sons Inc.: Hoboken, NJ, USA, 2014.

51. Trentelman, C.K.; Irwin, J.; Petersen, K.A.; Ruiz, N.; Szalay, C.S. The Case for Personal Interaction: Drop-Off/Pick-Up Methodology for Survey Research. *J. Rural Soc. Sci.* **2016**, *31*, 68–104.

52. Steele, J.; Bourke, L.; Luloff, A.E.; Liao, P.-S.; Theodori, G.L.; Krannich, R.S. The Drop-Off/Pick-Up Method for Household Survey Research. *Community Dev. Soc. J.* **2001**, *32*, 238–250. [CrossRef]

53. Fragkou, M.C.; McEvoy, J. Trust matters: Why augmenting water supplies via desalination may not overcome perceptual water scarcity. *Desalination* **2016**, *397*, 1–8. [CrossRef]

54. Stern, M.J.; Coleman, K.J. The Multidimensionality of Trust: Applications in Collaborative Natural Resource Management. *Soc. Nat. Resour.* **2015**, *28*, 117–132. [CrossRef]

55. Chandler, J.; Shapiro, D. Conducting Clinical Research Using Crowdsourced Convenience Samples. *Annu. Rev. Clin. Psychol.* **2016**, *12*, 53–81. [CrossRef]

56. Ward, J.H. Hierarchical Grouping to Optimize an Objective Function. *J. Am. Stat. Assoc.* **1963**, *58*, 236–244. [CrossRef]

57. Everitt, B.S.; Landau, S.; Leese, M. *Cluster Analysis*, 4th ed.; Oxford University Press Inc.: London, UK, 2001.

58. AAPOR Response Rate Calculator. Available online: https://www.aapor.org/Education-Resources/For-Researchers/Poll-Survey-FAQ/Response-Rates-An-Overview.aspx (accessed on 14 October 2019).

59. Ward, L.A.; Cain, O.L.; Mullally, R.A.; Holliday, K.S.; Wernham, A.G.; Baillie, P.D.; Greenfield, S.M. Health beliefs about bottled water: A qualitative study. *BMC Public Health* **2009**, *9*, 196. [CrossRef]
60. March, H.; Garcia, X.; Domene, E.; Sauri, D. Tap water, bottled water or in-home water treatment systems: Insights on household perceptions and choices. *Water* **2020**, *12*, 1310. [CrossRef]
61. York, A.M.; Barnett, A.; Wutich, A.; Crona, B.I. Household bottled water consumption in Phoenix: A lifestyle choice. *Water Int.* **2011**, *36*, 708–718. [CrossRef]
62. Johnstone, N.; Serret, Y. Determinants of bottled and purified water consumption: Results based on an OECD survey. *Water Policy* **2012**, *14*, 668–679. [CrossRef]

 *hydrology*

Article

Interdisciplinary Water Development in the Peruvian Highlands: The Case for Including the Coproduction of Knowledge in Socio-Hydrology

Jasper Oshun [1,*], Kristina Keating [2], Margaret Lang [3] and Yojana Miraya Oscco [4]

1 Department of Geology, Humboldt State University, Arcata, CA 95521, USA
2 Department of Earth and Environmental Sciences, Rutgers University, Newark, NJ 07102, USA; kmkeat@rutgers.edu
3 Department of Environmental Resources Engineering, Humboldt State University, Arcata, CA 95521, USA; margaret.lang@humboldt.edu
4 Department of Political Science, University of Toronto, Toronto, ON M5S3G3, Canada; yojana.mirayaoscco@mail.utoronto.ca
* Correspondence: oshun@humboldt.edu

Abstract: Agrarian communities in the Peruvian Andes depend on local water resources that are threatened by both a changing climate and changes in the socio-politics of water allocation. A community's local autonomy over water resources and its capacity to plan for a sustainable and secure water future depends, in part, on integrated local environmental knowledge (ILEK), which leverages and blends traditional and western scientific approaches to knowledge production. Over the course of a two-year collaborative water development project with the agrarian district of Zurite, we designed and implemented an applied model of socio-hydrology focused on the coproduction of knowledge among scientists, local knowledge-holders and students. Our approach leveraged knowledge across academic disciplines and cultures, trained students to be valued producers of knowledge, and, most importantly, integrated the needs and concerns of the community. The result is a community-based ILEK that informs sustainable land and water management and has the potential to increase local autonomy over water resources. Furthermore, the direct link between interdisciplinary water science and community benefits empowered students to pursue careers in water development. The long-term benefits of our approach support the inclusion of knowledge coproduction among scholars, students and, in particular, community members, in applied studies of socio-hydrology.

Keywords: socio-hydrology; knowledge coproduction; integrated local environmental knowledge; education and training; community-based water development

Citation: Oshun, J.; Keating, K.; Lang, M.; Miraya Oscco, Y. Interdisciplinary Water Development in the Peruvian Highlands: The Case for Including the Coproduction of Knowledge in Socio-Hydrology. *Hydrology* **2021**, *8*, 112. https://doi.org/10.3390/hydrology8030112

Academic Editors: Tamim Younos, Alaina J. Armel, Tammy Parece, Juneseok Lee and Jason Giovannettone

Received: 15 June 2021
Accepted: 23 July 2021
Published: 2 August 2021

1. Introduction

In the high-elevation Andes of Perú, agrarian communities, many of which are Indigenous, depend on local water resources for sustenance. Sustainable and lasting water management in the Peruvian Andes requires a complete perspective of water resources that incorporates physical science, social context, and the production of knowledge. Although human impacts often are not fully quantified in watershed studies, there is a growing need to integrate the combined effects of humans and hydrology to focus on "land-change science" [1,2] that integrates society with nature by way of socio-hydrology [3] or analyses of "waterscapes" shaped by both hydrological and social flows [4,5]. Here, we present a framework for explicitly including the coproduction of knowledge from scientific experts, students, and local knowledge-holders in the rural Andean village of Zurite and argue that such a framework is critical for sustainable water development.

Water resources in the Peruvian Andes are impacted by humans at global, regional, and local scales. Global-scale climate forcing accelerates the melting of glaciers, impacting

95

water supply (e.g., see [6,7]). Local and regional development projects, such as hydroelectric dams or mining operations, impact water use and drive laws affecting water accessibility (e.g., see [8,9]). Additionally, traditional Indigenous practices, dating to pre-Inca times, govern land management and aquifer recharge, and inform local water use and community water allocations (e.g., see [10]). These activities result in complex spatiotemporal effects on the water supply and create challenges for the study of water resources and the implementation of sustainable water management in the region. Carey et al. developed a socio-hydrology framework bridging natural and social sciences to model how downstream communities may adapt to changes in local water supply [11]. The focus of their study was the Santa River basin, a watershed in which communities, mining operations and hydroelectric dams compete for shrinking supplies of glacial meltwater. In addition to scientific data to understand water resources, Carey et al. incorporated five human variables—politics and economics, laws and institutions, technology, land and resource use, and societal response—into their framework. The framework provides a useful link between science and society, as these factors strongly influence communal water use and sustainable water management.

Here, we argue that, in order for socio-hydrology to not only link natural and social science but also to result in community-level adaptation to climate change, we must include a sixth factor: the collaborative coproduction of knowledge. Our approach to knowledge coproduction leverages community needs and knowledge with our scientific expertise, and trains a collaborative, interdisciplinary, and multinational cohort of students in community-minded approaches to water resources research. Including knowledge coproduction in the socio-hydrology framework, particularly when using an interdisciplinary, applied, and community-minded approach, provides a platform for knowledge transfer and mutual learning between all stakeholders in the project, and is critical for achieving community-level resiliency to climate change such as local sustainable land and water management.

2. Combined Socio-Hydrology Hydro-Social and Integrated Local Environmental Knowledge Conceptual Model

We used an integrated local environmental knowledge (ILEK) approach that focuses on the coproduction of knowledge to inform decision-making and action in Zurite (Figure 1). The ILEK approach was developed by Sato et al. [12–14] and is purported to increase a community's resiliency and ability to adopt environmentally sustainable changes. We used this approach to outline the knowledge actors who collectively produce the knowledge and inform sustainable environmental decision-making. All the knowledge actors in the ILEK approach contribute to three interconnected outcomes: knowledge coproduction, decision-making and action, and institutional change. In our framework, the knowledge actors were the local knowledge holders, i.e., the local stakeholders from the community of Zurite, the scientific experts (geologists, geophysicists, engineers, hydrologists, sociologists), and the U.S. and Peruvian students.

The knowledge coproduced by this set of actors applies the hydro-social framework [11], including "upstream" and "downstream" knowledge, to develop an ILEK linking upstream water resources to downstream water use and decision-making that, in turn, guides land use/management, resulting in potential changes to upstream water resources. Here, upstream knowledge focused on water resources originating in the Upper Ramuschaka Watershed (URW), the primary water source for the village of Zurite, and included the disciplines of hydrology, geology, geomorphology, and geophysics. Downstream knowledge focused on water use, local knowledge, and decision-making in Zurite, and includes the disciplines of engineering, sociology, history, and community studies, with a focus on canal reconstruction and the potential for adaptations to water use. Two major goals of the project were sustainable water management within the village of Zurite and developing a trained cohort of socio-hydrologists. By including knowledge coproduction and the training of an interdisciplinary and multinational cohort of students, the ILEK produced will benefit from the diversity of knowledge actors (in terms of age, subject area expertise, ethnic and cultural background, etc.) who, together, will be better at solving complex environmental problems [15,16]. All knowledge actors will benefit from a platform facilitating knowledge exchange and mutual learning. Additionally, students will benefit when trained to understand the complexities of upstream and downstream disciplines affecting sustainable water management and are empowered through their contributions to the project's success.

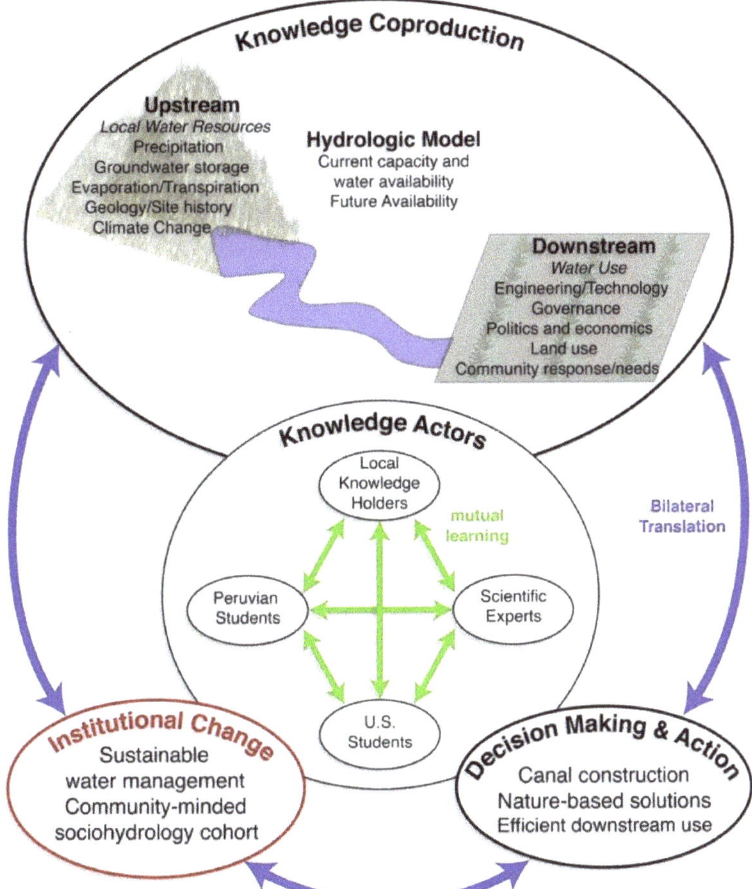

Figure 1. The conceptual model of knowledge coproduction, applied to a hydro-social framework [11]. scientific experts, local knowledge-holders, and U.S. and Peruvian students combine their forces, to learn from one another and coproduce upstream and downstream knowledge. The result is local environmental knowledge (ILEK) [12–14] that integrates upstream and downstream disciplines, and western and traditional science, to inform decision-making and promote institutional change, such as sustainable water management.

We applied this combined hydro-social/ILEK framework to a two-year water development project that took place in Zurite (population 3600), Peru, named Bonanza en los Andes (referred to as Bonanza). The structure of this paper closely follows the three interconnected themes of our project: science, society, and knowledge coproduction. First, we summarize the upstream knowledge relating to water resources in the Vilcanota Watershed, which includes Zurite, and the URW. Next, we summarize the downstream knowledge and discuss water and society in the Peruvian highlands. Third, we describe our approach to knowledge coproduction, which blends upstream and downstream disciplines and focuses on training cohorts of students who contribute to an ILEK. Finally, we report the knowledge produced, as well as the short-term and anticipated long-term impacts of our project on the community of Zurite and on participating students.

3. Upstream and Downstream Knowledge: Water Resources and Use in Andean Perú

3.1. Upstream Water Resources throughout the Vilcanota Watershed

Perú holds more freshwater resources per capita than any other South American country; however, the longitudinal barrier of the Andes forms a barrier between the moist eastern flanks, draining to the Amazon Basin, and the rain shadow desert along the populous Pacific coast [17]. High elevation peaks trap moisture and high-elevation, headwater streams coalesce to form large eastward-flowing rivers, such as the Vilcanota-Urubamba (Vilcanota) River.

The Vilcanota River originates in the Altiplano near Ausangate (6372 m.a.s.l.) and flows northwest through the Sacred Valley, north of Cusco (Figure 2). The Vilcanota Watershed totals 11048 km^2 at the confluence of the Urubamba and Lucumayo Rivers in Santa Maria (elevation 1200 m.a.s.l., [8]), and is the second most glaciated tropical watershed in the world. However, by 2016, only 1.28% of the watershed was glaciated, reflecting a 30% reduction since 1985 [18–21]. The high elevation portions of the watershed, including the village of Zurite, have a present-day Köppen–Geiger climate classification of CwB [22].

The hydrograph of the Vilcanota River is driven primarily by precipitation, which is concentrated between November and April (less than 4% of annual precipitation falls between May and August) [8]. Glacier meltwater contributions to the Vilcanota River are estimated to be less than 3% annually [23]. The vast majority of the Vilcanota Watershed (~78%) is part of the puna biome: a seasonally dry grass and shrubland environment existing above the tree line and below the permanent snow line, along the spine of the Andes from central Perú to central Chile and Argentina [24–27]. Groundwater contributions from the puna sustain perennial headwater streams and account for nearly all the annual discharge in the Vilcanota River [8,23]. The city of Cusco, the regional capital and home to over 420,000 people [28], draws approximately 90% of its water from the combined sources of Laguna Piuray—a lake fed by an approximately 26 km^2 catchment area draining the puna—and the Vilcanota River, which is primarily groundwater-fed [29]. Within the Department of Cusco, over 20% of its inhabitants are without access to a permanent water supply [30]. The percentage without access to safe water resources rises to 42% in the rural province of Anta (55,000 inhabitants), of which Zurite is one of 9 districts [31].

Much of the puna exists in landscapes that are sculpted by glaciers with distinctly glacial morphology, such as steep cirque walls, hanging valleys, moraine deposits, and, importantly, gently sloping valley floors in which peat-forming, seasonally saturated wetlands—known as bofedales—are often found [24,26,32]. Bofedales act as shallow aquifers, filling in the wet season, due primarily to groundwater contributions from upslope, and draining to streams throughout the dry season [33]. As temperatures warm and glaciers melt, the puna biome is predicted to expand to higher elevations [34], suggesting even greater proportions of puna-derived runoff from these watersheds. Seasonal variability in precipitation in the Vilcanota Watershed is projected to increase [35], which, importantly, could lead to a delay in the onset of precipitation in September through December, a time critical for planting the most widely grown crop in the Department of Cusco, corn [36]. The effect of changes in the timing and quantity of precipitation and the impacts on water resources, particularly in the dry season, are poorly understood.

3.2. Downstream Knowledge: Water and Society in Andean Peru

3.2.1. Water Rights in Perú

To understand the societal impacts on water use in Zurite, it is helpful to understand the historical context of water rights in Perú. Both the General Water Law of 1969 and agrarian reform in the 1970s had a positive impact on the ability of local communities to regulate their water use. These changes placed water management decisions in the hands of local entities, known as Comisiónes de Regantes (Water Users' Commissions), and redistributed land from large haciendas to indigenous families (see, e.g., [37]). The decentralization of power allowed for increased local autonomy over water and land management (see, e.g., [31]). However, in the 1990s, a wave of neoliberal economic policies resulted in expanded mining operations and hydroelectric dams, which increased water demand [38]. To support these operations, decision-making power was concentrated in the hands of companies, despite their small numbers compared to stakeholders within the watersheds [39,40]. More recently, there has been additional pressure on the limited water supplies, due to an increase in the amount of water exported, carried in the form of vegetables [6]. The increased demand for water has disproportionately affected small, rural, and predominantly indigenous communities and, in some cases, led to conflict over water resources between mining companies and local stakeholders (see, e.g., [5,6,41]).

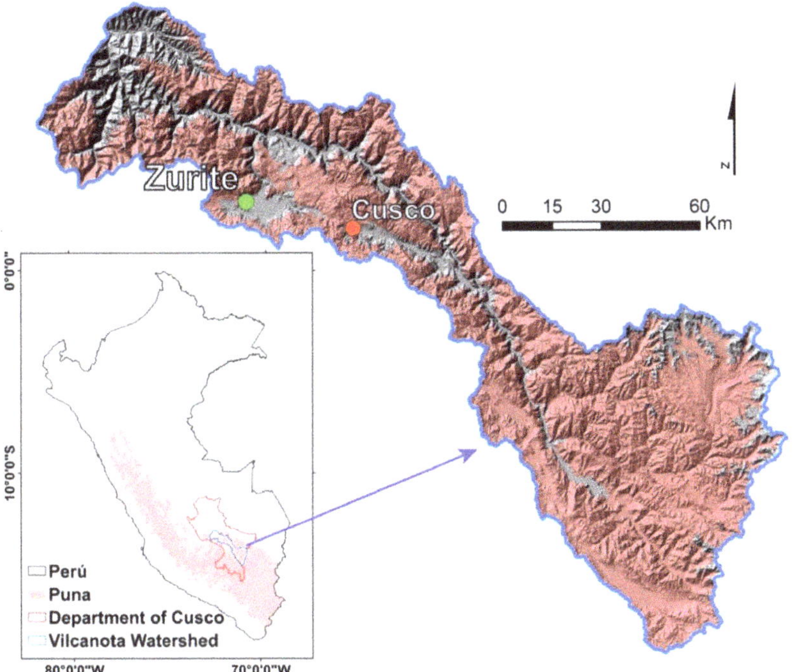

Figure 2. Zurite is located approximately 40 km to the northwest of Cusco, within the 11,048 km^2 Vilcanota Watershed [8]. The Vilcanota flows to the NW and is entirely within the Department of Cusco, outlined in red on the inset map. Red shading indicates the extent of the puna biome in Perú and within the Vilcanota Watershed [25]. The Sacred Valley is the gray region (lower elevation than the puna) directly north of Cusco and Zurite.

Over the final decades of the 20th century, water resources in Andean Perú were targeted for privatization by the World Bank, to (1) decentralize decision making, (2) create and protect private property rights for water, and (3) establish a market for water to improve water use efficiency [37,42]. In actuality, aging infrastructure and flood irrigation practices result in poor water-use efficiency and 86% of national freshwater use by the agricultural sector [17] and market forces resulting from the privatization of water are not likely to incentivize water conservation [37]. Alternative approaches for water management, in which decision-making power lies with the community, have been shown to increase efficiency and protect Indigenous communities' access to water by leveraging traditional water management practices (see, e.g., [37]). As presented in our conceptual model above (Figure 1), we argue that decision-making is best informed by the coproduction of knowledge from a diverse body, including scientific experts, local knowledge-holders, and students.

3.2.2. Indigenous Andean Communities and Allin Kawsay

A core principle of Allin Kawsay, the Indigenous Andean cosmovision, is to live in harmony with the environment and the community [43,44]; this principle guides the communal laws and customs of water allocation within Indigenous Andean communities [45]. Allin Kawsay exemplifies traditional ecological knowledge (TEK) and refers to the principles of "good living" in Quechua. Similar to the TEK of the Indigenous peoples of North America, in Allin Kawsay, all aspects of a physical space—mountains and other landforms, plants, animals—are interconnected and deserving of the same respect and treatment [46]. Furthermore, the health of the mountains, other landforms, plants and animals serve as a reflection of the health of the community. [47,48]. Mountains, in particular large peaks, are homes to powerful spirits, or apus (see, e.g., [45,49]). The apus provide water, but, when dissatisfied, can also cause lake outburst floods that are interpreted as ominous signs [50]. Allin Kawsay forms a holistic world view that emphasizes a collaborative relationship between humans and the physical space, rather than an extractive one.

Although Indigenous Andean communities have been resilient over generations and have historically adapted to changes in climate, in part due to the guidance of Allin Kawsay, the current accelerated rates of climate change, combined with a lack of government resources, endangers access to local water supplies, resulting in local hardship [51,52]. Many Indigenous communities are reliant on agricultural production for subsistence and possess limited alternatives for producing a living, yet they have typically been excluded from water resources allocation discussions (see, e.g., [8,53]). A 2009 law, which defined water as national property, sought to prevent the privatization of water and increase water use efficiency [54]. This law created a national administration for the use of water, as well as providing agency to local governments to manage local water resources [39]. Nevertheless, a frequent turnover in government leadership, particularly at the national level, may result in changes with detrimental effects on Indigenous communities that are best countered by longer-term planning (exceeding 10 years) within these communities [21].

Much of the past research has focused on examining the impact of melting glaciers on Indigenous communities' access to water (see, e.g., [8,11,41,52,53]). However, Indigenous Andean communities that are dependent on non-glacial water sources, such as the puna, are also vulnerable to changes in climate and/or policies that affect access to local water resources. To develop sustainable water management within the Indigenous community of Zurite, which depends on water originating from the puna for sustenance, we used an interdisciplinary approach to knowledge coproduction that integrates the perspective of Allin Kawsay with community needs, upstream and downstream disciplines, and student and local training to produce ILEK [13,55]. In addition to building local capacity and community resiliency within Zurite, this approach will train a culturally educated cohort of socio-hydrologists and provide a framework that can be adapted to other communities. Our approach focuses on issue-driven and solution-oriented science, and the critical role of the local community in transforming the relationship between the environment and society, to ultimately achieve sustainable futures [13].

4. Framework for the Coproduction of Knowledge

4.1. Project Goals and Approach to Knowledge Coproduction

We designed a year-long education program to support the project's objectives of water resources research spanning upstream water sources in the URW and downstream water uses in the community of Zurite. We leveraged Dr. Oshun's longstanding relationship with Zurite (he lived and volunteered in Zurite in 2003 and has since returned to support several development projects), such that we began our project with a collaborative spirit, uniting

U.S.-based scientists and students with Peruvian students and the leadership in Zurite. The municipal government, Farmer's Union, and Water Users' Commission expressed a need for canals, and we began our project with an agreement to collaborate on the design and construction of irrigation canals, which we incorporated into our upstream and downstream water resources research. Both the research and the irrigation canal collaboration would contribute to ILEK via a continuous exchange of ideas and knowledge coproduction among the diverse group of participants. A guiding principle of our approach was that all knowledge-holders (see Figure 1; local knowledge holders, U.S. students, Peruvian students, and scientific experts) are valuable assets and are thus critical to the overall project success.

We designed the curriculum of our project to guide the students through the application of research, to the benefit of the community of Zurite. The curriculum included topics in the upstream disciplines of hydrology, geology, geophysics, and ecology, and the downstream disciplines of engineering, sociology, history, and community studies. We leveraged our experiential and academic expertise to include TEK and Allin Kawsay within the curriculum, to provide context and a deeper and more holistic understanding of water resources, usage and the needs of the community.

As longer-term student participation in projects has been shown to maximize the positive and lasting impact on students [56,57], our education program was designed to span a full year, with three components each year: a preparatory course in the spring, an international immersive field experience in the summer, and an independent research project in the fall (summarized in Figure 3). U.S. students from the primary research institute (Humboldt State University, HSU) committed to the program for at least half the year, with the majority of students participating in all three components. Peruvian students and U.S. students who were not from HSU participated in the summer international research experience and, in some cases, also participated in the fall independent research. The project ran for two years (2018 and 2019). The students' backgrounds and each component of the program are summarized below.

Figure 3. The educational framework for our Bonanza project. We embedded the summer international research experience within coursework to maximize the impacts and foster sustained learning.

4.2. Student Participants

Over two years, 29 students participated in either the coursework, the international field experience, or both (12 students in 2018 and 17 students in 2019). Students came from undergraduate (18) and graduate (11) programs in the United States (23) and Perú (6). Students were pursuing degrees in geology (13), environmental resources engineering

(7), geophysics (2), hydrology (2), physics (1), geography (1), film (2), and environment and community (1). Two students participated in both years of the program. The 22 HSU students took 1–4 semesters of preparatory and research-focused course work, spanning the international field experience.

4.3. Spring Preparatory Course

We designed the spring preparatory course to build a common foundation of knowledge among students with diverse educational backgrounds, and to encourage students to apply their disciplinary skills to pursue individual research. We built foundational knowledge through introductory lectures focused on the disciplines necessary to achieve our objectives of water resources research in the URW and the construction of irrigation canals in Zurite: regional geology, the puna landscape, the Andean cosmovision, Incan and contemporary irrigation practices, and the community of Zurite. These lectures were accompanied by exercises and discussions that combined to form an interdisciplinary focus on water, which included physical hydrology, agricultural water needs, regional and local water management, and the issues of water quantity and quality encountered by local communities and community organizations. We fostered feelings of empathy and connectedness, developed communication skills, and engaged students in teamwork through exercises connecting our scientific objectives to the points of view and needs of the Indigenous and farming community of Zurite. These exercises were designed to instill a sense of respect for people of other cultures, which, through the international summer research, grew into intercultural competence or cultural humility (see, e.g., [58–60]).

Concurrently, we guided students through the research process by leveraging their disciplinary and newly acquired skill sets. Students worked independently to develop research questions, conduct literature reviews, generate hypotheses, and construct a plan for field research. The full list of projects and skills gained by the students associated with each project are provided in Table 1. Example projects included a simple hydrologic model to inform site selection for the installation of stream gages, a flight plan to collect drone data to generate a digital elevation model, an assessment of current and projected irrigation needs, a calculation of crop-specific irrigation needs and irrigation canal capacities in Zurite, and the hydraulic engineering of proposed sections of irrigation canals. These projects integrated our growing data set with the varied skillsets of the students in the class, and directly contributed to our objectives of understanding water resources in the Andean puna and informing sustainable water use in the community of Zurite. At the conclusion of the semester, students presented a progress report and a testable hypothesis, and identified their responsibilities and needs, to test their hypotheses in the summer field research.

Table 1. List of student projects, contributions to ILEK in Zurite, and the skillsets acquired by students. For the locations, U indicates upstream, and D indicates downstream.

Project	Location	Contributions to ILEK	Skill Sets
Geologic field mapping	U	Geologic map of URW, including identification of potentially active faults and landslide hazards	4-dimensional thinking, integration of Spanish language geology terms, cross-cultural collaborations in the field
Digitizing geologic map	U	Spatially oriented geologic map	GIS skills, interpreting field notes, 4-dimensional thinking

Table 1. *Cont.*

Project	Location	Contributions to ILEK	Skill Sets
Drone flight plan and data collection	U	1-m digital elevation model of URW, videos of groups working in the field and of the landscape	UAV pilot license, flight experience at high elevation, structure from motion data analyses
Geophysical Analysis	U	Application of geophysical methods to determine the subsurface structure and inform aquifer storage and hydrologic flow pathways	Field survey design, team management, data processing analysis, and interpretation, AGU poster presentations
Slope stability analysis	U	Identification of landslide hazards above Zurite	Application of published model, geospatial skills, interpretation of model sensitivity to model parameters
Installation of hydrologic monitoring equipment	U	Continuous recording of precipitation and temperature, discharge distributed throughout the URW	Identification of suitable sites, rating curve construction, cross-cultural learning, data analysis and quality control, AGU presentation
Installation of deep monitoring wells	U	Continuous measurements of groundwater resources beneath hillslope and in bofedal (2019–2021)	Contract work, interdisciplinary learning, cross-cultural learning, language practice
Distributed discharge measurements	U	Spatiotemporally distributed discharge at 1–3-month intervals (2019–2021) to identify connections between landscape structure and hydrologic productivity	Interdisciplinary learning, field skills and technology to apply to senior theses, contract-based employment, cross-cultural learning, language practice
Estimates of seasonally dynamic water storage in URW	U	Connection of geology and puna landscape structure, including geophysics, to rainfall-runoff metrics and total water yield	Application of mathematical analysis presented in recent literature to URW stream data, processing data in R, experience teaching other students
Soil characteristics and plant water status in the URW	U	Characterization of the URW within the puna biome, soil classification, plant water availability and source water identification	Application of research methods to a new environment, opportunity to teach others in the classroom and field

Table 1. *Cont.*

Project	Location	Contributions to ILEK	Skill Sets
Distributed evapotranspiration model Zurite	U	Development of model to predict ET	Geologic field mapping, ground-truthing of a remotely sensed model, cross-cultural learning, language practice
Hydrologic modeling using MODFLOW	U	Identification of wet, low gradient regions (bofedales) and springs, guided future hydrologic field measurements	MODFLOW, model sensitivity analyses, teamwork through the integration of datasets (geologic, hydrologic, seismic) through collaborations with student colleagues
Quantification of water storage in bofedales	U	Hydraulic properties of bofedales, estimation of dynamic storage and contributions to streamflow	Collection of data in the field over two seasons, interpretation of multiple methods, spatial analyses, and the connection of results to broader project and community needs
Stage-discharge rating curve for diversion weir at outlet of URW	D	Relationship to quantify continuous discharge from URW and connect to water demand downstream	Application of engineering skills for community benefit, analysis of existing data, connection to necessary field measurements
Irrigation needs: current and under future climate scenarios	D	Quantification of total irrigation water demand, projections of changes to supply (small) and demand (large increase under a warming climate water use and identification of opportunities to boost local resiliency	Application of engineering skills to the benefit of the local community, cultural humility, interpretation of risk to community, and potential resiliency
Hydraulic modeling of existing and proposed canal network HEC-RAS, and estimated cost	D	Design of proposed canal, including material and labor cost estimates	Application of engineering skills for community benefit, cross-cultural learning, and cultural humility
Water Quality in Zurite	D	Distributed tests in new potable water system showed good water quality in 2018	Application of engineering coursework, communication with the community, cultural humility

Table 1. *Cont.*

Project	Location	Contributions to ILEK	Skill Sets
Film—interviews of project participants and community	D	Zuriteños empowered to voice their concerns to outside scientists, opportunity to learn across the community and from the Bonanza group	Constructing narrative arc, developing questions, Spanish Quechua language, data management, empathy, cultural humility

4.4. Summer International Research Experience

We led month-long research and community development trips to Zurite in 2018 and 2019. Students stayed in the homes of three different host families, with shared meals and discussions at the home of a centrally located host family. The homestays and engagement with the community encouraged open-mindedness, and exposed students to a diversity of values and cultural backgrounds e.g., [61].

Much of the logistics and the execution of fieldwork were led by the students, with the program faculty providing guidance in the field and through evening discussions. In the URW, fieldwork included geologic field mapping, the installation of hydrologic monitoring equipment—including rain gages, subsurface moisture probes, and groundwater monitoring wells—soil substrate measurements, and intensive geophysical surveys. In Zurite, fieldwork included surveys of the irrigation water distribution infrastructure, measurements used to design the construction of new canals, and gathering of water use information from local water officials. U.S. students worked alongside Peruvian students, providing mutual learning opportunities for field techniques such as geologic mapping, hydrologic monitoring, geophysical surveys, drone flights, rotary drilling, and well installation, as well as sharing customs. The unique experience of working in Zurite, including the strong tropical sun, the exhilarating thin air, and unique smells, tastes and sounds, provided a learning experience that cannot be replicated in a classroom [62].

Members of the community contributed to our fieldwork as part of the organized faena, or work patronage paid to the community. We offered training focused on gaging rivers, learned local perspectives, and explained our scientific objectives. Incorporating local representatives in our fieldwork allowed for an exchange of ideas between the local knowledge-holders and the students, and also demonstrated the community's investment in our research and in the water development project. The shared experience between students and volunteers provided an opportunity to both gain linguistic and cultural capital (see, e.g., [59,63]) and carry forward an experience of seeing the world through someone else's eyes.

Our role in Zurite, beyond the scientific objectives in the URW, and our contributions to the design and construction of irrigation canals, was as facilitators of knowledge coproduction contributing to ILEK. Our approach has been inclusive and participatory. Rather than communicating in one direction from science to citizens, we followed the best practices of including stakeholders (see, e.g., [64]) to identify community needs to help inform best practices in society-environment interactions (see, e.g., [65]).

4.5. Fall Research Projects

In the fall, we advised independent student research, and coordinated discussions to encourage collaboration between students in weekly meetings. Embedding the students' international research experience within the coursework, and participation in the follow-up semester, instills lasting learning [56]. We introduced new methodological approaches through four introductory labs, in which students explored and processed drone-derived imagery, precipitation and runoff data to construct a water balance, seismic data to determine water-holding units in the subsurface, and canal hydraulics data to design irrigation canals. As the semester progressed, we loosened the structure, allowing more time for students to creatively explore their individual research topics. We met with students to review their progress, in a series of checkpoints designed to lead students through analyses, interpretation, and preparation for oral and written reports. Importantly, students were asked to include a "capacity-building statement", outlining a plan to integrate the knowledge they produced into the community sphere. By explicitly including the needs of the community, we guided students to apply their research, fostering a deeper appreciation for the host country and, importantly, an enhanced understanding of one's agency and responsibility as a local and global citizen (see, e.g., [66]). Table 1 presents a list of student-led projects, either in the classroom, in the field, or both, each project's contribution to ILEK, and the skillsets acquired by the students.

5. Knowledge Coproduced and Project Impacts

5.1. Upstream Knowledge: Upper Ramuschaka Watershed and Local Water Resources

The primary water source used for irrigation in Zurite is the 2.12 km^2 Upper Ramuschaka Watershed (Figure 4). The URW presents as a typical puna landscape showing glacial morphology, including steep headwalls of bedrock or grasslands and low gradient bofedales. The underlying geology is composed primarily of sandstones, conglomerates, quartzites and mudstones, with isolated carbonates in the western headwaters of the URW. The sedimentary units are part of the San Jeronimo Group, which is Eocene to Oligocene in age [67]. The sedimentary units typically dip to the southwest; however, there are a number of smaller-scale folds and faults in the URW. There is a large quartz monzodiorite intrusive complex, which is likely part of the Yauri-Andahuaylas Batholith (42–30 Ma) along the western boundary of the URW [67–69].

We measured annual precipitation in WY 2019 to be 752 mm, and in WY 2020 to be 825 mm (details found in [70]), which is similar to the 38-year average of 855 mm in Anta (10 km away, [71]). Over our study period, approximately 50% of precipitation occurred between December and February, and only 1–3% between June and August. Runoff accounted for 62–80% of annual precipitation and was also highly seasonal, with only 10–18% occurring in the dry season. Importantly, however, streamflow measured at the diversion weir located at the outlet of the URW never fell below 11 L/s. Streamflow in the URW is fed directly by groundwater draining from hillslopes or the slow drainage of bofedales, which hold large volumes of water and cover 11.5% of the URW. We estimate, conservatively, that seasonal storage in bofedales amounts to about one-half of dry season runoff in the URW. The presence of bofedales in the source watershed of Zurite, and their capacity to seasonally store and release groundwater to sustain perennial flow, represents a built-in source of natural resiliency and is the focus of ongoing research [70].

5.2. Downstream Knowledge: Water Use and Governance in Zurite

Through discussions with local leaders and water users, we produced the following summary of how water is used in Zurite within the context of water resources derived from the URW. Systems in Zurite can be described as 'Indigenous', meaning that irrigation is for the most part carried out in a manner similar to traditional Andean practices (see, e.g., [72–74]). Water draining from the URW is diverted via a concrete weir and inter-basin transfer canal, where it is routed through a system of canals to flood and irrigate crops to the east of Zurite (Figure 4). Corn, grown in the wet season, represents approximately 80% of the agricultural production across Zurite's 1200 ha of cropland, with lesser production of potatoes, quinoa, wheat, fava beans, and forage for animals. The irrigation need is most acute over the dry winter, from May to September (primarily to grow fava beans) and, crucially, to irrigate fields recently sowed with corn before the onset of the rainy season, which begins in September (Figure 5).

Figure 4. Zurite and the Upper Ramuschaka Watershed (URW). The URW is located 1300 m above the village, within the puna biome. Water is often diverted from the URW via an inter-basin canal for irrigation to the east of Zurite. Blue squares show the locations of 4 surface reservoirs, and the green-shaded area shows the area in which we collaborated with the local community to construct 1.3 km of irrigation canals (map and photo below).

Agricultural plots, known as chacras, are typically 0.25 to 1 ha and are owned and operated by individual families. Because the Ramuschaka Watershed is within the District of Zurite, the community of Zurite has the right to form and elect a local Water Users' Commission. Similar to other Andean communities, the Water Users' Commission manages turnos, or turns, in which users pay 15 soles/hectare (~USD 5/hectare) for the right to divert water from the extensive canal network to flood-irrigate their chacras. An aging canal infrastructure system and the nature of flood irrigation result in poor water-use efficiency. The president of the commission is elected every 4 years and is responsible for regulating and ensuring equitable water allocation to users in the community. In order to operate, the Water Users' Commission in Zurite must follow the regulations outlined by both the regional Water Administrative Authority in Cusco and the National Water Authority and collect and pay taxes to both administrative bodies. According to the Zuriteños, only rarely do these payments return to directly benefit the local community through irrigation system improvements. Instead of incorporating local best practices of land and water management, government institutions often seek to impose a homogeneous system of water infrastructure across diverse agrarian communities.

The efficient use of water resources in Zurite is challenged by limited storage, an aging infrastructure, and, in at least one case, a lack of community-based knowledge. The community operates five reservoirs with a total storage capacity of 10,500 m^3. This volume is substantially below dry-season irrigation demands and necessitates some fields being left fallow [75]. These reservoirs do not, nor were they intended to, function as long-term storage but instead operate as water elevation controls, allowing the multiple low flow-rate water sources to accumulate prior to delivery. The reservoirs are filled and drained approximately daily during the dry season, as water users schedule irrigation delivery on a rotating schedule to meet individual families' needs. Thus, the disruption of water delivered from both local and regional water supplies for even a short time can threaten a family's crops and livelihood. The construction of these systems across steep and, in some cases, unstable terrain makes irrigation disruptions extremely common. Figure 6a illustrates the construction variety throughout most of the system—concrete canals with control gates that minimize seepage water losses run parallel to and in series with hand-dug, earthen canals. Figure 6b shows a previously well-designed and controlled canal, constructed by the municipality, after it was damaged by a debris flow in March 2019. The debris flow was caused by a poorly designed and hydraulically mismanaged larger canal in the upper basin that was installed by regional authorities to transfer water from the outlet of the URW to an adjacent small drainage channel. Here, national water organizations proceeded without the involvement of local knowledge in the canal design, which led, in part, to this catastrophic debris flow and the destruction of the canal and surrounding crops of corn.

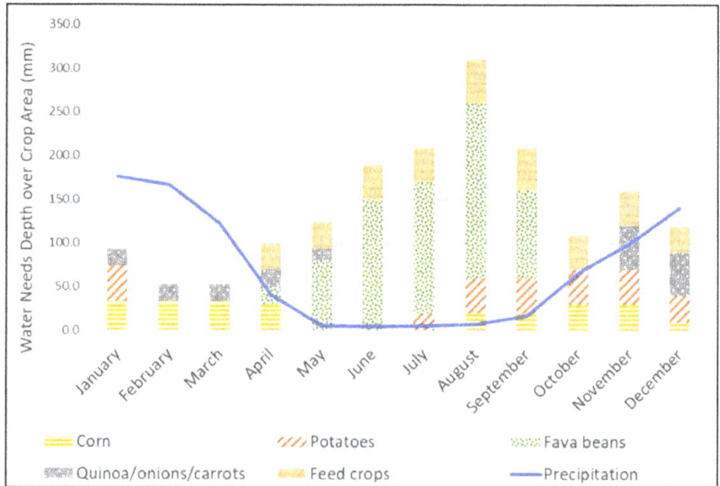

Figure 5. Average monthly precipitation (mm) and typical monthly irrigation needs (mm) for major crops grown in Zurite. The need for water is most acute in the dry winter to irrigate fields of fava beans, feed crops, and recently planted corn (August-November). Figure by W. Wunderlich and A. Virgil.

5.3. Project Impacts on the Community of Zurite

5.3.1. Immediate Infrastructure Benefit

The Bonanza project has resulted in immediate and longer-term benefits to the community of Zurite. Over the two years of the program, we collaborated with the municipal government, Farmer's Union, and Water Users' Commission in Zurite to negotiate, plan, and execute a USD 71,283 canal development project. Our monetary contribution to the project, funded by Geoscientists Without Borders, totaled USD 20,000. We advised student researchers, in collaboration with Zurite's engineers, in designing the canals. In March 2020, the community of Zurite finished building the 1.3 km of irrigation canals (Figure 7). These canals extended the irrigation system, to provide water to and boost the crop yield of land owned and farmed by over 100 families.

5.3.2. Benefits of Knowledge Coproduction: Identification of the Risks and Opportunities of Current Water Resources and Irrigation Practices

Two examples illustrate our inclusive approach to knowledge coproduction, the learning that occurs in both directions, and the initial steps we have taken to develop an ILEK with the community. First, we learned of the principal concerns within the community through discussions with Zuriteños. An elder related, "We are worried about climate change and we want our children to as well . . . we think the educational aspect is the only way that we will have sustainable development in our communities." We agree that education and local training can directly contribute to the ILEK necessary for sustainable water management. As the elder expressed, younger generations are viewed as the inheritors of the land who, one day, will assume positions of authority and responsibility. Through our conversations with the community, we have gained a deeper knowledge of the issues at hand, and we are better positioned to work collaboratively to produce, share, and use knowledge [64].

Figure 6. (**a**) A modern concrete canal with control gates adjacent to an earthen canal. (**b**) Destruction to concrete canal following March 2019 debris flow.

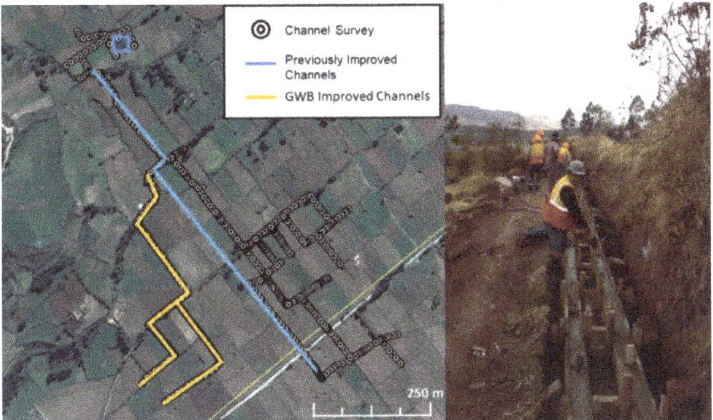

Figure 7. The left panel shows the area of canal improvements (green-shaded area of Figure 4). Circles show the extent of the irrigation canal network, with the blue square indicating the location of a surface reservoir that regulates flow to the canals. The blue line shows canals constructed by the community prior to 2018. The yellow line shows the extent of the 1.3 km of canals built through the Geoscientists Without Borders-funded Bonanza collaboration with the community of Zurite. The entire project totaled USD 71,283 and GWB's contribution was USD 20,000. The photograph on the right shows the early phases of construction in fall 2019.

We interpreted the elder's comment as a need for bottom-up approaches to building local capacity through training and the coproduction of knowledge to form an ILEK. Our response to this comment was to train locals in basic hydrological monitoring. Although in their early stages, these training initiatives have boosted the local capacity to monitor local water resources and inform water management.

Second, our fieldwork in the URW, one vertical kilometer above Zurite and a 50-minute car ride, or approximately a 4-h walk, provided an opportunity for local representatives to see their local water source. The remoteness of the URW disconnects Zurite's primary water source from the community. Tomás Ruiz López, the Community President of Zurite, remarked in 2019 that he had not seen the bofedales of the URW in twenty years. By working alongside our research team, Sr. Ruiz López and other representatives from the community have gained a new perspective and ILEK that will inform land management decisions in the community.

In order to disseminate our scientific findings on the importance of bofedales, and our recommendations for land conservation, we have written a progress report (2020), with a final report due in 2021, and we have presented our findings at two town hall meetings (2019 and 2020) and at a virtual symposium (2021). Our inclusion of community concerns in our reports, and our willingness to contribute our knowledge to ILEK in Zurite builds trust, and also leads to adaptive capacity and enhanced resiliency (see, e.g., [76,77]). The developing ILEK of our collaborative coproduction of knowledge identifies the risks of climate change-induced precipitation volatility, inefficient irrigation practices and limited water storage, frequent disruptions to infrastructure, and the opportunities of nature-based solutions (see, e.g., [10]), focusing on bofedal conservation or the introduction and expansion of more efficient irrigation practices, such as sprinkler irrigation.

We are currently working with the community to design and install up to ten sprinkler systems with flexible, replaceable, and relatively cheap parts. These sprinklers will connect to the existing canal network to increase water use efficiency, thus decreasing the total water demand on water resources from the URW building resiliency through adaptive water management. The presence of the sprinkler systems may also serve as a training model for neighboring communities and thus transfer resiliency beyond the community of Zurite.

5.4. Impacts on Student Participants

We used qualitative student commentaries to assess the benefits of knowledge coproduction on participating students. In particular, we focus on the high-impact practices of embedding our program in academic coursework [56], and increased self-confidence to pursue future research opportunities [57]. Below, we organize the impacts to students around three themes—the interdisciplinarity and applied nature of knowledge coproduction, the explicit inclusion of community knowledge and expressed needs, and the increased feelings of belonging and empowerment felt by students.

Bonanza leveraged the disciplinary knowledge of a diverse set of students and applied this knowledge to the objective of understanding water resources in the puna and building capacity and resilience in Zurite. Students cite the acquisitions of interdisciplinary perspectives as one of the primary benefits of research and study abroad [57]. A geology student, who participated in our program in 2018 wrote, "This project required me to utilize all of my past research and academic experience to work in a team environment with students, faculty and scientists from the U.S. and Perú to study the geology, geophysics and hydrology of a rural watershed in the Peruvian Andes." An environmental resources engineering (ERE) student from 2019 commented, "It was so valuable to get to work with multiple disciplines and gain new perspectives. Definitely life-changing." These sentiments were shared by a Peruvian student who participated from 2018 to 2020, "To be part of this research team was a great honor for me due to the learning that took place and, thanks to the members of this organization, I learned a great deal about geologic mapping, hydrogeology, hydrology, and geophysics."

Bonanza explicitly incorporated the knowledge and needs of the community of Zurite in the student research and learning experience. Whereas the experience of living and researching abroad is a high-impact practice that leads to greater student engagement [78,79] and builds a community that boosts student success (see, e.g., [80–82]), our program trained students not only to apply their knowledge for the betterment of the community but also to identify specific adaptations or actions that might be taken to build local capacity and contribute to sustainable water management. Students blended scientific knowledge with experiential learning to form their own ILEK. An ERE student who participated in both years wrote of what she gained from the cultural perspective: " . . . I appreciated [learning] about indigenous communities in the Andes . . . a Peruvian graduate student [co-author Yojana Miraya Oscco], who coincidentally was studying at Humboldt, discussed her research [on Indigenous community organizing]. It is super important to be informed about the communities you visit when doing research and something that is neglected far too often in the sciences . . . " This response provides evidence that our students understand the community perspective and leave our program with a more open mind, characteristic of those who practice cultural humility [58].

Bonanza inspired and empowered students, through the explicit integration of scientific and societal needs and leadership training. Research experience, and the satisfaction felt by students who contribute to shared successes, cultivate feelings of belonging, increased social and psychological engagement and ultimately increase interest in pursuing research careers [83–85]. In our case, students were motivated to pursue water resources careers. An ERE student from 2019 wrote, "This trip made me rethink what approach I want to take to grad school (location, concentration), and made me think more about pursuing a career in water resources." Another ERE student from 2019 cited a better understanding of their career goals and increased self-confidence: "This trip solidified my reason for becoming an engineer and broadened my perspective of what an engineer's job is." A geology student from 2019 echoed this increased self-confidence, "This trip broadened my horizons immensely. It made me realize that I can do meaningful work in almost any part of the world while still doing something I enjoy."

Bonanza has empowered students to be leaders in collaborative, community-based and applied research. Four geology and geophysics students presented at the American Geophysical Union Fall Conference in 2018 and 2019. Two geology students graduated from HSU to enter Ph.D. programs, and one geology and 3 engineering students have sought out and found employment in water resources management. The research generated in two seasons of fieldwork is the basis for a master's thesis at HSU [70], and the equipment purchased through this program supported the undergraduate senior theses of two Peruvian students.

6. Conclusions

The Bonanza project presents the benefits of including coproduction and application of knowledge within a hydro-social framework. Similar to past models, we explicitly connected upstream and downstream disciplines focused on water resources. What is novel, however, is our focus on the coproduction of knowledge and the application of this knowledge to the benefit of scientists, students, and the local community. Our project successfully addressed two major goals: to contribute directly to sustainable water management within the village of Zurite and to produce an ILEK while training a culturally educated cohort of socio-hydrologists. The resulting ILEK, which incorporates both Allin Kawsay and Western approaches to hydrologic science, has built the local capacity to measure, monitor and manage local water resources, and generated local resilience to natural and socio-political external forces.

Specifically, we explored water resources in the URW, and linked our results to the needs of the community of Zurite. We collaborated with the community to design and construct 1.3 km of irrigation canals, to bring water to the fields of over 100 families, and trained members of the community to monitor local water resources. Based on our study of the URW and the needs of Zurite, we identified key upstream and downstream measures to build ILEK and local resiliency. Upstream, we recommended the conservation of bofedales, to sustain their role as shallow surface reservoirs integral to perennial streamflow Downstream, we recommend expanding surface water storage, increasing water use efficiency, and examining the impacts of these on water quality.

Our model of knowledge coproduction has had positive and, importantly, lasting educational impacts. We designed and implemented a year-long program that trained 29 students from the U.S. and Perú to be interdisciplinary and community-minded researchers. These students report a change in their life and career outlook, citing increased cultural humility and understanding of local community needs, and have gone into graduate programs and jobs that are focused on water resources. The mutual learning that occurred during different elements of the program created a more complete understanding of the hydrological framework than would have been possible without input from all knowledge-holders.

Based on our work from the Bonanza project, we conclude that combining Western scientific approaches and Allin Kawsay within the ILEK framework can result in impactful changes for water sustainability and can have a lasting impact on all knowledge-holders in the program, including student participants. Our work placed a strong emphasis on local action and defining best practices for sustainable water management and was strengthened by building bridges across different scientific disciplines, and between western scientific approaches and Indigenous and local knowledge, within the community in which we worked. The resulting ILEK provides a framework that can be applied to water development to build local capacity and resiliency and train a culturally educated cohort of socio-hydrologists to work with communities beyond Zurite.

Author Contributions: Conceptualization, J.O., K.K. and M.L.; methodology, J.O., K.K., M.L. and Y.M.O.; investigation, J.O., K.K. and M.L; resources, J.O., K.K., M.L. and Y.M.O.; data curation, J.O., K.K., M.L. and Y.M.O.; writing—original draft preparation, J.O., K.K., M.L. and Y.M.O.; writing—review and editing, J.O., K.K., M.L. and Y.M.O.; supervision, J.O.; project administration, J.O.; funding acquisition, J.O., K.K. and M.L. All authors have read and agreed to the published version of the manuscript.

Funding: This project was funded by a two-year award from Geoscientists Without Borders (award number: 2017080009) via the Society of Exploration Geophysicists (Founding Supporter Schlumberger). Additional support from North Coast Rotary and the Sponsored Programs Foundation, Humboldt State University. Canal construction costs were supported by GWB, the Municipality of Zurite, the Community of Zurite (San Nicolas de Bari), and the Comisión de Regantes.

Institutional Review Board Statement: Not applicable.

Informed Consent Statement: Not applicable.

Data Availability Statement: No new data were created or analyzed in this study. Data sharing is not applicable to this article.

Acknowledgments: This project would not have been possible without the generous support of Tomás Ruiz López and Gladis Quispe. Thank you to the Municipality of Zurite, the Community of San Nicolas de Bari, and the Comisión de Regantes in Zurite. Uriel Ccopa Villena provided essential logistical support. Many undergraduate and graduate students from the United States and Perú contributed to this study. Z. Pearson read an earlier draft and provided many helpful comments that greatly improved the structure and overall quality.

Conflicts of Interest: The authors declare no conflict of interest. The funders had no role in the design of the study; in the collection, analyses, or interpretation of data; in the writing of the manuscript, or in the decision to publish the results.

References

1. De Fries, R.; Eshleman, K.N. Land-use change and hydrologic processes: A major focus for the future. *Hydrol. Process.* **2004**, *18*, 2183–2186. [CrossRef]
2. Turner, B.L.; Lambin, E.F.; Reenberg, A. The emergence of land change science for global environmental change and sustainability. *Proc. Natl. Acad. Sci. USA* **2007**, *104*, 20666–20671. [CrossRef] [PubMed]
3. Sivapalan, M.; Savenije, H.H.G.; Blöschl, G. Socio-hydrology: A new science of people and water: Invited Commentary. *Hydrol. Process.* **2012**, *26*, 1270–1276. [CrossRef]
4. Swyngedouw, E. Modernity and Hybridity: Nature, Regeneracionismo, and the Production of the Spanish Waterscape, 1890–1930. *Ann. Assoc. Am. Geogr.* **1999**, *89*, 443–465. [CrossRef]
5. French, A. Webs and Flows: Socionatural Networks and the Matter of Nature at Peru's Lake Parón. *Ann. Am. Assoc. Geogr.* **2019**, *109*, 142–160. [CrossRef]
6. Baraer, M.; McKenzie, J.M.; Mark, B.G.; Bury, J.; Knox, S. Characterizing contributions of glacier melt and groundwater during the dry season in a poorly gauged catchment of the Cordillera Blanca (Peru). *Adv. Geosci.* **2009**, *22*, 41–49. [CrossRef]
7. Rabatel, A.; Francou, B.; Soruco, A.; Gomez, J.; Cáceres, B.; Ceballos, J.L.; Basantes, R.; Vuille, M.; Sicart, J.-E.; Huggel, C.; et al. Current state of glaciers in the tropical Andes: A multi-century perspective on glacier evolution and climate change. *Cryosphere* **2013**, *7*, 81–102. [CrossRef]
8. Drenkhan, F.; Carey, M.; Huggel, C.; Seidel, J.; Oré, M.T. The changing water cycle: Climatic and socioeconomic drivers of water-related changes in the Andes of Peru. *WIREs Water* **2015**, *2*, 715–733. [CrossRef]
9. Oré, M.; Geng, D. Políticas públicas del agua en las regiones: Las viscisitudes para la creación del Consejo de Recursos Hídricos de la cuenca Ica-Huancavelica. In *¿Escasez de Agua? Retos Para La Gestión de La Cuenca Del Río Ica*; Fondo Editorial PUCP; San Miguel: Lima, Perú, 2014.
10. Ochoa-Tocachi, B.F.; Bardales, J.D.; Antiporta, J.; Pérez, K.; Acosta, L.; Mao, F.; Zulkafli, Z.; Gil-Ríos, J.; Angulo, O.; Grainger, S.; et al. Potential contributions of pre-Inca infiltration infrastructure to Andean water security. *Nat. Sustain.* **2019**, *2*, 584–593. [CrossRef]
11. Carey, M.; Baraer, M.; Mark, B.G.; French, A.; Bury, J.; Young, K.R.; McKenzie, J.M. Toward hydro-social modeling: Merging human variables and the social sciences with climate-glacier runoff models (Santa River, Peru). *J. Hydrol.* **2014**, *518*, 60–70. [CrossRef]
12. Sato, T.; Kikuchi, N. Creation and Sustainable Governance of New Commons through Formation of Integrated Local Environmental Knowledge (ILEK Project). 2013. Available online: https://www.chikyu.ac.jp/rihn_e/project/E-05.html (accessed on 15 June 2021).
13. Kitamura, K.; Nakagawa, C.; Sato, T. Formation of a Community of Practice in the Watershed Scale, with Integrated Local Environmental Knowledge. *Sustainability* **2018**, *10*, 404. [CrossRef]
14. Sato, T.; Chabay, I.; Helgeson, J. *Transformations of Social-Ecological Systems*; Springer: Berlin/Heidelberg, Germany, 2018. [CrossRef]
15. Hong, L.; Page, S.E. Groups of diverse problem solvers can outperform groups of high-ability problem solvers. *Proc. Natl. Acad. Sci. USA* **2004**, *101*, 16385–16389. [CrossRef]
16. Aminpour, P.; Gray, S.A.; Singer, A.; Scyphers, S.B.; Jetter, A.J.; Jordan, R.; Murphy, R.M., Jr.; Grabowski, J.H. The diversity bonus in pooling local knowledge about complex problems. *Proc. Natl. Acad. Sci. USA* **2021**, *118*, 17–21. [CrossRef]
17. ANA. Recursos Hídricos En El Perú. 2012. Available online: https://www.gob.pe/institucion/ana/noticias/139639-formulacion-del-plan-nacional-de-recursos-hidricos-se-realiza-de-manera-integrada-en-puno (accessed on 15 June 2021).
18. ANA. Inventario Nacional de Lagunas de Las Cordilleras Vilcanota Y Carabaya, San Isidro, Lima, Perú. 2014. Available online: https://repositorio.ana.gob.pe/bitstream/handle/20.500.12543/199/ANA0000015.pdf?sequence=4&isAllowed=y (accessed on 15 June 2021).
19. Salzmann, N.; Huggel, C.; Rohrer, M.; Silverio, W.; Mark, B.G.; Burns, P.; Portocarrero, C. Glacier changes and climate trends derived from multiple sources in the data scarce Cordillera Vilcanota region, southern Peruvian Andes. *Cryosphere* **2013**, *7*, 103–118. [CrossRef]
20. Kronenberg, M.; Schauwecker, S.; Huggel, C.; Salzmann, N.; Drenkhan, F.; Frey, H.; Giraáldez, C.; Gurgiser, W.; Kaser, G.; Juen, I.; et al. The Projected Precipitation Reduction over the Central Andes may Severely Affect Peruvian Glaciers and Hydropower Production. *Energy Procedia* **2016**, *97*, 270–277. [CrossRef]
21. Drenkhan, F.; Huggel, C.; Guardamino, L.; Haeberli, W. Managing risks and future options from new lakes in the deglaciating Andes of Peru: The example of the Vilcanota-Urubamba basin. *Sci. Total Environ.* **2019**, *665*, 465–483. [CrossRef]
22. Peel, M.C.; Finlayson, B.L.; McMahon, T.A. Updated world map of the Köppen Geiger climate classification. *Hydrol. Earth Syst. Sci.* **2007**, *11*, 1633–1644. [CrossRef]
23. Buytaert, W.; Moulds, S.; Acosta, L.; De Bièvre, B.; Olmos, C.; Villacis, M.; Tovar, C.; Verbist, K.M.J. Glacial melt content of water use in the tropical Andes. *Environ. Res. Lett.* **2017**, *12*, 114014. [CrossRef]
24. Squeo, F.A.; Warner, B.G.; Aravena, R.; Espinoza, D. Bofedales: High altitude peatlands of the central Andes. *Rev. Chil. Hist. Nat.* **2006**, *79*, 245–255. [CrossRef]
25. Josse, C.; Cuesta, F.; Navarro, G.; Barrena, V.; Cabrera, E.; Chacón-Moreno, E.; Ferreira, W.; Peralvo, M.; Saito, J.; Tovar, A. Atlas de los Andes del Norte y Centro. Bolivia, Colombia, Ecuador, Perú y Venezuela. In Secr Gen Comunidad Andina Programa Reg Ecobona Condesan-Proy Páramo Andino Programa BioAndes EcoCiencia NatureServe LTA-UNALM IAvH ICAEULA

CDC-UNALM RUMBOL SRL. Lima, Perú. 2009. Available online: http://www.saber.ula.ve/handle/123456789/39336 (accessed on 15 June 2021).

26. Fonkén, M.S.M. An introduction to the bofedales of the Peruvian High Andes. *Mires Peat* **2014**, *15*, 1–13.

27. Ochoa-Tocachi, B.F.; Buytaert, W.; De Bièvre, B.; Célleri, C.; Crespo, P.; Villacís, M.; Llerena, C.A.; Acosta, L.; Villazón, M.; Guallpa, M.; et al. Impacts of land use on the hydrological response of tropical Andean catchments. *Hydrol. Process.* **2016**, *30*, 4074–4089. [CrossRef]

28. INEI. Estimaciónes y Proyecciones de Población Total Por Sexo de Las Principales Ciudades (2012–2015). Jesus Maria, Lima, Perú. 2015. Available online: https://www.inei.gob.pe/estadisticas/indice-tematico/poblacion-y-vivienda/ (accessed on 15 June 2021).

29. SEDACUSCO. Memoría Annual, Cusco. 2019. Available online: https://www.sedacusco.com/transparencia/memoria/2019.pdf (accessed on 15 June 2021).

30. INEI. Censos Nacionales 2017: XII de Población, VII de Vivienda y III de Comunidades Indígenas. Jesus Maria, Lima, Perú. 2017. Available online: http://censo2017.inei.gob.pe/ (accessed on 15 June 2021).

31. Grille, R.G.; Remy, M.G. Building Democracy with Equality: The Participatory Experience in the Rural Province of Anta, Cusco, Peru. *IDS Bull.* **2009**, *40*, 22–30. [CrossRef]

32. Salvador, F.; Monerris, J.; Rochefort, L. Peatlands of the Peruvian Puna ecoregion: Types, characteristics and disturbance. *Mires Peat* **2014**, *15*, 1–17.

33. Cooper, D.J.; Sueltenfuss, J.; Oyague, E.; Yager, K.; Slayback, D.; Cabrero Caballero, E.M.; Argollo, J.; Mark, B.G. Drivers of peatland water table dynamics in the central Andes, Bolivia and Peru. *Hydrol. Process.* **2019**, *33*, 1913–1925. [CrossRef]

34. Tovar, C.; Arnillas, C.A.; Cuesta, F.; Buytaert, W. Diverging Responses of Tropical Andean Biomes under Future Climate Conditions. *PLoS ONE* **2013**, *8*, e63634. [CrossRef]

35. Vergara, W.; Kondo, H.; Méndez Pérez, J.; Palacios, E. *Visualizing Future Climate in Latin America: Results from the Application of the Earth Simulator*; World Bank: Washington, DC, USA, 2007; pp. 1–173.

36. Banco Nacional de Reserva de Perú. *Cusco: Sintesis de Actividad Económica*; Departamento de Estudios Económicos Sucursal Cusco: Cusco, Peru, 2020.

37. Tarwick, P.B. *The Struggle for Water in Peru: Comedy and Tragedy in the Andean Commons*; Stanford University Press: Palo Alto, CA, USA, 2003.

38. Bebbington, A.; Bury, J. Institutional challenges for mining and sustainability in Peru. *Proc. Natl. Acad. Sci. USA* **2009**, *106*, 17296. [CrossRef]

39. Oré, M.; Rap, E. Políticas neoliberales de agua en el Perú. Antecedentes y entretelones de la ley de recursos hídricos. *Debates Sociol.* **2009**, *34*. Available online: https://revistas.pucp.edu.pe/index.php/debatesensociologia/article/view/2533 (accessed on 15 June 2021).

40. Vos, J.M.C. Metric Matters: The Performance and Organisation of Volumetric Water Control. in Large-Scale Irrigation in the North Coast of Peru. 2002. Available online: https://edepot.wur.nl/198423 (accessed on 15 June 2021).

41. Carey, M.; French, A.; O'Brien, E. Unintended effects of technology on climate change adaptation: An historical analysis of water conflicts below Andean Glaciers. *J. Hist. Geogr.* **2012**, *38*, 181–191. [CrossRef]

42. Boelens, R.; Zwarteveen, M. Prices and Politics in Andean Water Reforms. *Dev. Chang.* **2005**, *36*, 735–758. [CrossRef]

43. Zimmerer, K. The Indigenous Andean Concept of "Kawsay", the Politics of Knowledge and Development, and the Borderlands of Environmental Sustainability in Latin America. *PMLA* **2012**, *127*, 600–606. [CrossRef]

44. Lajo, J. Qhapaq Ñan: La Ruta Inka de Sabiduría. Editorial Abya Yala. 2006. Available online: https://digitalrepository.unm.edu/abya_yala/358 (accessed on 15 June 2021).

45. Gelles, P.H. *Water and Power in Highland Peru: The Cultural Politics of Irrigation and Development*; Rutgers University Press: New Brunswick, NJ, USA, 2000.

46. Pierotti, R.; Wildcat, D. Traditional ecological knowledge: The third alternative (commentary). *Ecol. Appl.* **2000**, *10*, 1333–1340. [CrossRef]

47. Deloria, V., Jr. Traditional Technology. *Winds Chang.* **1990**, *5*, 12–17.

48. Salmón, E. Kincentric ecology: Indigenous perceptions of the human–nature relationship. *Ecol. Appl.* **2000**, *10*, 1327–1332.

49. Gose, P. *Deathly Waters and Hungry Mountains: Agrarian Ritual and Class. Formation in an Andean Town*, 4th ed.; University of Toronto Press: Toronto, ON, Canada, 1994.

50. Bolin, I. The glaciers of the Andes are melting: Indigenous and anthropological knowledge merge in restoring water resources. In *Anthropology and Climate Change: From Encounters to Actions*; Leftcoast Press: Walnut Creek, CA, USA, 2009; pp. 228–239.

51. Painter, J. Deglaciation in the Andean Region. UNDP—Human Development Report Office. 2007. Available online: http://hdr.undp.org/en/content/deglaciation-andean-region (accessed on 20 May 2021).

52. Perez, C.; Nicklin, C.; Dangles, O.; Vanek, S.; Sherwood, S.G.; Halloy, S.; Garrett, K.A.; Forbes, G.A. Climate Change in the High Andes: Implications and adaptation strategies for small-scale farmers. *Int. J. Environ. Cult. Econ. Soc. Sustain.* **2010**, *6*, 71–88. [CrossRef]

53. Lynch, B.D. Vulnerabilities, competition and rights in a context of climate change toward equitable water governance in Peru's Rio Santa Valley. *Glob. Environ. Chang.* **2012**, *22*, 364–373. [CrossRef]

54. French, A. ¿Una nueva cultura de agua?: Inercia institucional y la gestión tecnocrática de los recursos hídricos en el Perú. *Anthropologica* **2016**, *34*, 61–86. [CrossRef]

55. Ostrom, E. A general framework for analyzing sustainability of social-ecological systems. *Science* **2009**, *325*, 419–422. [CrossRef]

56. Rausch, K. Embedded Programs as a Model for Increased Study Abroad Access. *Inst. Int. Educ.* 2019. Available online: www.iie.org/en/Learn/Blog/2019/07/Embedded-Programs-as-a-Modelfor-Increased-Study-Abroad-Access (accessed on 17 September 2020).

57. Taboada, H.; Ferregut, C.; Flores, B.; Bililign, S.; Engel, L.; Rathbun, L.; Ramos, I.; Winter, R. Broadening Participation of Underrepresented Minorities in STEM Research Abroad. Public Report. 2020. Available online: www.utep.edu/engineering/imse/stemresearchabroad/_Files/docs/TABOADA---%20OISE1848137-report.pdf (accessed on 20 September 2020).

58. Tervalon, M.; Murray-Garcia, J. Cultural humility versus cultural competence: A critical distinction in defi. *J. Health Care Poor Underserved* **1998**, *9*, 117–125. [CrossRef]

59. Yosso, T.J. Whose culture has capital? A critical race theory discussion of community cultural wealth. *Race Ethn. Educ.* **2005**, *8*, 69–91. [CrossRef]

60. Bok, D.C. *The SAGE Handbook of Intercultural Competence*; Sage: Thousand Oaks, CA, USA, 2009.

61. Deardorff, D.K. Identification and Assessment of Intercultural Competence as a Student Outcome of Internationalization. *J. Stud. Int. Educ.* **2006**, *10*, 241–266. [CrossRef]

62. Mankiewicz, C. Essay: Student Science Research in Education Abroad. *Front. Interdiscip. J. Study Abroad* **2005**, *12*, 123–125. [CrossRef]

63. Orellana, M.F.; Bowman, P.; Orellana, M.F.; Bowman, P. Cultural diversity research on learning and development: Conceptual, methodological, and strategic considerations. *Educ. Res.* **2003**, *32*, 26–32. [CrossRef]

64. Vogel, C.; Moser, S.C.; Kasperson, R.E.; Dabelko, G.D. Linking vulnerability, adaptation and resilience science to practice: Pathways, players and partnerships. *Glob. Environ. Chang.* **2007**, *17*, 349–364. [CrossRef]

65. Cash, D.W.; Moser, S.C. Linking global and local scales: Designing dynamic assessment and management processes. *Glob. Environ. Chang.* **2000**, *10*, 109–120. [CrossRef]

66. Bringle, R.G.; Hatcher, J.A. A Service-Learning Curriculum for Faculty. *Mich. J. Community Serv. Learn.* **1995**.

67. Carlotto, V.; Concha, R.; Cárdenes, J.; Garcia, B.; Villafuerte, C. *Geología y Geodinámica En La Quebrada Qenqo: Aluviones Que Afectaron Zurite-Cusco (2010)*; INGEMMET: Lima, Peru, 2010.

68. Bonhomme, M.; Fornari, M.; Laubacher, G.; Sébrier, M.; Vivier, G. New Cenozoic K-Ar ages on volcanic rocks from the eastern High Andes, southern Peru. *J. S. Am. Earth Sci.* **1988**, *1*, 179–183. [CrossRef]

69. Carlotto, V.; Carlier, G.; Jaillard, E.; Sempéré, T.; Mascle, G. Sedimentary and structural evolution of the Eocene-Oligocene Capas Rojas basin: Evidence for a late Eocene lithospheric delamination event in the southern Peruvian Altiplano. *Andean Geodyn. Ext. Abstr.* **1999**, 141–146.

70. Wunderlich, W. Spatiotemporal Patterns in Water Yield from the Humid Puna: A Case Study in the Agrarian District of Zurite, Perú. 2021. Available online: https://digitalcommons.humboldt.edu/etd/481/ (accessed on 31 May 2021).

71. Aybar, C.; Fernández, C.; Huerta, A.; Lavado, W.; Vega, F.; Felipe-Obando, O. Construction of a high-resolution gridded rainfall dataset for Peru from 1981 to the present day. *Hydrol. Sci. J.* **2020**, *65*, 770–785. [CrossRef]

72. Coward, E. Principles of Social Organization in an Indigenous Irrigation System. *Hum. Organ.* **1979**, *38*, 28–36. [CrossRef]

73. Coward, E.W. Indigenous organisation, bureaucracy and development: The case of irrigation. *J. Dev. Stud.* **1976**, *13*, 92–105. [CrossRef]

74. Mabry, J.B.; Cleveland, D.A. *The Relevance of Indigenous Irrigation. Canals and Communities: Small-Scale Irrigation Systems*; University of Arizona Press: Tucson, AZ, USA, 1996; pp. 227–259.

75. Comisión de Regantes, Z.C. Sistema de Regación, Zurite, Perú. 2020. Available online: https://es.slideshare.net/belubel83/05-sistema-de-rotacin (accessed on 15 June 2021).

76. Adger, W.N.; Hughes, T.P.; Folke, C.; Carpenter, S.R.; Rockström, J. Social-ecological resilience to coastal disasters. *Science* **2005**, *309*, 1039. [CrossRef]

77. Moser, S.C.; Luganda, P. Talk for a change: Communication in support of societal response to climate change. *IHDP Newsl.* **2006**, *1*, 17–20.

78. Kinzie, J.; Gonyea, R.; Shoup, R.; Kuh, G.D. Promoting persistence and success of underrepresented students: Lessons for teaching and learning. *New Dir. Teach. Learn.* **2008**, *115*, 21–38. [CrossRef]

79. Kuh, G.D. Excerpt from high-impact educational practices: What they are, who has access to them, and why they matter. *Assoc. Am. Coll. Univ.* **2008**, *14*, 28–29.

80. Tinto, V. Learning better together: The impact of learning communities on student success. *High. Educ. Monogr. Ser.* **2003**, *1*, 1–8.

81. Museus, S.D. The culturally engaging campus environments (CECE) model: A new theory of success among racially diverse college student populations. *High. Educ. Handb. Theory Res.* **2014**, *29*, 189–227.

82. Johnson, M.D.; Sprowles, A.E.; Goldenberg, K.R.; Margell, S.T.; Castellino, L. Effect of a Place-Based Learning Community on Belonging, Persistence, and Equity Gaps for First-Year STEM Students. *Innov. High. Educ.* **2020**, *45*, 509–531. [CrossRef]

83. Villarejo, M.; Barlow, A.E.; Kogan, D.; Veazey, B.D.; Sweeney, J.K. Encouraging Minority Undergraduates to Choose Science Careers: Career Paths Survey Results. *CBE—Life Sci. Educ.* **2008**, *7*, 394–409. [CrossRef]

84. Lopatto, D. Undergraduate Research Experiences Support Science Career Decisions and Active Learning. *CBE—Life Sci. Educ.* **2007**, *6*, 297–306. [CrossRef]
85. Thoman, D.B.; Brown, E.R.; Mason, A.Z.; Harmsen, A.G.; Smith, J.L. The role of altruistic values in motivating underrepresented minority students for biomedicine. *BioScience* **2015**, *65*, 183–188. [CrossRef]

 *hydrology*

Article

Regional Climate Change Impact on Coastal Tourism: A Case Study for the Black Sea Coast of Russia

Evgeniia A. Kostianaia [1] and Andrey G. Kostianoy [1,2,*]

[1] P.P. Shirshov Institute of Oceanology, Russian Academy of Sciences, 36, Nakhimovsky Pr., 117997 Moscow, Russia; evgeniia.kostianaia@ocean.ru

[2] Laboratory for Integrated Research of Water Resources, S.Yu. Witte Moscow University, 12, Build. 1, Second Kozhukhovsky Pr., 115432 Moscow, Russia

* Correspondence: kostianoy@ocean.ru

Abstract: Regional climate change is one of the key factors that should be taken into account when planning the development of the coastal tourism, including investments and construction of tourism-related infrastructure. A case study for the Black Sea coast of Russia shows a series of potential negative hydrological, meteorological, and biological factors that accompany regional warming of the Black Sea Region, that can impede the development of coastal tourism and devalue billions of dollars in investments by the State, private companies, and individuals. We discuss such natural phenomena as air and sea warming, extreme weather events, coastal upwelling, heavy rains, river plumes, wind and waves, tornado, rip currents, sea-level rise, algal bloom, introduced species, and other features characteristic for the region that seriously impact coastal tourism today, and may intensify in the nearest future. Sporadic occurrence of extreme weather events, unpleasant and sometimes dangerous sea and atmosphere phenomena during the summer tourist season, and from year to year can be of critical psychological importance when choosing your next vacation and tourism destination. The research does not include anthropogenic factors, geopolitical, and socio-economic processes, and the COVID-19 pandemic that play an important role in the sustainable development of coastal tourism as well.

Keywords: Black Sea; coastal tourism; regional climate change; warming; wind; waves; sea level rise; upwelling; heavy rain; river plume; algal bloom; introduced species

check for
updates

Citation: Kostianaia, E.A.; Kostianoy, A.G. Regional Climate Change Impact on Coastal Tourism: A Case Study for the Black Sea Coast of Russia. *Hydrology* **2021**, *8*, 133. https://doi.org/10.3390/hydrology8030133

Academic Editors: Tamim Younos, Tammy Parece, Juneseok Lee, Jason Giovannettone and Alaina J. Armel

Received: 20 July 2021
Accepted: 31 August 2021
Published: 6 September 2021

Publisher's Note: MDPI stays neutral with regard to jurisdictional claims in published maps and institutional affiliations.

1. Introduction

According to the UN Atlas of the Oceans [1], coastal tourism and recreation include the full range of tourism, leisure, and recreationally oriented activities that take place in the coastal zone and the offshore coastal waters such as recreational boating, cruises, swimming, recreational fishing, surfing, windsurfing, supping, snorkeling, diving, etc. These include the infrastructure supporting coastal development (e.g., hotels, resorts, restaurants, marinas, beaches, recreational fishing facilities, shops, roads, railways, airports, transportation). However, according to ECORYS [2], differentiation should be made between coastal and maritime tourism. For the purpose of this study, we shall use the definitions from [2], where coastal tourism would imply beach-based recreation activities (such as, for example, sunbathing, swimming, surfing, etc.), as well as non-beach related land-based tourism in the coastal area (this covers any recreation or tourism activities in the coastal area that require the sea to be in the proximity). Associated manufacturing industries and supplies also refer to this type. At the same time, maritime tourism refers to tourism that is mostly water-based than land-based, such as, for example, cruising, yachting, boating. This type also covers manufacturing of equipment, landside facilities, and necessary services.

The primary focus of this study is the impact of climate change on coastal tourism rather than maritime tourism. The main purpose of tourists coming to the Black Sea coast

117

of Russia is to sunbathe, swim, and snorkel, which falls under the coastal tourism category. Climate change effects seem to be able to impact coastal tourism in a more pronounced form as air and sea temperature warming, heavy precipitation and flooding, algal bloom, and introduction of alien species are expected to affect the possibility to lie on the beach, get into the water, and swim. These parameters seem to have a lesser impact on such maritime tourism activities as cruising, yachting, and boating.

Coastal tourism plays an important role in the socio-economic development of many countries. For some of the countries, it was an engine for economic development, other countries successfully used coastal tourism to overcome economic crises, and for others, it is the main source of their budget revenue. We have plenty of examples in Europe: the MENA region, Southeast Asia, the Caribbean, and other parts of the world. Every country which has an exit to a warm sea invests in the development of coastal tourism, because it brings an important income into state and regional budgets, provides employment and jobs to the local population. According to different sources, in 2006, the coastal tourism industry has contributed about 10% of the total global Gross Domestic Product (GDP) [1,3,4]. In fact, in 2019, prior to the COVID-19 pandemic, the travel and tourism industry accounted for 25% of all new jobs in the world, 10.6% of all jobs (334 million), and 10.4% of global GDP (USD 9.2 trillion). At the same time, international visitor spending totaled USD 1.7 trillion in 2019 (6.8% of total exports, 27.4% of global services exports) [3]. Most of these values belong to coastal tourism because this is the preferred destination for summer vacation.

The development of coastal tourism seriously impacts the coastal zone, terrestrial and marine environment, water and energy resources. Fresh and drinking water consumption, seawater, ground, air and noise pollution, plastic and microplastic pollution, untreated sewage, wastewater discharge into the sea, accumulation of waste and garbage, land degradation and land-use, coastal erosion, habitat and biodiversity loss, destruction of aesthetic value, and physical beauty of the coast are among the main threats caused by coastal tourism [4,5]. Uncontrolled tourism development in some cases can lead to environmental damage that can cost local populations and governments more than they would gain from the coastal tourism industry [4]. On the other hand, sustainable development of coastal tourism has the potential to create beneficial effects on the environment by raising awareness of environmental values, contributing to environmental protection and conservation, thus resulting in an increase in their economic importance [1]. The environment also can have both positive and negative impacts on the coastal tourism development, for example, via extreme weather events; physical, chemical, and biological processes in the sea; earthquakes and volcano eruption, and regional climate change which has been of great importance during the past 40 years, but usually ignored in strategies of coastal tourism development. All this leads to the necessity of sustainable coastal tourism management and development, which can be done, for example, in the framework of Integrated Coastal Zone Management [6], Marine Spatial Planning [7], and Maritime Clusters [8–11], which have been increasingly acknowledged as essential boosters for innovation and diversification of the Blue Economy [11].

In the Russian Federation, the Black Sea coast of Russia is the major resort area for Russian citizens, because the Arctic Seas are not taken into consideration, the Baltic Sea and the Sea of Japan are much colder, and the Sea of Azov and the Caspian Sea have a little resort area infrastructure in comparison with the Black Sea. Thus, the major coastal tourism flow is directed to the Krasnodar Krai (Region) and the Republic of Crimea (Figure 1). In 2019, 17.3 million tourists visited the Krasnodar Krai, 60% of them during the summer season, and 7.43 million tourists visited Crimea. These numbers include both foreign and domestic visitors [12].

Figure 1. Geographical map of the Northeastern Black Sea based on Maps-For-Free (https://maps-for-free.com/#close, accessed on 30 August 2021). Blue line is the Kuban River. The Greater Caucasus stretches along the Black Sea coast.

The region under consideration is located between the Kerch Strait in the north and the state border with Abkhazia in the south, i.e., in the latitudinal band between 43°23′ and 45°30′ N (Figure 1). For comparison, this geographical band corresponds to the Northern Adriatic Sea between Split in Croatia and Venice in Italy, or the northern part of the Ligurian Sea between Cannes in France and Genoa in Italy. Both regions in the Mediterranean Sea are well known as the best places for coastal tourism and summer vacation in the Mediterranean due to a warm climate and warm sea. The same is true for the Russian coast of the Black Sea.

The climate in most of the territory of the Krasnodar Krai is moderately continental, on the Black Sea coast from Anapa to Tuapse—a semi-dry Mediterranean climate, south of Tuapse—humid subtropical. The high-altitude climatic zoning is characteristic of the mountains. In general, the region is characterized by hot summers and mild winters. The average January temperature on the Black Sea coast is 0 … +6 °C, in Sochi +6 °C. The average July temperature is +22 … +24 °C. Annual precipitation is from 400 to 600 mm in the flat part, and up to 3250 mm in the mountains [13].

The development of resorts on the Black Sea coast of Russia in the Crimea and the Krasnodar Krai began at the end of the 19th century. Today, this is a major resort and coastal tourism area for Russian citizens who spend their summer holidays on numerous beaches of the Black Sea. In December 2015, the Krasnodar Krai was named the most attractive tourist region of Russia in the "National Tourism Rating" (second and third places—St. Petersburg and Moscow). It includes resorts of federal significance Sochi, Gelendzhik, and Anapa, as well as numerous small villages and resorts of regional significance located between Anapa and Adler at the border with Abkhazia (from north to south—Vityazevo, Anapa, Dyurso, Yuzhnaya Ozereevka, Shirokaya Balka, Myskhako, Novorossiysk, Kabardinka, Gelendzhik, Divnomorskoe, Dzhankhot, Krinitsa, Betta, Arkhipo-Osipovka, Dzhubga, Lermontovo, Novomikhailovskiy, Olginka, Nebug, Agoy, Tuapse, Shepsi, Lazarevskiy, Vardane, Loo, Dagomys, Sochi, Matsesta, Khosta, Kudepsta, Adler) [14]. For instance, in Sochi, there are around 1000 certified hotels from 5-star (192) to mini-hotels (414) [15]. In Crimea, the most well-known resorts are located in (from west to east): Evpatoriya, Sevastopol, Balaklava, Simeiz, Alupka, Yalta, Gurzuf, Alushta, Sudak, Koktebel, Feodosiya, and Kerch [14].

In recent years, the most attention in the Krasnodar Krai has been paid to the development of a health resort complex, which makes it possible to increase the load of the health resorts in the off-season. The health resort sphere of the Krasnodar Krai includes more than 200 organizations, their total capacity is about 100 thousand places. This is about 21% of the bed capacity of all Russian health resorts. On average, up to 1.5 million people annually rest and recover in health resorts, thus, the Krasnodar Krai occupies more than 25% of the Russian market of such services [12].

During the past 10 years, the Russian Government has invested significant funds in the development of tourism on the coasts of the Black Sea, including large infrastructure projects—reconstruction of airports in Sochi (Adler), Gelendzhik, Anapa, Krasnodar, and Simferopol; construction of the Crimean Bridge across the Kerch Strait and federal highway "Tavrida" from Kerch to Sevastopol; construction of Imeretinsky Resort and Olympic Park south of Sochi; building of new hotels and reconstruction of old ones; construction of the offshore gas pipeline "Dzhubga–Lazarevskoye–Sochi", and many others. All these projects have been performed with an Environmental Impact Assessment (EIA) which, according to the International Association for Impact Assessment (https://www.iaia.org, accessed on 30 August 2021), is "the process of identifying, predicting, evaluating and mitigating the biophysical, social, and other relevant effects of development proposals prior to major decisions being taken and commitments made".

Usually, the development of tourism and infrastructure on the Russian coasts of the Black Sea is carried out without assessment of the regional climate change impact in the nearest future and in the long-term perspective. To our knowledge, the same is typical for other resort areas in the world. This is not an obligatory procedure such as the EIA, but the ongoing climate change in the Black Sea Region forces such an analysis to be done to be sure that the investment will be as efficient as possible. The main feature of regional climate change in the Black Sea Region is the warming of the climate and seawater, which are regarded as evident and positive consequences of climate change for coastal tourism development. However, along with the warming of the region, climate change can lead to a series of negative processes (in relation to coastal tourism development) in the atmosphere and the sea which should be carefully studied before adopting long-term regional development programs.

Regional climate change in the Black Sea is accompanied by the intensification of extreme weather events. The Fourth [16] and Fifth [17] Climate Change Assessment Reports of the Intergovernmental Panel on Climate Change (IPCC) indicate that in the 21st century, climate change will be accompanied by an increase in the frequency, intensity, and duration of events with extremely high or low air temperatures, extreme precipitation or drought. All this may lead to floods, droughts, fires, shallowing of rivers, lakes, and water reservoirs, desertification, dust storms, algal bloom in the seas, and freshwater reservoirs. In turn, these phenomena in many cases may lead to chemical and biological pollution of water, land, and air, as well as to deterioration in the quality of life of the population, significant financial losses associated with damage to housing, businesses, roads, and railways, agriculture and forestry, coastal tourism, and in many cases even to human losses. The First [18] and Second [19] Roshydromet assessment reports on climate change and the consequences on the territory of the Russian Federation confirm these forecasts. Forecasts of independent groups of scientists have been repeatedly confirmed over the past 20 years—heavy rains, floods, droughts, and fires in various regions of the Russian Federation, including the coastal zone of the Krasnodar Krai, were yearly observed. In this regard, the analysis and forecasting of extreme weather events associated with regional climate change in the coastal zone of the Krasnodar Krai and Crimea are extremely important tasks, given the importance of coastal tourism for these regions.

Research interest towards the relationship between climate and weather, on the one hand, and tourism and recreation, on the other hand, became evident already in the 1950s [20]. A lot of attention has been given to evaluate predicted climate change by certain touristic regions, as consequences of such climatic changes will vary across regions and might mean different implications, increasing or decreasing the touristic flow. Such effects are manifold and have regional specifics. Detailed discussions can be found in publications from some of the major touristic coastal regions of the world, such as, for example, published by: Becken [21] for Australia; Grimm et al. [22] for Brazil; Layne [23] for Caribbean SIDS; Becken [24] for Fiji; Friedrich et al. [25] for South Africa. The main discussed impacts for these regions are the same: higher temperatures, increased frequency, and intensity of extreme storms, increased precipitation, sea-level rise, sea temperature rise, change in ecosystems. Some other specific impacts are ocean acidification, coral bleaching, migration of species, the appearance of illnesses, change in the appearance of insects and organisms. Layne [23] and Lincoln [26] also point to the impact of climate change on human health and its implications for the tourist sector, citing such issues as extreme sunburn, dehydration, heatstroke, damage of public health facilities, water shortage during droughts, increase in skin diseases, cardio-respiratory conditions, and heat-related illnesses, spread of diseases through stagnant contaminated water. Some other indirect impacts are presented by Santos-Lacueva et al. [27]: increased energy consumption for cooling, increased water price due to scarcity, increased water consumption for recreation and comfort of tourists, artificialization of beaches due to sea-level rise, and the need for pest fumigation.

Klueva et al. [28] evaluated summer tourism for several regions in Russia, using the "tourism climate index"—TCI. For coastal tourism in the south of Russia, the authors came to the conclusion that climatic resources will increase towards the middle of the 21st century. Towards the end of the century, the TCI will lower, however, it will still remain within the range of comfortability for coastal tourism.

Despite the increasing significance of continuous research on climate change impacts on coastal tourism for the Black Sea coast of Russia, there are very few studies on this subject for this specific region. Extreme precipitation in summers, including this summer of 2021, has already led to significant disruption of transport services, evacuation of people, economic losses, and death tolls. Therefore, it is important to continue such research for this specific region to assess not only direct impacts on coastal tourism but also indirect consequences, including economic implications. A good example of such detailed research is a study by Arabadzhyan et al. [29], where the authors identified the following nine impact chains:

1. Loss of tourist experience value in the destination due to changes in environmental attributes.
    1.1. Loss of attractiveness of marine environments due to loss of species, increase in exotic invasive species or degradation of landscape.
    1.2. Loss of attractiveness and comfort due to beach availability reduction.
    1.3. Loss of attractiveness due to increased danger of forest fires in tourism areas.
    1.4. Loss of attractiveness of land environments due to loss of species, increase in exotic invasive species or degradation of landscape.
2. Loss of tourist experience value in the destination due to changes in human being comfort (or health).
    2.1. Loss of comfort due to increase in thermal stress and heat waves.
    2.2. Increase in health issues due to emergent diseases.
3. Loss of tourist experience value in the destination due to the change in the quality of infrastructure and facilities.
    3.1. Increase in damages to infrastructures and facilities (accommodation, promenades, water treatment system, etc.).
    3.2. Decrease in available domestic water for the tourism industry.
    3.3. Loss of attractiveness due to loss of cultural heritage (monuments, gastronomy, etc.).

According to this breakdown, our research concerns impact chains 1.1. (partially), 1.2., 2.1., and 3.1. Further directions would also need to concern other impact chains, for which it would be essential to cooperate and receive information from the tourist industry specialists, marine biologists, health workers, public utilities and cultural tourism representatives.

The aim of this paper is to review a series of potential negative hydrological, meteorological, and biological factors for the Black Sea coast of Russia that accompany regional warming of the Black Sea Region and can impede the development of coastal tourism and devalue billions of dollars of investments by the State, private companies, and individuals. We discuss such natural phenomena as air and sea warming, extreme weather events, coastal upwelling, heavy rains, river plumes, wind and waves, tornado, rip currents, sea level rise, algal bloom, introduced species, and other features characteristic for the region that seriously impacts coastal tourism today, and may intensify in the nearest future. In the paper, we do not investigate the impact of anthropogenic factors, geopolitical and socio-economic processes, and the COVID-19 pandemic that plays an important role in the sustainable development of coastal tourism as well.

2. Materials and Methods

The review of different meteorological, hydrological, and biological features and processes is made on the basis of scientific publications, our own research conducted for the Black Sea coast of Russia during the past 30 years, and our own experience as "tourists" coming to Novorossiysk for summer vacation every year.

The analyses and visualizations of data for the air temperature and atmospheric precipitation used in this study were produced with the Giovanni online data system v.4.32, developed and maintained by the NASA Goddard Earth Sciences Data and Information Services Center (GES DISC) [30]. As for the air temperature, we used the area-averaged 2 m air temperature (monthly mean) with a spatial resolution of $0.50 \times 0.625°$ for the region 43.4–45.2° N; 36.6–40.0° E, which covers the coastal zone of Russia from the Kerch Strait to the border with Abkhazia (Figure 1), and the time period from 1981 to 2020 (MERRA-2 Model M2IMNXASM_5.12.4) [31]. For the sea surface temperature, we used the area-averaged (monthly mean) skin water temperature produced by MERRA-2 M2TMNXOCN V5.12.4 Model with the same spatial resolution and for the same region [32]. For these parameters, we investigated interannual variability of monthly averaged data for the summer period only (June–August) as well as for May and September, separately.

For atmospheric precipitation (mm/month) we used an area-averaged "Merged satellite-gauge precipitation estimate final run" (GPM_3IMERGM v06) (monthly mean) model with a spatial resolution of $0.1 \times 0.1°$ for the same region and for the time period 2000–2020 [33]. For the wind speed (m/s), we used area-averaged monthly mean values of the wind speed from the MERRA-2 M2TMNXFLX v5.12.4 Model with a spatial resolution of $0.50 \times 0.625°$ for the same region and for the time period from 1981 to 2020 [34].

Satellite Synthetic Aperture Radar (SAR) image was downloaded from the ESA Copernicus Open Access Hub (https://scihub.copernicus.eu, accessed on 30 August 2021). For optical imagery we used NASA Worldview Snapshots (https://wvs.earthdata.nasa.gov, accessed on 30 August 2021).

3. Results

Regional climate change leads to a number of changes in the atmosphere and the sea, which potentially can be both positive and negative to socio-economic development. The most known consequence of climate change is warming, but this is not a single effect. Climate change is accompanied by changes in atmospheric circulation, the position of atmospheric fronts, trajectories of cyclones and anticyclones, atmospheric pressure, wind speed and direction, precipitation intensity and location, intensification of extreme weather events, and other processes and phenomena. In the sub-sections below, we try to describe the major natural features which are observed regularly during the past two decades at the Russian coast of the Black Sea, and we focus especially on the negative ones in relation to coastal tourism.

3.1. Warming of the Region and Extreme Events

Warming of the air and the sea is an evident consequence of global and regional climate change in the Black Sea Region. Ginzburg et al. [35] showed that the air temperature in the Black Sea Region is rising with a rate of +0.053 °C/year for 1980–2020, which is three times faster than for the 1935–2017 time period. The highest rate of change of +0.06 °C/year is located along the coastal zone of the northeastern Black Sea. Since the late 1990s, the maximum monthly mean values of the near-surface air temperature during summer on average increased considerably, with an extreme value of 27.2 °C in 2010, when a blocking anticyclone stood over the central part of European Russia for 55 days since the end of June [36]. The above-mentioned analysis, as well as previous publications on regional climate change in the Black Sea Region mentioned in the review [37], showed air temperature trends for all 12 months of the year during a certain time period. In the present study, we are interested in what is happening during summertime, which is a tourist season. This is why we calculated the interannual variability of the air temperature for the summer season (June, July, August) for a 40 years-long time period from 1981–2020 (Figure 2).

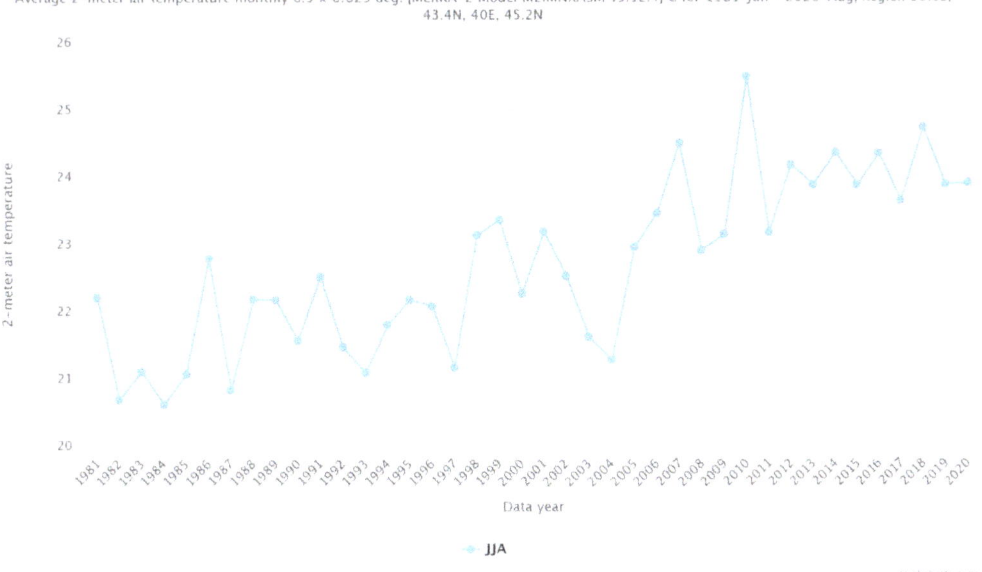

Figure 2. Interannual variability of the near-surface air temperature over the northeastern Black Sea during summer from 1981–2020 based on the MERRA-2 M2IMNXASM_5.12.4 Model.

Figure 2 confirms that, on average, the air temperature during summer has been progressively warming over the past 40 years from around 21 °C in the early 1980s to around 24 °C in the 2010s. The transition between the period of a slight increase in the air temperature to a more stable period occurred between 2004 (21.3 °C) and 2012 (24.2 °C). The hottest summer, 25.5 °C, was registered in 2010 due to a long-standing blocking of an anticyclone over the central part of European Russia. Year-to-year variability of the air temperature varied from 0.5 to 2.5 °C from 1981 to 2011, but since 2012 the air temperature variability has become more stable and has not exceeded 1 °C.

These data show that the air temperature, on average, gets warmer. The air temperature is becoming more comfortable for coastal tourism for a longer period of time, and the tourist season slowly expands for May and September. The same analysis for

May (Figure 3) showed that the air temperature rose from around 15 °C at the end of the 20th century to 16 °C in the first decade of the 21st century, and 17 °C in the 2010s. First, the average air temperature increase in May is lower than the same parameter for the average value for the summertime. Second, 17 °C does not seem to be comfortable for opening the tourist season. Third, the weather in May is very unstable, because from year to year the air temperature has varied from 15 to 19 °C over the past 10 years. Thus, it is evident that during the coming decades, May will not be a month from which the tourist season will start. At the same time, September shows to be a better candidate for expansion of the summer season because its average air temperature is 3–4 °C higher than in May, and it is already comparable with June (Figure 4). Since 2005, its average temperature almost yearly has been over 20 °C, which can be regarded as a psychological mark for summertime, and in 2015, 2017, and 2020 it was over 22 °C.

Global and regional warming is accompanied by extreme weather events such as heat and cold waves, draughts and frosts, heavy rains and snowfalls, etc. This effect is also observed in the eastern part of the Black Sea which was especially investigated for the period 1950–2015 based on daily air temperature data [38]. Kostianoy et al. [38] showed an increase in the amplitude of air temperature extremes with positive anomalies from 3.6 °C to 3.9 °C for phenomena exceeding one standard deviation (1 SD), and from 5.5 °C to 6 °C for phenomena exceeding 2 SD. At the same time, the amplitude of extreme events with negative anomalies remained practically unchanged: 3.9 °C and 7.2 °C, respectively. The number of extreme events with positive anomalies exceeding 1 SD increased from 10–14 to 28–32 events per year, and exceeding 2 SD—from 1–2 to 12–14 events per year. At the same time, the number of extreme events with negative anomalies exceeding 1 SD decreased from 22–24 to 8–10 events per year, and exceeding 2 SD—from 5–6 to 2–3 events per year. The average duration of extreme events with positive anomalies exceeding 1 SD increased from 2.5 to 3.5 days, but with negative anomalies, practically remained unchanged—3 days. The duration of extreme events with anomalies exceeding 2 SD increased from 1 to 2 days for positive anomalies events and remained the same (2 days) for negative anomalies [38].

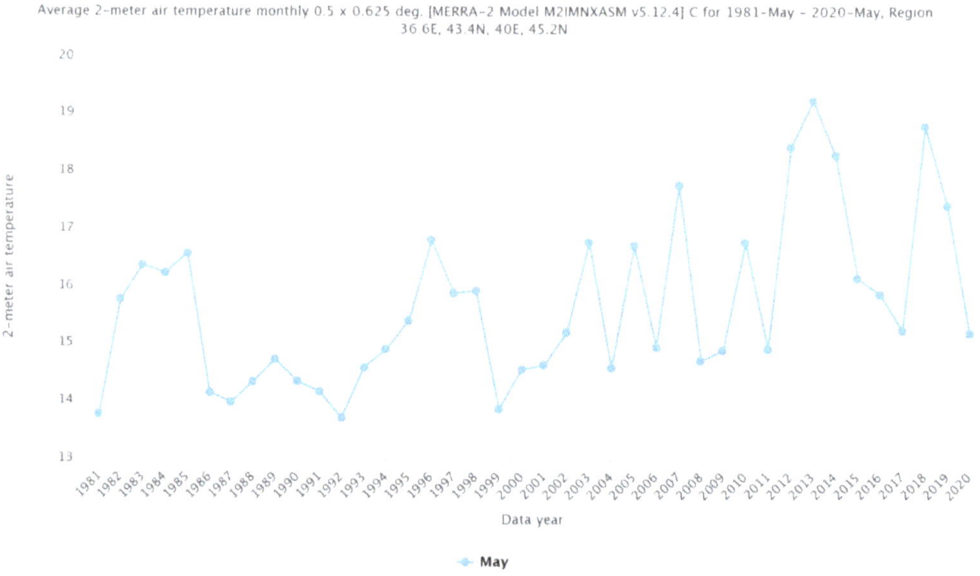

Figure 3. Interannual variability of the near-surface air temperature over the northeastern Black Sea in May from 1981–2020 based on the MERRA-2 M2IMNXASM_5.12.4 Model.

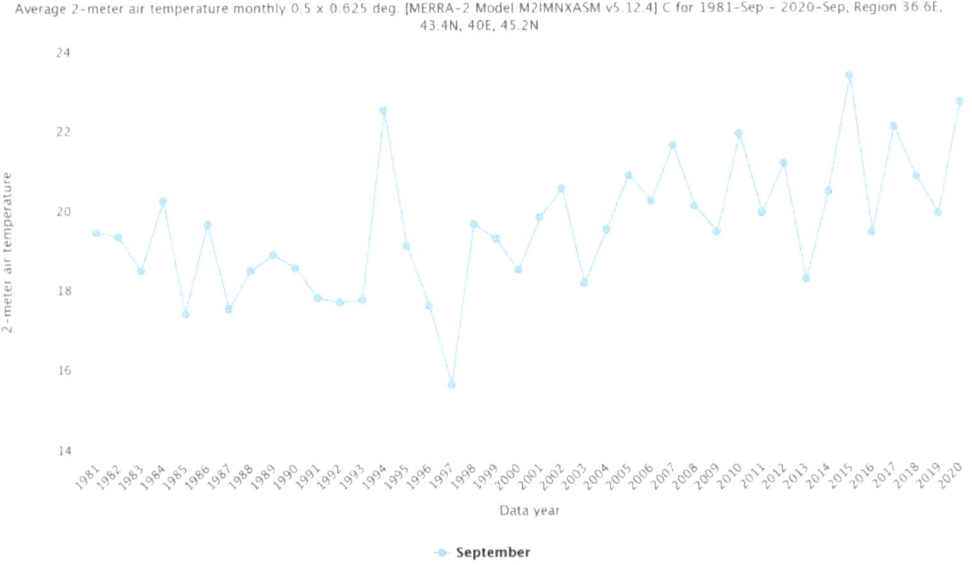

Figure 4. Interannual variability of the near-surface air temperature over the northeastern Black Sea in September from 1981–2020 based on the MERRA-2 M2IMNXASM_5.12.4 Model.

It means that, on average, heat waves in the Eastern Black Sea became a bit stronger, their frequency doubled for anomalous events exceeding 1 SD and reached 28–32 events per year, while strong events exceeding 2 SD increased 10 times from 1–2 to 12–14 per year, and the duration of both types of extreme events increased as well.

3.2. Sea Warming

The sea surface temperature (SST) is a second important factor for coastal tourism. The SST in the Black Sea, like the air temperature, has significantly increased over the past 40 years. Since the end of the 1990s, the maximum monthly average summer SST values have increased. Summer highs in most cases exceeded 25 °C with an extreme of 26.94 °C in August 2010. The maximum/minimum summer SST values correspond to approximately the same years as for the air temperature. The SST linear trend for the whole Black Sea for 1982–2020 is equal to + 0.052 °C/year, which is identical to the air temperature rate of change. The SST trends are unevenly distributed over the Black Sea area: the highest values (+0.058–0.060 °C/year) are observed along the northeastern coast of the sea (Russian coast), the minimum (less than + 0.044 °C/year)—in the center of the Western Black Sea (in the area of the western cyclonic gyre) and at the northwestern shelf of the sea [35]. Roughly the same picture of the distribution of SST trends over the sea area for the period 1981–2015 was presented in [39]. In the northeastern part of the sea from 1983–2015, the SST trends were recorded within + 0.075–0.084 °C/year [40].

In the present study, we are interested in the SST variability during summertime, this is why we calculated the interannual variability of SST for the summer season (June, July, August) only for a 40-years-long time period (1981–2020) and for the northeastern part of the Black Sea (Figure 5). We found that, on average, in the 1980s, the SST was around 22 °C, in the 1990s—23 °C, in the 2000s there was a sharp rise of the SST from 23 °C to 26 °C in 2010, and in the 2010s the SST stabilized around 25 °C. We have to note that features of interannual variability of the SST are similar to those characteristics of the air temperature (see Figure 2).

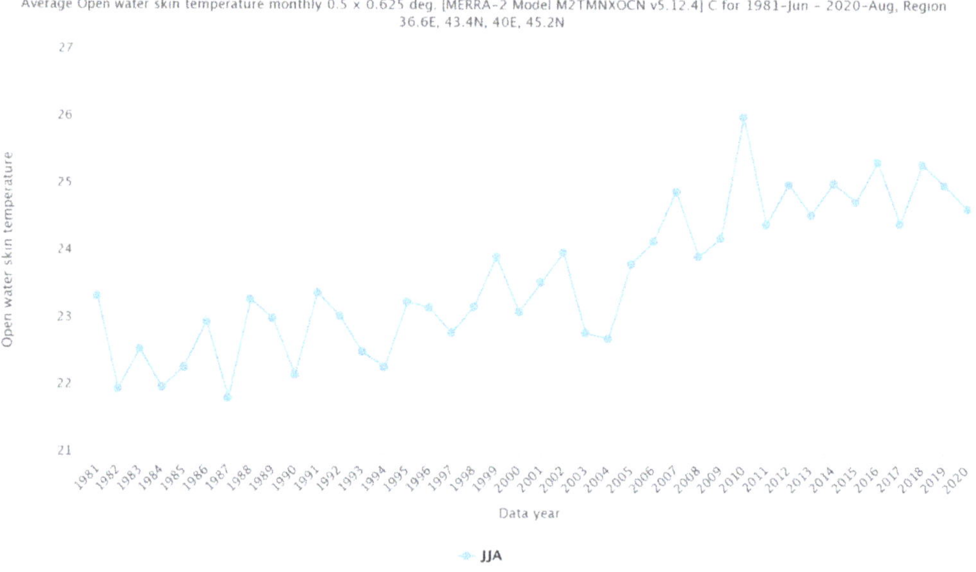

Figure 5. Interannual variability of the sea surface temperature in the northeastern Black Sea during summer from 1981–2020 based on the MERRA-2 M2TMNXOCN V5.12.4 Model.

We also repeated the same analysis for May and September (Figures 6 and 7) to understand in what direction the summer hydrological season is expanding. May shows a slowly warming, but only in 2012, 2013. and 2018 the SST was over 18 °C, and in the past decade, the SST varied from 14.76 °C in 2011 to 18.76 °C in 2018. Thus, the SST in May became much more unstable in comparison with the previous three decades and the SST remains too cold to open the swimming season (Figure 6). Similar to the case with the air temperature, September is already becoming a fully-fledged tourist month suitable for comfortable swimming because the SST during the past two decades has varied between 21.75 °C in 2003 and 24.87 °C in 2020 with a steady tendency to increase (Figure 7). September and October of 2020 were exceptionally warm, and it was possible to swim in the coastal zone of Russia till the end of October.

We also calculated the relationship between the air temperature and SST for the region under investigation and the same time period in the form of the scatter plot shown in Figure 8. It shows a relationship between the monthly mean 2m air temperature and the monthly mean SST. It is interesting that in the middle of the graph, i.e., between 5 and 20 °C of the air temperature, there is a hysteresis curve where there are different values of SST depending on the direction of change of the air temperature during a change of seasons. This is an interesting and evident result which is explained by the fact that in spring the air temperature is rising much faster than warming of the sea surface, and in autumn with a decrease in the air temperature, SST remains high for a longer period of time. Thus, at the same values of the air temperature, SST is higher in autumn than in spring.

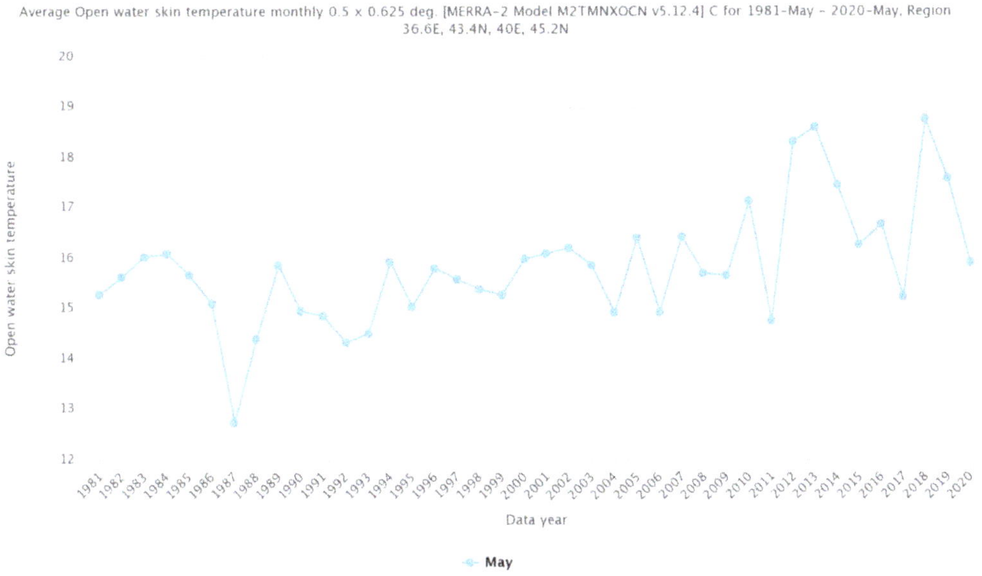

Figure 6. Interannual variability of the sea surface temperature in the northeastern Black Sea in May from 1981–2020 based on the MERRA-2 M2TMNXOCN V5.12.4 Model.

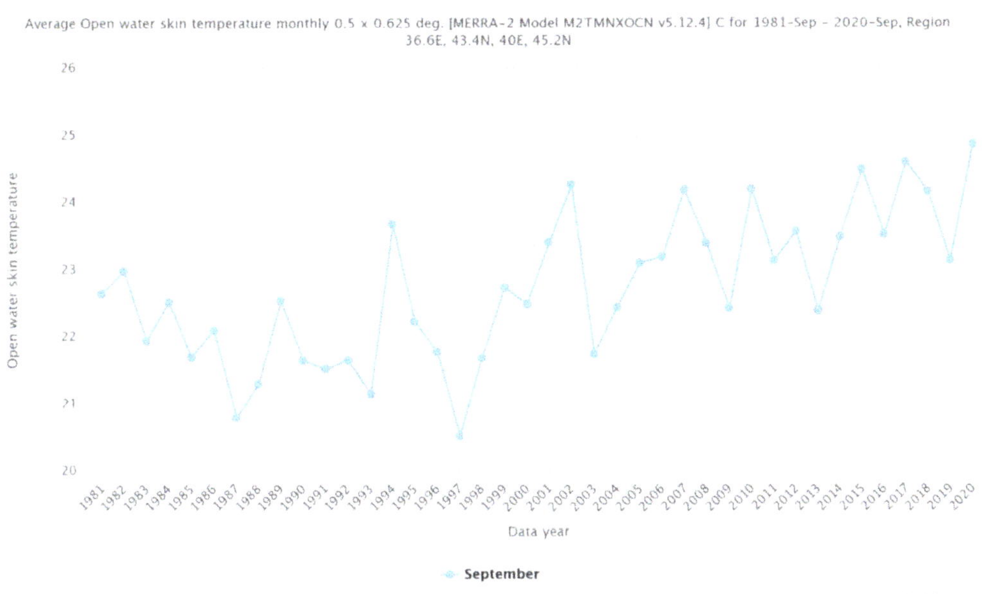

Figure 7. Interannual variability of the sea surface temperature in the northeastern Black Sea in September from 1981–2020 based on the MERRA-2 M2TMNXOCN V5.12.4 Model.

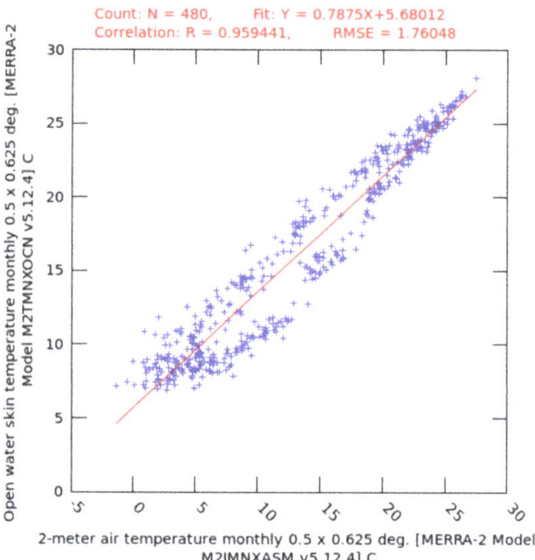

Figure 8. Relationship between the monthly mean 2−meter air temperature and the monthly mean SST for 1981–2020.

3.3. Heavy Rains and River Plumes

Heavy rains have become a serious problem for the coastal zone of the Krasnodar Krai of the Russian Federation and the Black Sea coast of Turkey. They lead to a significant increase in the water level in rivers, flooding of villages and even large cities, damage to urban and tourist infrastructure, roads and railways, bridges, beaches, and the washout of garbage and sewage into rivers. This leads to environmental problems and even human casualties.

For example, on 8 August 2002, the city of Novorossiysk and the nearby resort area Shirokaya Balka were flooded as a result of heavy rain, where, in less than a day there fell an equivalent of six-month precipitation (Figure 9). The tourist infrastructure was destroyed in Shirokaya Balka. According to official figures, almost 20,000 people were affected. More than 100 people died in the flooded area [41]. On 6–7 July 2012, according to the Hydrometeorological Center of Russia, the equivalent of five-month precipitation (275 mm) fell overnight in Krymsk, Novorossiysk, and Gelendzhik. One hundred and seventy-one people died in the flood, which damaged nearly 13,000 homes and affected nearly 30,000 more. The flooding was part of the aftermath of a severe storm that hit the Krasnodar Krai, which resulted in precipitation of the equivalent of almost six months in two days, and it was the largest in the last 70 years. Heavy rains caused dangerous flooding on rivers, including a catastrophic flood on the Adagum River near the city of Krymsk. The total damage is estimated at RUB 20 billion (USD 0.3 billion). Such torrential rains occur annually, for example: 7–8 September 2018, when torrential rains hit the entire coastal zone from Adler to the Kerch Strait; 24–25 October 2018, when torrential rains destroyed roads, railways, and bridges in the Tuapse region; 24 June 2019, when torrential rains caused landslides and damaged roads and tourist infrastructure in the mountain resort area of Krasnaya Polyana near Adler; 29 June and 16 July 2019, when torrential rains hit the city of Sochi [42].

 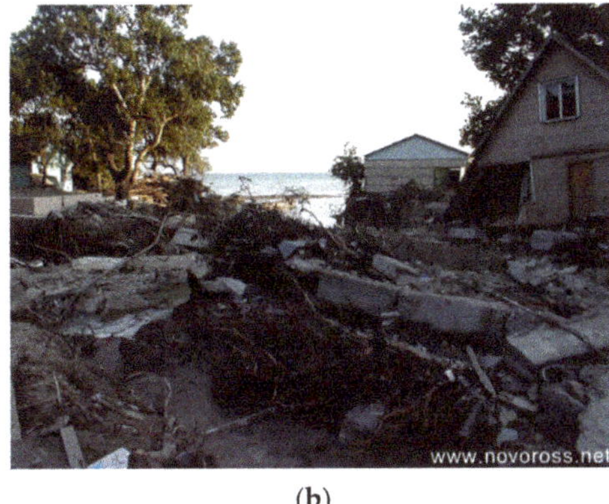

(a)   (b)

Figure 9. (**a**) Flooded streets and embankment in the center of Novorossiysk on 8 August 2002; (**b**) damaged tourist infrastructure in Shirokaya Balka on 8 August 2002.

The summer of 2020 was rather dry, however the summer of 2021 again witnessed catastrophic rains and flooding. On 17 and 18 June 2021, heavy rains hit the coast of the Krasnodar Krai and of Crimea. In Yalta, the monthly norm of precipitation fell on the first night, in Kerch—2 norms of precipitation. Houses, streets, roads, the federal highway "Tavrida" turned out to be flooded. The hailstorm broke through the windows of cars and roofs of houses in some districts of the Krasnodar Krai. Following this, on June 18 in Yalta, another three monthly precipitation rates fell. For three weeks, the forces of the Ministry of the Russian Federation for Civil Defence, Emergencies, and Elimination of Consequences of Natural Disasters were engaged in dismantling debris, clearing streets, pumping out water, restoring housing, flooded and damaged infrastructure, and cleaning streets. The summer vacation was completely ruined, and the beaches were closed for three weeks. People returned tickets, canceled hotel reservations, and returned home.

On 26 June 2021, heavy rains in Gelendzhik and Anapa (50 mm for one hour) flooded the centers of cities, as well as damaged embankments and beaches. On 4 and 5 July 2021, the cities of Adler, Sochi, Lazarevskoye, Tuapse, Lermontovo, Dzhubga, Novorossiysk, and Anapa were flooded again as a result of heavy rains and the storm surge that occurred along the whole Black Sea coast of Russia from Adler to Kerch (Figure 1). In the Khosta Region of Sochi, one monthly norm of precipitation fell during the day, in the Bakhchisaray Region of the Crimea—2–3 norms of precipitation. More than 300 houses, roads, streets, embankments, beaches, gas and water supply, and hundreds of cars were flooded and broken. The only road leading from Dzhubga to Sochi along the entire Black Sea coast was closed for several days due to landslides and destructions by mudflows in several places. Eight people died, and several dozens were injured.

More heavy rains followed on 23 July 2021 (floods in Sochi and Khosta), 8 August 2021 (floods in Sochi, Tuapse, Novorossiysk, and Kerch), 10 August 2021 (flood in Novorossiysk, 4.5 mm during 1 h, Anapa, Vityazevo), 12–13 August 2021 (floods in Novorossiysk, Anapa, Kerch—2–4 monthly precipitation norm or 40–200 mm during 1 day), 15–16 August 2021 (floods in Gelendzhik, Novorossiysk (100 mm), Anapa (219 mm), Kerch (176 mm), 2000 houses were flooded. Water in Anapa was present till 20 August, 500 houses were still flooded, as well as a dozen power electricity stations which were switched off); 18–19 August 2021 (floods in Anapa—monthly norm).

We calculated monthly average precipitation over the northeastern Black Sea for June–August from 2000 to 2020 and found that since 2007 there was a steady rise of precipitation from 39.2 mm/month in 2007 to 86.8 in 2014 and 88.3 mm/month in 2017 (Figure 10). A peak in 2002 was partially related to the catastrophic rains on 21–22 June and 8 August in the area of Novorossiysk. A relative minimum of 2020 (58.3 mm/month) was related to the drought which was observed from summer 2019 till December 2020.

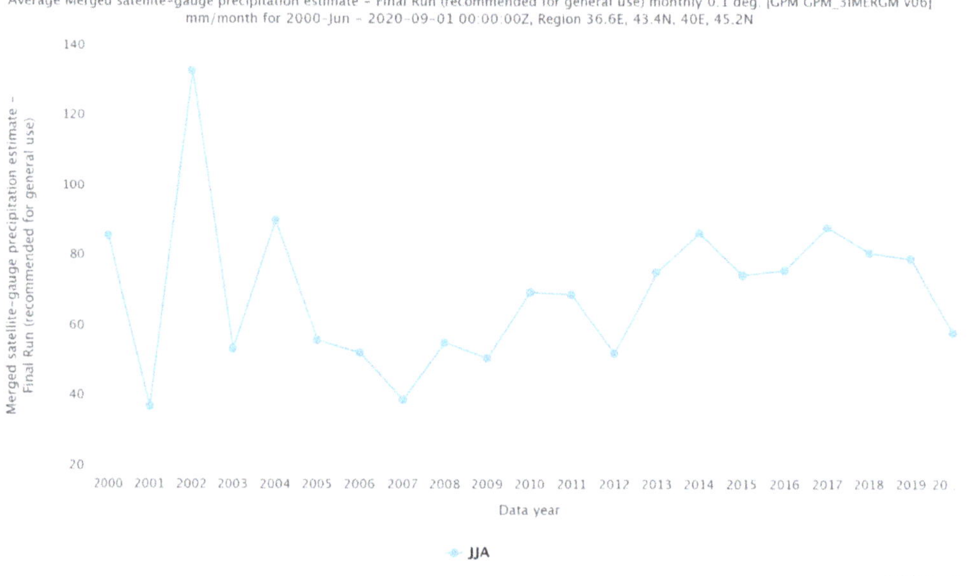

Figure 10. Interannual variability of atmospheric precipitation (mm/month) in the northeastern Black Sea in summer from 2000–2020 based on the GPM_3IMERGM v06 Model.

Looking at Figure 10, we cannot say that there are changes in heavy rains because: (1) the graph shows average values of total precipitation (mm/month) for three summer months; (2) the shown parameter is average monthly precipitation measured in mm/month but not in mm/hr which should be taken from records at meteorological stations; (3) the values are averaged over a large area. Zolina et al. [43,44], Zolina and Bulygina [45], based on records from meteorological stations, showed that in Europe, including the Black Sea coast of Russia, the structure of precipitation has changed to more abundant rainfalls. During 1950–2008, the occurrence of the association of heavy rainfall with longer WP (wet periods defined as consecutive days with significant precipitation > 1 mm/day) was increasing by 3–4% per decade. The occurrence of intense precipitation associated with longer WPs increased from 40% in the 1950s–1960s to 55% in the 1990s–2000s. Heavy rainfalls associated with longer WPs intensified over Europe with upward trends of 2–3% per decade in Western Europe and >5% per decade in European Russia, implying actual changes from 4 to 9 mm/day over the 60-year period. The changing character of WPs (short rain events have been regrouped into prolonged wet spells) may significantly alter the frequency and strength of floods. The highest values of the thresholds for absolute extreme precipitation (more than 26 mm/day) are observed in the North Caucasian Federal District (including the Caucasian Black Sea coast), which is characterized by a complex orography and the presence of mountain systems [45]. They found that for all seasons of the year, practically throughout the entire territory of Russia, there is a steady tendency towards an increase in the values of extreme precipitation up to 8% per decade (values for 1966–2012), which corresponds to an increase in the absolute values of extreme precipitation by about 4 mm/day over the last 50 years in winter and about 7 mm/day in summer.

Besides damage to coastal infrastructure, numerous rains lead to the cooling of seawater in the coastal zone and may delay the establishment of the comfortable conditions for swimming in the tourist season. Such a case was observed in June 2021, when seawater was colder than usual by about 5 °C, while the air temperature was quite typical for this month. The other problem related to heavy rains and river plumes concerns seawater quality, which is very dirty, with a very high concentration of suspended matter and all kinds of pollutants gathered on land. Usually, after heavy rains, people do not swim in the sea for 1–2 days, in case of repeated heavy rains, and beaches can be officially closed for swimming for a week or more. This was the case in Yalta in the second half of June 2021.

River runoff forms river plumes in the coastal zone of the sea (Figure 11), which, depending on the speed and direction of the wind, the velocity, and direction of the coastal current, the magnitude of the runoff, and the difference in density with seawater, can have different forms and spread in different directions from the river mouth, as well as participate in the mesoscale and sub-mesoscale circulation of coastal waters [42,46–50]. River plumes significantly affect the quality of seawater and the sanitary and epidemiological situation on the beaches of the resort area of the Krasnodar Krai. The situation is exacerbated by the fact that sewerage systems are in poor condition in cities, or there are no such systems at all in most small villages along rivers and coastal areas. After heavy rains, wastewater very often reaches the sea and poses a serious threat to human health.

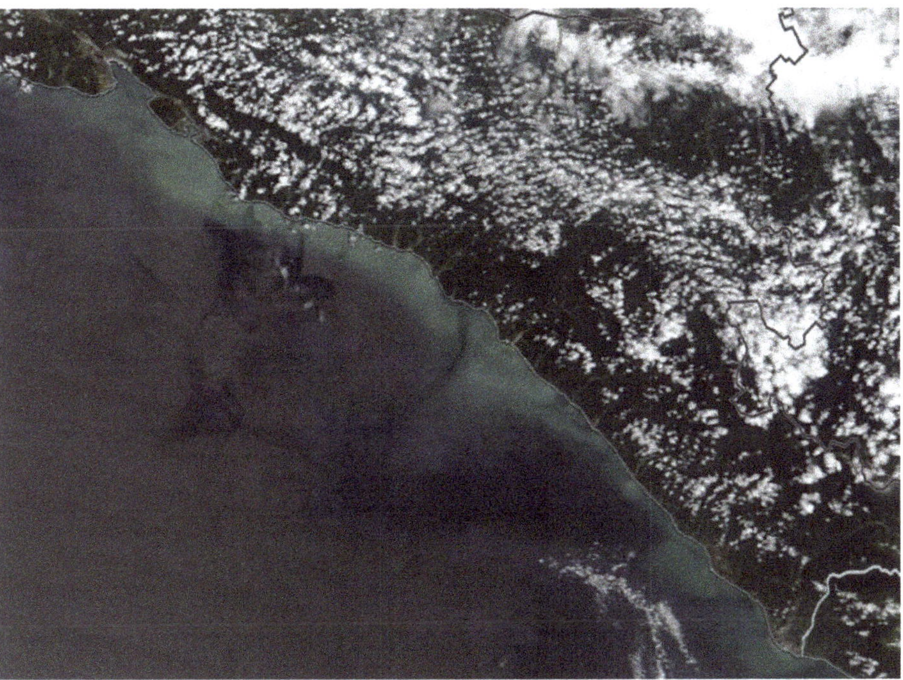

Figure 11. Joint river plumes (light green colors) along the Russian coast of the Black Sea from Novorossiysk to Adler and the border with Abkhazia on 7 July 2021 (Terra MODIS True Color Corrected Reflectance, https://wvs.earthdata.nasa.gov/api/v1/snapshot?REQUEST=GetSnapshot&LAYERS=MODIS_Terra_CorrectedReflectance_TrueColor,Coastlines_15m, Reference_Features_15m&CRS=EPSG:4326&TIME=2021-07-07&WRAP=DAY,X,X&BBOX=43.335571,37.41211,44.84436, 40.327148&FORMAT=image/jpeg&WIDTH=1327&HEIGHT=687&AUTOSCALE=TRUE&ts=1626260065393, accessed on 14 July 2021).

Figure 11 shows a satellite view of the coastal zone of the northeastern Black Sea on 7 July 2021 after heavy rains occurred along the whole coast on 4–5 July. Light green colors show numerous river plumes generated at the mouths of small rivers located along the coast. Heavy rains generate huge river runoff which results in merging of separate river plumes into a single turbid water mass which is characterized by a high concentration of suspended matter and all kinds of pollution. River plumes propagate northwestward along the coast with a general current. Such a situation persists for several days until the flood flow of rivers stops and coastal waters mix with those around them.

3.4. Wind and Waves

Regional climate change-induced coastal processes such as sea-level rise, coastal flooding and erosion, and storm surge are the main reasons for coastal infrastructure damage and vulnerability [51–55].

Wind speed is one of the main climatic parameters. The wind has a significant effect on the exchange of momentum, heat, moisture, and trace particles between the atmosphere and the underlying ocean and land. It causes waves in the oceans and seas, storm surges, and has a huge impact on sea ice. It plays a key role in ocean circulation, which is responsible for the global transport of significant amounts of heat and carbon. Wind speed is a sensitive indicator of the state of the global climate system. Surface wind data are directly applied to industries such as transportation, construction, power generation, agriculture, human health, maritime safety, and emergency management [56].

In general, the Black Sea coast of Russia is not a windy region in summer. Strong winds and stormy conditions are usually observed in late autumn and winter. The most known natural phenomenon is Novorossiysk bora, a strong wind blowing from the northeast, with maximum occurrence and force in winter. Variability of the main directions of winds above the Black Sea is determined by the seasonal variability of the distribution of atmospheric pressure. In winter, north-easterly, northerly, and north-westerly winds prevail over the sea, and easterly winds prevail in its southeastern part. In summer, north-westerly, westerly, and south-westerly winds blow most often. In the cold half-year in the western part of the sea, the average wind speed reaches 7–8 m/s, and in the coastal zone—less than 7 m/s. In the southeastern part of the sea, the wind speed is 5–6 m/s, in the northeastern part—6–7 m/s. In the warm season, the average wind speeds are 1–1.5 m/s less, but they also increase from east to west. The lowest wind speed throughout the year is observed near the southern coast of Crimea and in the southeastern part of the sea [14].

Strengthening of the wind over the sea is most often due to the passage of atmospheric cyclones. The strongest wind in the Black Sea is Novorossiysk bora (or Nord-Ost, northeasterly wind), which is observed several times annually in autumn and winter (40–50 days) (Figure 12). The speed of the northeastern wind reaches 40 m/s, and with gusts—up to 80 m/s. Bora is often accompanied by a temperature drop of 10–20 °C. Its duration ranges from 1–3 days to a week and can affect from several dozens to 150 km offshore. The appearance of bora is limited to the area from Anapa to Tuapse, where the height of the mountains is not a barrier for such winds (Figure 1) [14].

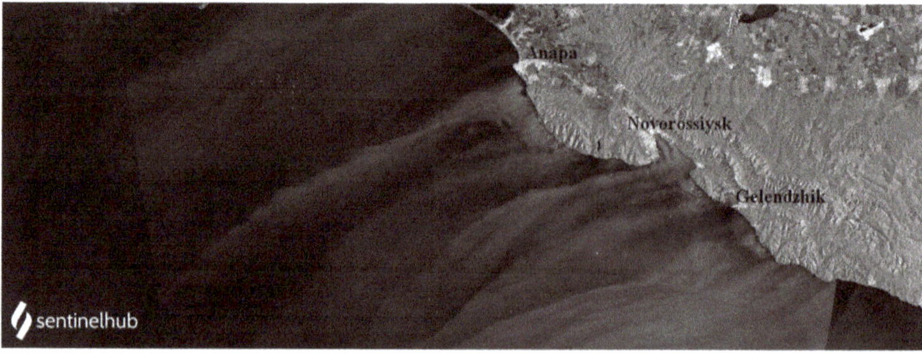

Figure 12. The impact of Novorossiysk bora on the coastal zone of the northeastern part of the Black Sea as revealed by Sentinel-1 SAR image on 5 August 2020. Grey bands propagating from the shoreline are a rough sea generated by the northeasterly wind. Ground length between Novorossiysk and Gelendzhik is 30 km.

During the past decades, the northeasterly wind starts to appear more often during the summertime when it is hot and dry. Its direction is almost perpendicular to the shore in the Krasnodar Krai, thus, even in summer, it is dangerous for swimmers as it generates strong offshore currents directly from the shoreline (Figure 12). In those places where the beaches are located under the cliffs, swimming is even more dangerous, since the cliffs shade the coastal strip 50–100 m wide from the wind, and then the waves and currents begin to accelerate, directed into the open sea, which swimmers cannot resist. Annually, several people are carried out to sea on inflatable mattresses at distances from 1 to 10 km, who have to be rescued on motorboats and even helicopters.

Kostianaia et al. [56] analyzed the seasonal and interannual variability of the wind speed in the eastern part of the Black Sea for the period from 1980 to 2013. In addition, a detailed analysis of extreme wind phenomena for the coastal regions of the Krasnodar Krai and Abkhazia was carried out, interannual variability of the amplitude, frequency, and duration of extreme wind events were studied. Changes in the monthly mean values of the wind speed module, calculated for the coastal region of the Krasnodar Krai (43.5–45° N; 37–40° E), show zero change in the average monthly value of the wind speed module. The average value for this period was approximately 2.3 m/s. The maximum monthly mean value of the wind speed module was observed in 1994 and amounted to approximately 9.2 m/s. On the coast of the Krasnodar Krai, there were a great number of maximum monthly mean values of the wind speed module: there were six cases when the monthly average wind speed was more than 6.5 m/s [56].

The analysis of extreme wind events has shown that there was an increase in the amplitude of extreme events with positive anomalies of the wind speed module from 4.35 m/s to 4.7 m/s for phenomena exceeding 1 SD and for phenomena exceeding 2 SD, there was a slight decrease in the amplitude of extreme events from 7 m/s to 6.8 m/s. The number of extreme events with positive anomalies exceeding 1 SD decreased from 33 to 32 events per year, and those exceeding 2 SD increased from 8 to 12 events per year. The average duration of extreme events with positive anomalies exceeding 1 SD increased from 1.65 to 1.75 days, and those exceeding 2 SD remained the same—1.38 days. In 1995, the maximum number of positive anomalies (44 events) in the wind speed module exceeding 1 SD was recorded [56].

We calculated interannual variability of the monthly averaged wind speed for June–August for 1981–2020 and found great variability from year to year from 5.6 in 1994 to 4.4 m/s in 1997. Wind conditions became more stable in the past decade with variations between 4.75 and 5.1 m/s (Figure 13).

A tornado is a very intense rotating column of air that contacts the cumulus clouds and the underlying surface. In the coastal zones of the seas and oceans, tornadoes are called waterspouts. They rotate with a velocity up to 180 km/h, have a diameter of several dozen meters, and can travel several kilometers before dissipating, sometimes coming to the shore [57]. This is a typical meteorological feature for tropical and subtropical waters and coastal zones of the USA, South America, Southeastern Europe, Southern Africa, India and Bangladesh, Australia, and New Zealand. Several decades ago, a tornado was a very exotic event in the Black Sea. Today, with progressive warming of the sea, in the northeastern part of the Black Sea, there are several cases of tornadoes every year with a maximum occurrence between June and September. Their coming ashore is sometimes accompanied by catastrophic consequences [57]. For example, in June 1991, residents of the city of Tuapse were seriously affected, and in August 2002, the city of Novorossiysk and its suburbs (Shirokaya Balka) were hit with a tornado and heavy rains (Figure 9). Houses, mini-hotels, power lines, roads, and railways were damaged, and several villages around were destroyed. Dozens of people died and went missing. More often, tornadoes occur along the beaches of Big Sochi where the sea surface temperature is the highest along the Black Sea coast of Russia. However, in recent years, waterspouts were registered in Gelendzhik, Novorossiysk, and even in Rostov-on-Don (2017 and 2019) located more northward, in the northeastern extremity of the Sea of Azov.

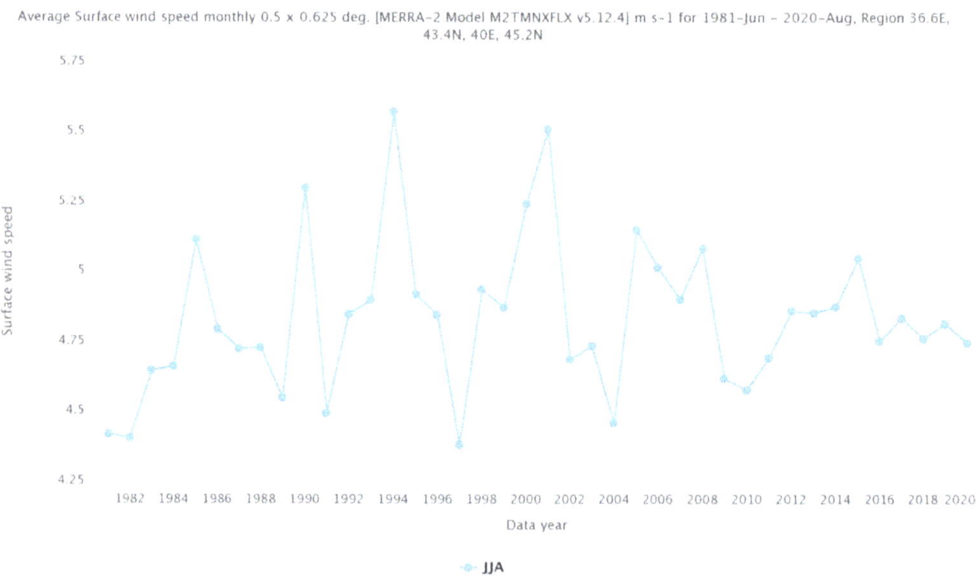

Figure 13. Interannual variability of the wind speed (m/s) in the northeastern Black Sea in summer from 1981–2020 based on the MERRA-2 M2TMNXFLX v5.12.4 Model.

The coastal zone is one of the most dynamic parts of the seas and oceans. It is here that the powerful wave energy transmitted to them by the wind is extinguished. The result of energy dissipation is the formation of strong currents and complex systems of intensive water exchange, the formation and movement of underwater bars, suspending and transfer of huge masses of sediments, and all these processes occur in a complex and in a variety of combinations. The state of the shores, especially beaches and coastal bottom relief, is largely determined by the nature of sediment movement in the coastal zone under the influence of wind waves, swells, and currents, therefore, in the context of the expansion of economic activity on the coast, scientific knowledge about the hydro- and litho-dynamic processes of the coastal zone and the transfer of huge masses of coastal sediments is extremely important [58,59].

The power and impact of winter and summer storms on the shores and beaches in Novorossiysk are shown in Figures 14 and 15a. In both cases, wind gusts were between 15 and 20 m/s. Strong winds in the summertime produce high waves along the shore which is dangerous for swimming, as well as water becomes turbid due to a high concentration of suspended matter even at a pebble beach. Winter storms, which usually are more severe, have also a negative impact on coastal tourism in summer because they destroy beaches (Figure 15a), which requires considerable funding for the reconstruction of beaches and infrastructure. One of the specific examples is the progressive destruction of the 1200 m long beach constructed in 2013 at the Olympic Park in Sochi before the opening of the Olympic Winter Games in 2014. During the past eight years, half of this beach was destroyed due to anthropogenic impacts (construction of a small port, which prevents natural movement of pebble and sediment from Mzymta Mouth along the coast) and the impact of storm surges. The authorities of the City of Sochi are forced to dump 25,000 tons of pebbles and rubble on this beach annually so that it does not erode.

Figure 14. (**a**) Winter storm in Novorossiysk on 22 January 2021 caused by southerly winds; (**b**) summer storm in Novorossiysk on 9 July 2021 caused by northeasterly winds (Nord-Ost).

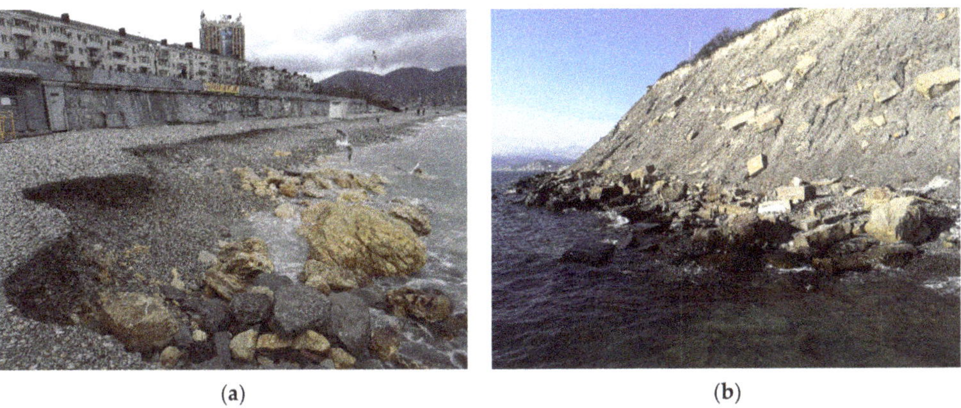

Figure 15. (**a**) Beach erosion in the City of Novorossiysk after a winter storm on 28 January 2021; (**b**) coast erosion in Shirokaya Balka near Novorossiysk due to landslide and talus processes, 4 February 2021.

Coastal and beach erosion is one of the negative processes that occur in the coastal zone due to wind, waves, and swells forcing and substantially impacting coastal infrastructure [54]. In any part of the coastal zone erosion can occur, since the resistance to erosion is not constant and this process depends on the ratio of the wave height, wave period, bottom slope and sediment structure, direction, and frequency of storm surges. Climatic changes in wind speed and direction not only near the coast (generation of wind waves) but also at a distance of tens and hundreds of kilometers (generation of swell) can affect the erosion of the coast even in those places where it was not previously observed [60].

For the Black Sea coast of Russia, the greatest influence on the formation of storm waves is exerted by the winds from the south, south-west, and west directions, and the prevalence of storm waves is from west, south-west, and south directions. Wave heights from these directions can reach 6.0 m and more with a maximum recorded 12.3 m. At such a wave height, the longest waves of 120–200 m are observed [14,61].

The ongoing degradation of the natural coastal zone and beaches on the Russian Black Sea coast is registered for several decades (Figure 15b). The tendencies of storm destruction of the coast may increase in connection with the ongoing rise in the level of the Black Sea, and the shortage of solid material in the coastal zone, which is caused by the fact that under intense hydrodynamic impact its entrainment exceeds its input [58,59,61]. For sandy beaches northward of Anapa (from Anapa to Veselovka) 47 km long, it was shown that the destructive effects of wind waves and swell can be comparable, and the role of swell has been increasing over the past 40 years. As a result, from the mid-1960s to the 2010s 20 to 80 m of beaches have already eroded [58,59]. Terrestrial runoff caused by heavy rains also leads to abrasion-landslide processes in the coastal zone (Figure 15b). In this relation, vulnerability maps for coast and beach erosion for the whole coastal zone of the Russian part of the Black Sea should be calculated on the basis of different scenarios of regional climate change causing changes in the wind–wave and precipitation regime.

The practical experience of coastal protection, as well as the results of hydro- and litho-dynamic studies, shows that the most effective way of coastal protection is the preservation and creation of wide beaches which play a role as smooth wave energy absorbers. It has been established that for the complete damping of wave energy of even a strong storm which may occur once every 25 years, the width of the beach on the Black Sea coast should be on average 30–35 m. With a smaller width, the beaches are subject to erosion, and the coast will be destroyed [61].

Rip current is a very dangerous natural hydrodynamic phenomenon typical for some beaches with shallow water. It occurs when wind and breaking waves push surface water towards the beach, which causes a local slight rise in the water level along the shore. Then, this excess of seawater tends to flow back to the open sea perpendicular to the shore via the route of least resistance, such as slightly deeper parts or breaks in the bottom topography of a sand bar. Gravity initiates a rip current through this gap tens meters wide with a velocity up to 3 m/s. Swimmers caught by a fast rip current begin to panic, try to swim against the water flow back to the shore, and finally exhaust themselves and drown. Rip current is a horizontal current that does not pull people under the water. Very often, rip currents are referred to incorrect terms such as "rip tides" and "undertow". For instance, the undertow occurs everywhere beneath shore-approaching waves, whereas rip currents are localized narrow offshore currents occurring at certain locations along the beach. Rip currents have become a serious problem for the beaches from Anapa to Vityazevo, which kill several people yearly. Every year, these shallow sandy beaches, which were known for decades as the safest beaches for children, are periodically closed due to the sporadic occurrence of rip currents. Tragic news arrived when this paper was under preparation. On 5 July 2021, during a storm at Vityazevo beach near Anapa, three men drowned while trying to save a child who was being carried away by a rip current. Rip currents may intensify due to an increase in wave energy (both wind waves and swell) and a related change in the bottom topography near the shoreline.

3.5. Coastal Upwelling

Coastal upwelling is a natural hydrodynamic phenomenon, which is caused by favorable alongshore (the coast should be on the left in the Northern Hemisphere) and offshore winds. It leads to the upwelling of relatively cold waters from deep layers right in the coastal zone with a temperature minimum near the shore. This is a large-scale process, which in the coastal zones of North-West and South-West Africa, and North-West and South-West America are thousands of km long, a hundred km wide, and in some places are observed all year round with seasonal variability of its intensity at northern and southern limits of the upwelling zone. In the Black Sea, there are no permanent upwelling zones, upwelling can arise sporadically at all coastal zones when favorable winds start to blow [62–66]. The main problem for beach tourism concerns the fact that during the upwelling event, the SST may drop by 10–15 °C from about 25 °C, which is a typical and comfortable temperature at the coasts of the Black Sea, and last from several days to several weeks. As a result, people cannot enter water even at an air temperature of 30 °C. The most known analog of this problem is the western coast of Portugal, where during summer coastal upwelling occurs every year, and it is the main cause why, at these coasts of Portugal, there are no resort areas while at the same latitudes in Spain, along the

Mediterranean coast, there are dozens of well-known resort areas and beach tourism is very popular.

Silvestrova [65] reported that from 1979 to 2016 120 cases of coastal upwelling were registered near Gelendzhik, which is one of the most important resort areas on the Black Sea coastal zone of Russia. In 51 cases, the SST drop was greater than 5 °C. According to this research, strong upwelling near Gelendzhik occurs every year with no evident tendency to increase or decrease in the frequency of appearance. In 1987, eight upwelling events were recorded, in 1995, 1996, 2004, 2009, and 2013—5–6 events. Stanichnaya and Stanichny [66] investigated the occurrence of coastal upwelling in the summer season between 15 May and 15 September. They showed that the average time of one upwelling event is of 5–7 days, while in some cases it can last 14–15 days. For the coastal zone of Crimea, from 1997 to 2011, the number of upwelling events varies from 5 to 12 per year (summer), the number of strong upwelling events with an SST drop of more than 5 °C—from 1 to 5 events, and the total duration—from 20 (2002) to 70 days (2011). Minimal SST values reached 8–9 °C, and the maximum difference with the surrounding waters was 15–16 °C. For instance, only during summer 2003, coastal upwelling occurred on 22–23 May, 7–11 June, 13–20 June, 22–25 June, 27 June–2 July, 7–21 July, 22–25 July, 13–16 August, and the SST near the shoreline varied from 10 to 17.5 °C instead of the typical 20–24 °C in-between the upwelling events [66].

Two examples of these upwelling events that have occurred along the coast of the Crimea and the Krasnodar Krai right during the summer tourist season are shown in Figure 16. On 30 July 2017, a strong upwelling was observed between Sevastopol and Yalta, where the SST was as low as 11–12 °C, while in the surrounding waters it was as high as 24–25 °C (Figure 16a). On 2 July 2018, the upwelling was detected along the Caucasian coast from Abkhazia in the south to Novorossiysk in the north (Figure 16b). Then, from 2 to 10 July it propagated from south to north reaching the area between Anapa and the Kerch Strait. The SST along the shore was 20–21 °C, while in the offshore waters it was 27 °C.

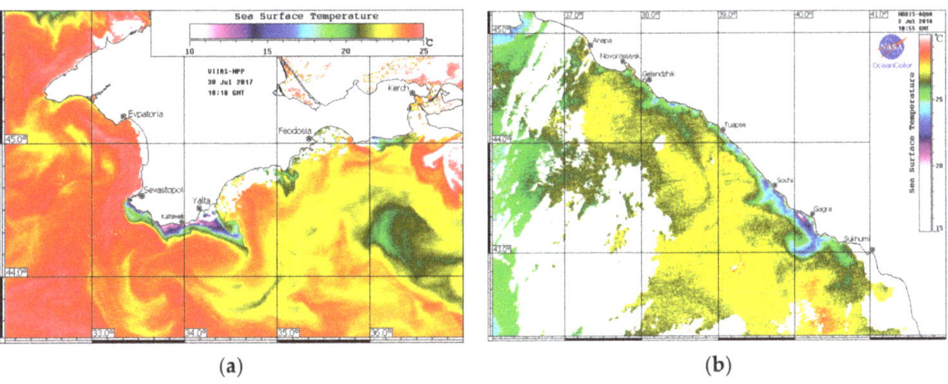

(a) (b)

Figure 16. (**a**) Coastal upwelling (shown in SST) at the southern coast of the Crimea on 30 July 2017 (VIIRS-NPP, 10:18 GMT); (**b**) coastal upwelling (shown in SST) along the coast of the Krasnodar Krai on 2 July 2018 (MODIS-Aqua, 10:55 GMT) (courtesy by D.M. Soloviev, Marine Hydrophysical Institute).

3.6. Sea Level Rise

Today, sea-level rise is not a serious problem for most of the shores of the Black Sea, but a combined effect of the sea-level rise and wave impact on the shores and beaches have already led to the destruction of beaches and shore tourism infrastructure in many places of the Russian coastal zone in the Black Sea and the Sea of Azov [61].

Ginzburg et al. [35] analyzed interannual changes in the level anomalies of the Black and Azov Seas according to the data of the along-track altimetry measurements of Topex/Poseidon and Jason-1, Jason-2, and Jason-3 satellites in the period of 1993–2020. Long-term variability of the Black Sea level is characterized by alternating periods of its rise and fall. At different time intervals, the rate of the level rise varied from the minimum value of +2.64 cm/year from January 1993 to June 1999 to the maximum +25.74 cm/year from August 2012 to July 2013; the rate of the level drop varied from −1.23 cm/year from June 1999 to April 2003 to −8.59 cm/year from June 2004 to February 2008. The average for the 1993–2020 period linear trend of the Black Sea level is of +0.32 ± 0.16 cm/year, which was found to be approximately 2.5 times lower than in 1993–2012 (+0.82 ± 0.18 cm/year), although 1.5 times more than from the 1920s to the mid-1990s (0.17–0.18 cm/year). For the northeastern part of the Black Sea, Lebedev et al. [40] showed that for the time interval of 1993–2015, the Black Sea level off the coast of the Krasnodar Krai grew at an average rate of 0.29 ± 0.03 cm/year, with an increase in rates from the Kerch Strait (0.28 cm/year) to Adler (0.31 cm/year). An average increase in the Black Sea level of 3.2 mm/year is consistent with the average world ocean sea-level rise [17].

3.7. Algal Bloom and Introduced Species

Algal bloom may be regarded as biological pollution of the coastal zone, which occurs as a result of an excess supply of nutrients (mainly nitrates and phosphates) in the coastal zone with river and land runoff, and high values of the SST. In the Black Sea, this problem occurs almost every year in the northwestern part of the Black Sea in the region of Odessa, and on shallow water sandy beaches of Anapa on the Russian coast [50,67]. The abundance of algae in the upper layers of the water makes the water green, not transparent, and entering into the water is unpleasant, and in some cases even dangerous (when swallowed), as it may contain toxic algae. In addition, the algae washed ashore starts to rot and the beach becomes unpleasant until cleaning (Figure 17).

(a)   (b)

Figure 17. (**a**) Typical summer green algal bloom on beaches of Anapa (https://turvopros.com/kogda-cvetet-more-v-anape, accessed on 16 July 2021); (**b**) concentration of brown algae in water and dead algae washed ashore with a storm near Dyurso on 11 October 2020.

Introduced species is yet another problem for the Black Sea and for the development of coastal tourism in particular. Many alien species have become established in the Black Sea since the middle of the 20th century due to a discharge of ballast water from ships coming from adjacent seas and even oceans, as well as due to regional warming which facilitates northward expansion of species from the Mediterranean Sea. The most dramatic example of alien species introduction to the Black Sea was the invasion of a gelatinous predator, the polymorphic ctenophore *Mnemiopsis leidyi* (Figure 18a), and later on, the accidental invasion of ctenophore *Beroe ovata* (Figure 18b) [68]. On the international level, it is suggested that *Mnemiopsis* was brought to the Black Sea with ballast waters from the coastal regions of North America at the beginning of the 1980s. By 1988, *Mnemiposis* had spread over the entire Black Sea area and showed an enormous abundance outburst in the fall of 1989 [68–71]. This led to a decrease in the biomass, abundance, and species diversity of edible zooplankton, fish larvae, and eggs, which are the principal food objects of *Mnemiopsis*. As a result, by 1991, the commercial fish catches had decreased 10 times; this especially referred to the anchovy, Mediterranean horse mackerel, sprat, as well as higher trophic levels—predator fishes and dolphins, who feed mainly on anchovies and sprats [68]. From the Black Sea, *Mnemiopsis* spread to the seas of Azov and Marmara, and in 1999—to the Caspian Sea. Then, in 1997, a new invader—ctenophore *Beroe ovata*—was first found in the northwestern part of the Black Sea, and in August 1999, the first outburst in the *Beroea* development over the entire Black Sea was observed. This predator feeds, first of all, on *Mnemiopsis*, thus, *Beroe* radically reduced the *Mnemiopsis* population and this resulted in the restoration of the Black Sea ecosystem within several years [68].

(**a**)

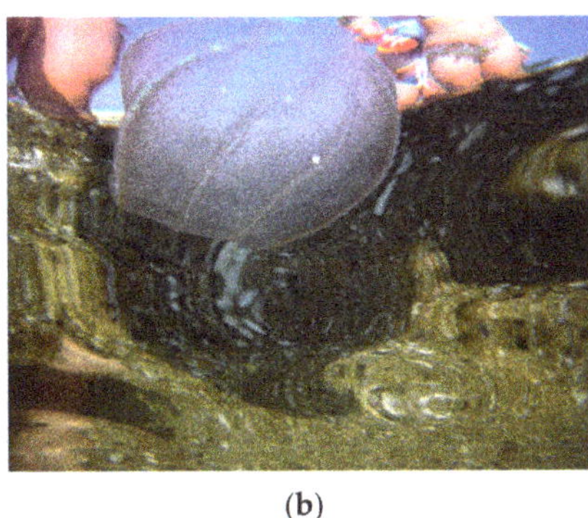

(**b**)

Figure 18. (**a**) *Mnemiopsis leidyi* (Courtesy by Bruno C. Vellutini—Own work, CC BY-SA 3.0, https://commons.wikimedia.org/w/index.php?curid=30155106, accessed on 18 July 2021); (**b**) *Beroe ovata* (Courtesy by Cristian Chirita—Own work, GFDL, https://commons.wikimedia.org/w/index.php?curid=10994369, accessed on 18 July 2021).

Both jellies are placed in the palm of your hand and do not pose any threat to adults or children, however, their number in different years reached such values when there were several dozens of them in one cubic meter of water. Swimming in such water is extremely unpleasant.

4. Discussion

Warming of the Black Sea coast of Russia, which includes both air temperature and sea surface temperature, is an evident positive factor that will favor the development of coastal tourism. The observed warming is accompanied by an expansion of the tourist season from June–August to September as well, which is already comparable with June by the air temperature. This will lead to a significant increase in water and electricity consumption related to an increase in the need for air conditioning and lighting. The lighting issue comes from the fact that time in the European part of Russia was established in such a way that, for example, in Novorossiysk on 22 June the sunrise is at 04:43 (local time) and sunset is at 20:18, and on 22 September, the sunrise is at 06:16 and sunset is at 18:26. It means that there is a lot of sun in the early morning and darkness in the early evening. Firstly, it is inconvenient for tourists who usually sleep in the morning, and secondly in the evening, when the air and water are still warm, swimming or walking in the dark is not very pleasant. This is an additional problem concerning the energy supply for lighting in the evening.

Regional climate change in the Black Sea is accompanied by the intensification of extreme weather events. The First [18] and Second [19] Roshydromet assessment reports on climate change and their consequences on the territory of the Russian Federation confirm these forecasts. Kostianoy et al. [38] showed that from 1950 to 2015, on average, heat waves in the Eastern Black Sea became a bit stronger (up to 4–6 °C hotter than the norm), their frequency doubled for anomalous events exceeding 1 SD and reached 28–32 events per year, while strong events exceeding 2 SD increased 10 times from 1–2 to 12–14 per year. The average duration of extreme events with positive anomalies exceeding 1 SD increased from 2.5 to 3.5 days, and with anomalies exceeding 2 SD increased from 1 to 2 days.

These facts may result in health issues for the local population and tourists due to heatstroke from overheating from the sun, as well as in problems with water and energy supply. One such example is a draught on the coasts of the Black Sea, which occurred in all the Black Sea countries and lasted from summer 2019 to December 2020. The draught was caused by a lack of rainfall everywhere. The draught resulted in shallowing of rivers, freshwater reservoirs, and a significant decrease in the groundwater level around the Black Sea (Figure 19). For example, in Turkey, the water storage level in water reservoirs was the lowest during the past 15 years, and the year 2020 was the driest for the past 5 years [72]. As a result, during two tourist seasons in several coastal cities of the Krasnodar Krai and Crimea, there were restrictions in the water supply to housing when water was given to houses during 3 h in the morning and 3 h in the evening. The situation was aggravated by a large number of tourists in the summer of 2020 because of restrictions to travel to resorts of Turkey, Egypt, Greece, Cyprus, and other countries related to the COVID-19 pandemic.

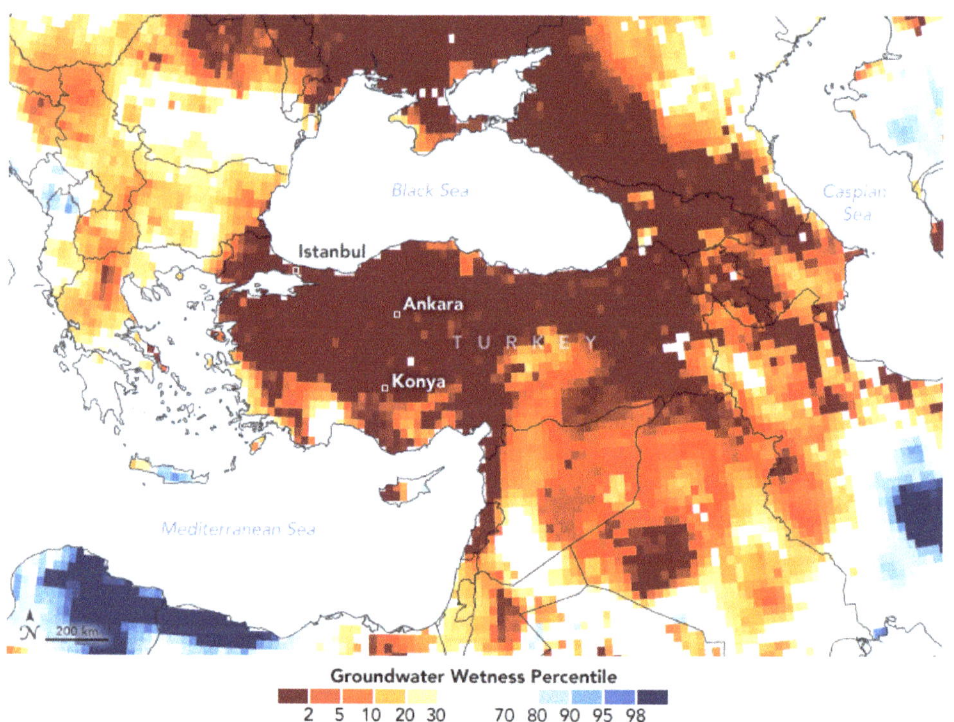

Figure 19. Shallow groundwater storage as of 11 January 2021 in the Black Sea Region and the Middle East, as measured by the Gravity Recovery and Climate Experiment Follow On (GRACE-FO) satellites. The color palette depicts the groundwater wetness percentile, i.e., how the amount of groundwater compares to long-term records (1948–2010). Blue areas have more water than usual, and orange and red areas have less groundwater. Credit: NASA's Earth Observatory [72].

Further increase in air temperature and intensity of drought periods will likely lead to more frequent forest and wildfires. This, in turn, poses a threat to the safety and security of the local population, will impact the attractiveness of the landscape, and the ultimate decision for tourists to choose this given location. This is already seen in such countries as Portugal, Spain, France, Italy, Montenegro, Greece, and Turkey [73].

We can also mention the recent paper by Klueva et al. [28] dealing with a forecast of the Tourism Climate Index (TCI), which merges tourism-related climate factors into a single index, for different regions of Russia, including the Black Sea coast of Russia. TCI is based on the air temperature, humidity, wind speed, atm precipitation, and time of sunshine. This is a very interesting study, which, for the coast of the Black Sea, shows that periods with TCI > 70 (which is characteristic for tourist comfort) will decrease from 90 days in 1981–2010 to about 30 days by 2050, and will tend to zero by the end of the 21st century. The main cause of this catastrophic decrease in TCI is a sharp rise in the expected air temperature which will be uncomfortable for people. Our research shows a set of additional factors like extreme weather events (heat waves, heavy rains, tornado), hydrological features like upwellings, river plumes, rip currents, wave impact, coastal erosion, biological factors, such as algal bloom, and introduced species. Both studies complement each other.

In recent years, torrential rains have become a real disaster for the resorts of the Krasnodar Krai and the Crimea, and although, on average, the amount of precipitation in summer has not increased over the past decade, the scale of the disaster caused by almost every rainstorm and flash flooding is enormous. We can recall heavy rains in the region on 17–18 June, 26 June, and 4–5 July 2021 that in some places were comparable by their disastrous impact to the heavy rain and flash flood in Western Germany, Southern Netherlands, and Southeastern Belgium on 15 July 2021. Even if it is one case for the entire summer holiday season, it can have disastrous consequences for the resort infrastructure, beaches, roads, and railways. For the coastal zone of the Krasnodar Krai, flooded railway stations and roads, damage to roads and railways bridges, as well as landslides and mudflows on these transport lines may cut all kinds of transport from several hours to several days because this is the only road and railways going along the coast from Dzhubga to Tuapse, and then to Sochi [74].

Change in the atmospheric circulation may lead to a change in cloudiness which can have a considerable effect on the attractiveness of a location for tourists. Such an example already exists on the Black Sea—the eastern part of the northern coastal zone of Turkey where the air temperature and sea surface temperature are even warmer than in the Krasnodar Krai, but which does not have sea resorts because this zone is almost always covered by clouds. This eastern part receives up to 2540 mm of rainfall per year, which makes it the wettest region in the country [75].

We did not find an increase in average wind speed during summer, but even at the present wind force, its direction can change in a way that will lead to serious erosion of the coasts and beaches due to the wave impact, to the appearance of coastal upwelling in the locations of the coastal zone of the Krasnodar Krai and the Crimea, where before it was not registered. This impact will be intensified with a progressive sea-level rise in the Black Sea. Upwelling may last for 1–2 weeks and the SST may drop by 10–15 °C. We consider upwelling as a serious threat to the development of coastal and beach tourism because there are many examples in the world where large-scale upwelling systems prevent its development even if the appropriate infrastructure exists. In Europe, the best example is the western coast of Portugal, and in North America—the coast of California. Today, the impact of algal bloom can be regarded as a local problem for the region of Anapa shallow water sandy beaches and along the coasts of the shallow Sea of Azov. The appearance of newly introduced species in the Black Sea is quite possible which is explained now by the warming of the sea and the establishment of favorable conditions for new species originated from the Mediterranean and Red Seas.

As we have shown, the frequency of extreme events is growing, so there is the variety of unpleasant events that make coastal tourism activities unpleasant or uncomfortable. Tourists might well fall into the situation where, if they visit the Black Sea coast of Russia every year, they will repeatedly catch one or another such event and will decide to go to another tourist destination the following year. Imagine that during one vacation, the duration of which for most vacationers is 1–2 weeks, tourists experience abnormal heat, lack of fresh water, and problems with electricity, in another year—heavy rain and flash flood accompanied by large river plumes, in the third year—strong wind and waves, in the fourth year—swell from distant storms, in the fifth year—an upwelling event for several days or even weeks, in the sixth year—algal bloom of waters or the invasion of jellyfish, and so on. These events might as well happen all in one summer. The likelihood of ruining the only vacation of the year is growing. Some of the unlucky people will next time think about where to go on vacation—to the Crimea, the Caucasus, Turkey, Egypt, or other countries. Until now, the decision was mainly dependent on the financial capabilities of tourists and on the price/quality ratio of the hotel and tourist services provided. At present, the weather factor also interferes in the decision-making process, which is a direct consequence of regional climate change. It will have to be taken into account not only by vacationers but also by the authorities and the tourism industry.

5. Conclusions

In the paper, we did not investigate the impact of anthropogenic factors, geopolitical and socio-economic processes, and the COVID-19 pandemic that also plays an important role in the sustainable development of coastal tourism on the Black Sea coasts of Russia. For instance, according to World Travel and Tourism Council [3], in 2020, due to the consequences of the COVID-19 pandemic, the total GDP contribution dropped from 10.4% to 5.5% (by 49.1% from USD 9.17 to 4.67 trillion), 62 million jobs were lost, representing a drop of 18.5%, leaving just 272 million employed across the tourism sector globally, compared to 334 million in 2019. The Council says: "The threat of job losses persists as many jobs are currently supported by government retention schemes and reduced hours, which without a full recovery of Travel and Tourism could be lost. Domestic visitor spending decreased by 45%, while international visitor spending declined by an unprecedented 69.4%" [3].

In Russia, according to World Travel and Tourism Council [3], in 2020, the total contribution of travel and tourism to GDP dropped by 47% from USD 75.5 to 40.1 billion, 205.7 thousand people (-5.1%) lost their jobs in the tourist sector, international visitor spending dropped by 69.6% from USD 14.8 to 4.5 billion, and domestic visitor spending dropped by 43.9% from USD 37.9 to 21.3 billion. The share of international visitors concerns the main tourist destinations in Russia—Moscow and St.-Petersburg, while the domestic visitor share reflects mainly coastal tourism in the Black Sea. New COVID-19-related restrictions for tourists on the coasts of the Black Sea for the summer season of 2021 will continue to considerably impact socio-economic conditions in the Krasnodar Krai and Crimea.

The above-mentioned example shows that external factors like geopolitical and socio-economic processes, global or domestic pandemics can suddenly cut one of the most important sectors of the national and global economy by around 50%. Regional climate change is one of these factors which affects the sustainable development of coastal tourism but is usually ignored. It would be wrong to believe that climate change is happening very slowly, so its effects can be felt in decades. The intensification of extreme weather conditions and negative natural processes in the atmosphere and the sea already today presents tourists with a serious choice—where is the best place to spend their next summer vacation? The answer to this question is not as obvious as it was in the previous decades.

The Russian Federation continues to invest significant funding in the development of infrastructure related to coastal tourism. On 2–5 June 2021, at the St.-Petersburg Economic Forum, a decision was taken to invest around USD 1.12 billion in the development of resorts and tourism in the Krasnodar Krai, in particular, in the construction of new resorts in Sochi and Anapa [76]. At the beginning of July 2021, the Russian Government announced a plan for the construction of a new tourist complex near Anapa called "New Anapa", which envisages the construction of 50 3–5 stars hotels, as well as restaurants, shops, roads, and other infrastructure which has to provide year-round tourist services. It is expected that it will provide 25,000 new jobs and will attract an additional 2–3 million tourists per year. This ambitious plan should be performed in three years and will cost around USD 3.5 billion [77]. The Russian Ministry of Transport and Russian Railways are developing plans to construct a new railroad from Goryachy Klyuch to Sochi (which presently extends along the shoreline from Tuapse to Sochi) through mountains and a new automobile road from Dzhubga to Sochi (which also extends along the coast) several kilometers from the beach and urban infrastructure. The railway project will cost around USD 19 billion and will be completed by 2030, the automobile road project is estimated to be USD 32 billion and will be accomplished by 2035 [78]. As usual, all these plans did not take into account the negative factors that can arise in the near future from regional climate change. The comprehensive analysis of these factors would make it possible to make plans scientifically sound and investments much more effective.

We believe that this current research will serve as the first step in the multifactor complex study of this important matter. The present and future research are funded within the framework of the Russian Ministry of Science and Education Project "Comprehensive studies of the ecological state of waters of the coastal zone of the northeastern shelf of the Black Sea in the framework of participation in the international project DOORS" related to collaboration with the DOORS (Developing Optimal and Open Research Support system to unlock the potential for blue growth in the Black Sea) Project—a large international project started on 1 June 2021 within the framework of the Blue Growth thematic competition of the HORIZON-2020 Program of the European Union. The DOORS consortium includes 37 organizations from 17 countries, including the P.P. Shirshov Institute of Oceanology of the Russian Academy of Sciences. It is expected that based on the results of the project, specific practical recommendations will be prepared for optimization of maritime economic activity and the Blue Growth in all the Black Sea countries. We hope that our research related to climate change impacts on coastal tourism will be a base for such recommendations because coastal tourism is a major economic sector for all the Black Sea countries.

In this study, we focused on the assessment of climatic conditions for beach-based coastal tourism which makes it comfortable for tourists to spend time on the beach or in the water. This research has limitations as it has not covered many related aspects such as, for example: impact of climate change on wildlife in the region, detailed analysis of the impact of climate change on railway and road infrastructure, airports, and consequences for the touristic attractiveness of the region; the impact of climate change on yachting, cruising, boating, surfing, sup surfing, windsurfing, sailing, etc. This research also does not suggest any management strategies as this was out of the scope of our research. This paper has not either discussed the economic consequences of such climate change impacts on the region as this merits a separate study.

For future research, the authors are planning to develop the following directions of research: (1) detailed investigation of individual factors and their interannual variability; (2) investigation of local peculiarities for different parts of the Black Sea coast of Russia; (3) collaboration with other Black Sea countries in the framework of the DOORS Project to investigate the impact of climate change on their coastal zone; (4) elaboration of recommendations for local authorities and tourism business to overcome potential negative consequences of regional climate change.

Author Contributions: Conceptualization, E.A.K. and A.G.K.; methodology, A.G.K.; formal analysis, E.A.K. and A.G.K.; investigation, E.A.K. and A.G.K.; resources, A.G.K.; data curation, A.G.K.; writing—original draft preparation, E.A.K.; writing—review and editing, A.G.K.; visualization, E.A.K.; supervision, A.G.K.; project administration, A.G.K.; funding acquisition, A.G.K. All authors have read and agreed to the published version of the manuscript.

Funding: This research was funded in the framework of the Russian Ministry of Science and Education Project "Comprehensive studies of the ecological state of waters of the coastal zone of the northeastern shelf of the Black Sea in the framework of participation in the international project DOORS" performed under the Agreement on the provision of a grant from the federal budget in the form of a subsidy 13.2251.21.0008. The APC was covered by the MDPI.

Institutional Review Board Statement: Not applicable.

Informed Consent Statement: Not applicable.

Data Availability Statement: All data used in the present analysis are freely available from the databases and websites mentioned in Materials and Methods Section.

Acknowledgments: We acknowledge collaboration with the DOORS Project (Grant Agreement ID 101000518) performed in the framework of the EU HORIZON-2020 Programme. We acknowledge the use of imagery from the Worldview Snapshots application (https://wvs.earthdata.nasa.gov), part of the Earth Observing System Data and Information System (EOSDIS). We are thankful to Dmitry M Soloviev from Marine Hydrophysical Institute, who provided us with thermal imagery presenting recent upwelling events.

Conflicts of Interest: The authors declare no conflict of interest. The funders had no role in the design of the study; in the collection, analyses, or interpretation of data; in the writing of the manuscript, or in the decision to publish the results.

References

1. UN Atlas of the Oceans. Available online: http://www.oceansatlas.org/subtopic/en/c/615/ (accessed on 2 July 2021).
2. ECORYS. *Study in Support of Policy Measures for Maritime and Coastal Tourism at EU Level*; ECORYS: Rotterdam, Brussels, 2013.
3. World Travel & Tourism Council. Available online: https://wttc.org/Research/Economic-Impact (accessed on 2 July 2021).
4. Ghosh, T. Coastal tourism: Opportunity and sustainability. *J. Sustain. Dev.* **2011**, *4*, 67–71. [CrossRef]
5. Gössling, S. Global environmental consequences of tourism. *Glob. Environ. Chang.* **2002**, *12*, 283–302. [CrossRef]
6. Mikhaylichenko, Y.G. Development of an integrated coastal zone management system for the Black and Caspian seas. In *Mitigation and Adaptation Strategies for Global Change*; Springer: Berlin/Heidelberg, Germany, 2006; Volume 11, pp. 521–537.
7. Schubert, M. Marine Spatial Planning. In *Handbook on Marine Environment Protection*; Salomon, M., Markus, T., Eds.; Springer: Berlin/Heidelberg, Germany, 2018; pp. 1013–1024.
8. De Vet, J.M.; Bocci, M.; Zonta, D.; Patrini, V.; Beyer, C.; Shipman, B.; Lampridi, V.; Triantaphyllidis, G. Support Activities for the Development of Maritime Clusters in the Mediterranean and Black Sea Areas. Final Report under FWC MARE/2012/06—SC D1/2013/01. Brussels. Berlin. Athens. 2014, 111p. Available online: https://webgate.ec.europa.eu/maritimeforum/system/files/Med%20clusters%20-%20Annexes%20def_0.pdf (accessed on 3 July 2021).
9. Hansen, E.R.; Holthus, P.; Allen, C.L.; Bae, J.; Goh, J.; Mihailescu, C.; Pedregon, C. OCEAN/MARITIME CLUSTERS: Leadership and Collaboration for Ocean Sustainable Development and Implementing the Sustainable Development Goals. World Ocean Council White Paper. 2018. Available online: http://www.bluemed-initiative.eu/wp-content/uploads/2018/03/Ocean-Maritime-Clusters-and-Sustainable-Development-WHITE-PAPER-FINAL-2018-logo.pdf (accessed on 3 July 2021).
10. Kostianaia, E.A.; Shapovalov, S.M.; Kostianoy, A.G. Marine clusters as a tool to achieve SDG14 in the Southern Seas of the Russian Federation. *J. Oceanol. Res.* **2018**, *46*, 234–247. [CrossRef]
11. Maritime Clusters in the Mediterranean Region. Union for Mediterranean. December 2019. Available online: https://ufmsecretariat.org/wp-content/uploads/2019/12/Maritime-Clusters-in-the-Mediterranean-Region_Dec-2019.pdf (accessed on 3 July 2021).
12. Kuban News. 20 August 2020. Available online: https://kubnews.ru/turizm/2020/08/20/kurorty-krasnodarskogo-kraya-dostizheniya-i-perspektivy-razvitiya/ (accessed on 4 July 2021).
13. Kostianoy, A.G.; Kosarev, A.N. (Eds.) The Black Sea Environment. In *The Handbook of Environmental Chemistry*; Springer: Berlin/Heidelberg, Germany, 2008; 457p.
14. Grinevetskiy, S.R.; Zonn, I.S.; Zhiltsov, S.S.; Kosarev, A.N.; Kostianoy, A.G. *The Black Sea Encyclopedia*; Springer: Berlin/Heidelberg, Germany, 2015; 889p.
15. Federal Tourism Agency. Available online: https://tourism.gov.ru/contents/turism_v_rossii/regions/yuzhnyyfo/krasnodarskiy-kray/klassifikatsiya-gostinits-v-krasnodarskom-krae (accessed on 4 July 2021).
16. Solomon, S.; Qin, D.; Manning, M.; Chen, Z.; Marquis, M.; Averyt, K.B.; Tignor, M.; Miller, H.L. (Eds.) *Contribution of Working Group I to the Fourth Assessment Report of the Intergovernmental Panel on Climate Change, 2007*; Cambridge University Press: Cambridge, UK, 2007.
17. IPCC. *Climate Change 2013: The Physical Science Basis. Contribution of Working Group I to the Fifth Assessment Report of the Intergovernmental Panel on Climate Change*; Stocker, T.F., Qin, D., Plattner, G.-K., Tignor, M., Allen, S.K., Boschung, J., Nauels, A., Xia, Y., Bex, V., Midgley, P.M., Eds.; Cambridge University Press: Cambridge, UK, 2013; 1535p.
18. *An Assessment Report on Climate Change and Its Consequences on the Territory of the Russian Federation*; V.1. Climate Change; Roshydromet: Moscow, Russia, 2008; 227p.
19. *The Second Assessment Report on Climate Change and Its Consequences on the Territory of the Russian Federation*; Roshydromet: Moscow, Russia, 2014; 1008p.
20. Scott, D.; Jones, B.; Mcboyle, G. *Climate, Tourism & Recreation: A Bibliography—1936 to 2006*; University of Waterloo: Waterloo, ON, Canada, 2006.
21. Becken, S. *Climate Change Impacts on Coastal Tourism*; CoastAdapt Impact Sheet 6; National Climate Change Adaptation Research Facility: Southport, Australia, 2016.

22. Grimm, I.J.; Alcântara, L.; Sampaio, C.A.C. Tourism under climate change scenarios: Impacts, possibilities, and challenges. *Rev. Bras. De Pesqui. Em Tur.* **2018**, *12*, 1–22.
23. Layne, D. Impacts of Climate Change on Tourism in the Coastal and Marine Environments of Caribbean Small Island Developing States (SIDS). *Caribb. Mar. Clim. Chang. Rep. Card Sci. Rev.* **2017**, *2017*, 174–184.
24. Becken, S. Harmonising climate change adaptation and mitigation: The case of tourist resorts in Fiji. *Glob. Environ. Chang.* **2005**, *15*, 381–393. [CrossRef]
25. Friedrich, J.; Stahl, J.; Hoogendoorn, G.; Fitchett, J.M. Exploring Climate Change Threats to Beach Tourism Destinations: Application of the Hazard—Activity Pairs Methodology to South Africa. *Weather. Clim. Soc.* **2020**, *12*, 529–544. [CrossRef]
26. Lincoln, S. Impacts of Climate Change on Society in the Coastal and Marine Environments of Caribbean Small Island Developing States (SIDS). *Caribb. Mar. Clim. Chang. Rep. Card Sci. Rev.* **2017**, *2017*, 115–123.
27. Santos-Lacueva, R.; Clavé, S.A.; Saladié, Ò. The vulnerability of coastal tourism destinations to climate change: The usefulness of policy analysis. *Sustainability* **2017**, *9*, 2062. [CrossRef]
28. Klueva, M.V.; Shkolnik, I.M.; Rudakova, Y.L.; Pavlova, T.V.; Kattsov, V.M. Summer tourism in the context of future climate change in Russia: Assessment based on a wide range of high-resolution conditional forecasts. *Russ. Meteorol. Hydrol.* **2020**, *6*, 47–59. (In Russian)
29. Arabadzhyan, A.; Figini, P.; García, C.; González, M.M.; Lam-González, Y.E.; León, C.J. Climate change, coastal tourism, and impact chains—A literature review. *Curr. Issues Tour.* **2020**, *1*, 1–36. [CrossRef]
30. Acker, J.G.; Leptoukh, G. Online Analysis Enhances Use of NASA Earth Science Data. *Eos Trans. AGU* **2007**, *88*, 14–17. [CrossRef]
31. Global Modeling and Assimilation Office (GMAO). *MERRA-2 instM_2d_asm_Nx: 2d, Monthly Mean, Single-Level, Assimilation, Single-Level Diagnostics V5.12.4*; Goddard Earth Sciences Data and Information Services Center (GES DISC): Greenbelt, MD, USA, 2015. [CrossRef]
32. Global Modeling and Assimilation Office (GMAO). *MERRA-2 tavgM_2d_ocn_Nx: 2d, Monthly Mean, Time-Averaged, Single-Level, Assimilation, Ocean Surface Diagnostics V5.12.4*; Goddard Earth Sciences Data and Information Services Center (GES DISC): Greenbelt, MD, USA, 2015. [CrossRef]
33. Huffman, G.J.; Stocker, E.F.; Bolvin, D.T.; Nelkin, E.J.; Tan, J. *GPM IMERG Final Precipitation L3 1 Month 0.1 Degree x 0.1 Degree V06*; Goddard Earth Sciences Data and Information Services Center (GES DISC): Greenbelt, MD, USA, 2019. [CrossRef]
34. Global Modeling and Assimilation Office (GMAO). *MERRA-2 tavgM_2d_flx_Nx: 2d, Monthly Mean, Time-Averaged, Single-Level, Assimilation, Surface Flux Diagnostics V5.12.4*; Goddard Earth Sciences Data and Information Services Center (GES DISC): Greenbelt, MD, USA, 2015. [CrossRef]
35. Ginzburg, A.I.; Kostianoy, A.G.; Serykh, I.V.; Lebedev, S.A. Climatic changes in hydrometeorological parameters of the Black and Azov seas (1980–2020). *Okeanologiya* **2021**, *61*, in press.
36. Arpe, K.; Leroy, S.A.G.; Lahijani, H.; Khan, V. Impact of the European Russia drought in 2010 on the Caspian Sea level. *Hydrol. Earth Syst. Sci.* **2012**, *16*, 19–27. [CrossRef]
37. Kostianoy, A.G.; Ginzburg, A.I.; Lebedev, S.A.; Sheremet, N.A. Southern seas of Russia. In *The Second Assessment Report of Roshydromet on Climate Changes and their Consequences on the Territory of the Russian Federation*; Federal Service for Hydrometeorology and Environmental Monitoring (Roshydromet), Kattsov, V.M., Semenov, S.M., Eds.; IGKE Roshydromet and RAS: Moscow, Russia, 2014; pp. 644–683.
38. Kostianoy, A.G.; Serykh, I.V.; Ekba, Y.A.; Kravchenko, P.N. Climate variability of extreme air temperature events in the Eastern Black Sea. *Ecol. Montenegrina* **2017**, *14*, 21–29. [CrossRef]
39. Avsar, N.B.; Jin, S.; Kutoglu, S.H. Interannual variations of sea surface temperature in the Black Sea. In Proceedings of the 2018 IEEE International Geoscience and Remote Sensing Symposium (IGARSS 2018), Valencia, Spain, 22–27 July 2018; pp. 5617–5620. [CrossRef]
40. Lebedev, S.A.; Kostianoy, A.G.; Bedanokov, M.K.; Akhsalba, A.K.; Berzegova, R.B.; Kravchenko, P.N. Climate changes of the temperature of the surface and level of the Black Sea by the data of remote sensing at the coast of the Krasnodar Krai and Republic of Abkhazia. *Ecol. Montenegrina* **2017**, *14*, 14–20. [CrossRef]
41. Komsomolskaya Pravda. Available online: https://www.kuban.kp.ru/daily/25929.3/2878361 (accessed on 20 July 2021).
42. Lebedev, S.A.; Kostianoy, A.G.; Soloviev, D.M.; Kostianaia, E.A.; Ekba, Y.A. On a relationship between the river runoff and the river plume area in the northeastern Black Sea. *Int. J. Remote Sens.* **2020**, *41*, 5806–5818. [CrossRef]
43. Zolina, O.; Simmer, C.; Gulev, S.K.; Kollet, S. Changing structure of European precipitation: Longer wet periods leading to more abundant rainfalls. *Geophys. Res. Lett.* **2010**, *37*, L06704. [CrossRef]
44. Zolina, O.; Simmer, C.; Belyaev, K.; Gulev, S.K.; Koltermann, P. Changes in the duration of European wet and dry spells during the last 60 years. *J. Clim.* **2013**, *26*, 2022–2047. [CrossRef]
45. Zolina, O.G.; Bulygina, O.N. Current climatic variability of extreme precipitation in Russia. *Fundam. Appl. Climatol.* **2016**, *1*, 84–103. [CrossRef]
46. Osadchiev, A.A. Dynamics of Propagation and Variability of Propagation of River Plumes in the Coastal Zone of the Sea. Ph.D. Thesis, P.P. Shirshov Institute of Oceanology, Moscow, Russia, 2013.
47. Osadchiev, A.A.; Zavialov, P.O. Lagrangian model for surface-advected river plume. *Cont. Shelf Res.* **2013**, *58*, 96–106. [CrossRef]
48. Zavialov, P.O.; Makkaveev, P.N. River plumes in the aquatoria of Sochi. *Sci. Russ.* **2014**, *2*, 4–13.

49. Kostianoy, A.G.; Kostianaia, E.A.; Soloviev, D.M. Satellite monitoring of Dzhubga-Lazarevskoye-Sochi offshore gas pipeline construction. In *Oil and Gas Pipelines in the Black-Caspian Seas Region*; Zhiltsov, S.S., Zonn, I.S., Kostianoy, A.G., Eds.; Springer International Publishing AG: Cham, Switzerland, 2016; pp. 225–260. [CrossRef]

50. Lavrova, O.Y.; Mityagina, M.I.; Kostianoy, A.G. *Satellite Methods of Detection and Monitoring of Marine Zones of Ecological Risks*; Space Research Institute: Moscow, Russia, 2016; 336p.

51. Dolan, A.H.; Walker, I. Understanding Vulnerability of Coastal Communities to Climate Change Related Risks. *J. Coastal Res.* **2004**, *39*, 1316–1323. Available online: https://www.biofund.org.mz/wp-content/uploads/2018/10/F2356.2003Walker-Understanding-Vulnerability-Of-Coastal-Communities-To-Climate-Change-Related-Risks.pdf (accessed on 4 July 2021).

52. Phillips, M.R.; Jones, A.L. Erosion and tourism infrastructure in the coastal zone: Problems, consequences and management. *Tour. Manag.* **2006**, *27*, 517–524. [CrossRef]

53. Bosello, F.; De Cian, E. Climate change, sea level rise, and coastal disasters. A review of modeling practices. *Energy Econ.* **2014**, *46*, 593–605. [CrossRef]

54. Kantamaneni, K. Coastal infrastructure vulnerability: An integrated assessment model. *Nat. Hazards* **2016**, *84*, 139–154. [CrossRef]

55. Phillips, M.R.; Jones, A.; Thomas, T. Climate Change, Coastal Management and Acceptable Risk: Consequences for Tourism. *J. Coast. Res.* **2018**, *85*, 1411–1415. [CrossRef]

56. Kostianaia, E.A.; Serykh, I.V.; Kostianoy, A.G.; Lebedev, S.A.; Akhsalba, A.K. Climate changes of the wind module in the region of the eastern coast of the Black Sea. *Vestn. Tver State Univ. Ser. Geogr. Geoecology* **2018**, *3*, 79–98.

57. Chernokulsky, A.; Kurgansky, M.; Mokhov, I.; Shikhov, A.; Azhigov, I.; Selezneva, E.; Zakharchenko, D.; Antonescu, B.; Kühne, T. Tornadoes in Northern Eurasia: From the Middle Age to the Information Era. *Mon. Weather. Rev.* **2020**, *148*, 3081–3110. [CrossRef]

58. Divinsky, B.V.; Kosyan, R.D. Influence of the climatic variations in the wind waves parameters on the alongshore sediment transport. *Oceanologia* **2020**, *62*, 190–199. [CrossRef]

59. Kosyan, R.D.; Divinsky, B.V. Problems of sediments transport modelling in the coastal area. *J. Oceanol. Res.* **2021**, *49*, 93–141. [CrossRef]

60. Shtremel, M.N. Classification of Sea Coastal Zones by the Type of Reaction of Sandy Bottom on Wave Impact (on the Example of Bulgarian Black Sea Coast). Ph.D. Thesis, P.P. Shirshov Institute of Oceanology, Moscow, Russia, 29 June 2021; 97p.

61. Krylenko, V.V.; Kosyan, R.D.; Krylenko, M.V. The coasts of the Caucasian north-eastern part of the Black Sea at the beginning of the XXI century. *J. Oceanol. Res.* **2021**, *49*, 68–92. [CrossRef]

62. Ginzburg, A.I.; Kostianoy, A.G.; Soloviev, D.M.; Stanichny, S.V. Coastal upwelling in the North-West Black Sea. *Issled. Zemli Iz Kosm.* **1997**, *6*, 61–72.

63. Ginzburg, A.I.; Kostianoy, A.G.; Soloviev, D.M.; Stanichny, S.V. Upwelling cyclonic eddies off south-western tip of the Crimea. *Issled. Zemli Iz Kosm.* **1998**, *3*, 83–88.

64. Silvestrova, K.P.; Zatsepin, A.G.; Myslenkov, S.A. Coastal upwelling in Gelendzhik region of the Black Sea and its relation to wind forcing and current. *Okeanologiya* **2017**, *57*, 521–530.

65. Silvestrova, K.P. Coastal Upwelling in the Northeastern Part of the Black Sea: Relationship with Wind and Current. Ph.D. Thesis, P.P. Shirshov Institute of Oceanology, Moscow, Russia, 2019; 141p.

66. Stanichnaya, R.R.; Stanichny, S.V. The Black Sea upwelling. *Curr. Probl. Remote. Sens. Earth Space* **2021**, *18*, 4.

67. Lavrova, O.Y.; Mityagina, M.I.; Kostianoy, A.G.; Strochkov, M. Satellite monitoring of the Black Sea ecological risk areas. *Ecol. Montenegrina* **2017**, *14*, 1–13. [CrossRef]

68. Shiganova, T. Introduced species. In *The Black Sea Environment*; Kostianoy, A.G., Kosarev, A.N., Eds.; Springer: Berlin/Heidelberg, Germany, 2008; pp. 375–406.

69. Vinogradov, M.E.; Shushkina, E.A.; Musaeva, E.I.; Sorokin, P.Y. A new intruder in the Black Sea—The ctenophore *Mnemiopsis*. *Okeanologiya* **1989**, *29*, 293–299.

70. Shiganova, T.A. Invasion of the Black Sea by the ctenophore *Mnemiopsis leidyi* and recent changes in pelagic community structure. *J. Fish. Oceanogr.* **1998**, *7*, 305–310. [CrossRef]

71. Shiganova, T.A.; Mirzoyan, Z.A.; Studenikina, E.A.; Volovik, S.P.; Siokou-Frangou, I.; Zervoudaki, S.; Christou, E.D.; Skirta, A.Y.; Dumont, H. Population development of the invader ctenophore *Mnemiopsis leidyi* in the Black Sea and other seas of the Mediterranean basin. *J. Mar. Biol.* **2001**, *139*, 431–445.

72. Turkey Experiences Intense Drought. Available online: https://earthobservatory.nasa.gov/images/147811/turkey-experiences-intense-drought?src=eoa-iotd (accessed on 13 July 2021).

73. Pešić, V.; Kostianoy, A.G.; Soloviev, D.M. The impact of wildfires on the Lake Skadar/Shkodra environment. *Ecol. Montenegrina* **2020**, *37*, 51–65. [CrossRef]

74. Kostianaia, E.A.; Kostianoy, A.G.; Scheglov, M.A.; Karelov, A.I.; Vasileisky, A.S. Impact of regional climate change on the infrastructure and operability of railway transport. *Transport. Telecommun.* **2021**, *22*, 183–195. [CrossRef]

75. Weather Atlas. Available online: https://www.weather-atlas.com/en/turkey-climate (accessed on 20 July 2021).

76. Rosbusinessconsulting. Available online: https://kuban.rbc.ru/krasnodar/07/06/2021/60bde2b19a7947f6a4a0fe50 (accessed on 20 July 2021).

77. Kommersant. Available online: https://www.kommersant.ru/doc/4887878 (accessed on 20 July 2021).

78. Rosbusinessconsulting. Available online: https://www.rbc.ru/business/20/11/2020/5fb756859a79478009000f03 (accessed on 20 July 2021).

Review

Social Barriers and the Hiatus from Successful Green Stormwater Infrastructure Implementation across the US

Jingyi Qi [1,*] and Nicole Barclay [2]

[1] Infrastructure and Environmental Systems Program, University of North Carolina at Charlotte, Charlotte, NC 28262, USA

[2] Department of Engineering Technology and Construction Management, University of North Carolina at Charlotte, Charlotte, NC 28262, USA; nbarclay@uncc.edu

* Correspondence: jqi1@uncc.edu

Abstract: Green stormwater infrastructure (GSI), a nature-inspired, engineered stormwater management approach, has been increasingly implemented and studied especially over the last two decades. Though recent studies have elucidated the social benefits of GSI implementation in addition to its environmental and economic benefits, the social factors that influence its implementation remain under-explored thus, there remains a need to understand social barriers on decisions for GSI. This review draws interdisciplinary research attention to the connections between such social barriers and the potentially underlying cognitive biases that can influence rational decision making. Subsequently, this study reviewed the agent-based modeling (ABM) approach in decision support for promoting innovative strategies in water management for long-term resilience at an individual level. It is suggested that a collaborative and simultaneous effort in governance transitioning, public engagement, and adequate considerations of demographic constraints are crucial to successful GSI acceptance and implementation in the US.

Keywords: stormwater management; social factors; green stormwater infrastructure

Citation: Qi, J.; Barclay, N. Social Barriers and the Hiatus from Successful Green Stormwater Infrastructure Implementation across the US. *Hydrology* **2021**, *8*, 10. https://doi.org/10.3390/hydrology8010010

Received: 17 December 2020
Accepted: 12 January 2021
Published: 15 January 2021

Publisher's Note: MDPI stays neutral with regard to jurisdictional claims in published maps and institutional affiliations.

1. Introduction

Urbanization can affect the hydrologic functions of urban watersheds and precipitation patterns [1–5]. The consequential increased use of impervious surfaces results in substantial increments of stormwater runoff volume and peak flow [6]. Thus, the transition from the conventional approach into a more sustainable stormwater management paradigm which includes green stormwater infrastructure (GSI), is indispensable to reducing substantial environmental, economic, and social damage [7–9]. Hence, there is also a need to understand the hindrances and limitations in GSI implementation.

GSI offers a promising solution to stormwater management by mimicking natural hydrological processes to reduce localized flooding events and water quality improvement through decentralized natural or engineered processes to treat stormwater runoff at its source [10]. In the US (United States), awareness of GSI has slowly increased over the past two decades. Its historical progress in stormwater management and background knowledge is documented in several in-depth publications [11–14]. Research teams across nations have developed various GSI practices and in addition, retrofits and hybrid measures on different spatial scales (such as watershed scale and site scale, etc.) with diverse primary purposes have been developed [15–20]. The details on these practices are well documented in the literature [21–28].

Numerous studies have evaluated the performance of GSI, particularly in economic and technical aspects [14,29–32]. GSI provides extra benefits to the community, such as raising property values, enriching life quality, and providing adaptable climate resilience [33–35]. Urban stormwater management has advanced gradually over the last two decades, thus various terminologies are used to define new principles and practices,

147

where the concepts behind them often overlap [14,36]. Using these different terms may reduce effective communication in certain circumstances, such as when documenting all the alternative stormwater practices used in the US to assess their performance in general [36]. To avoid confusion, the term GSI was used throughout this work in referring to all types of multi-purpose structural stormwater management practices that involve natural processes for runoff volume and water quality control.

Despite the progress, there are limited study efforts on non-technical factors, such as public perceptions and knowledge, that could explain the slow advancement in the wide adaptation of GSI to the desired level for stormwater management and sustainability capacity building [37]. The contradiction between the low implementation rate of GSI in major regions of the US and the actual demand to address climate change impacts suggests that certain factors are hindering the relevant decision-making processes [38,39]. Furthermore, a study discovered the mismatch in the percentage of their survey participants that expressed an intention to support GSI and the number of those who actually adopted GSI [40]. This result is in agreement with the findings in an exhaustive review [41]. Irrational decision-making behaviors in energy-related decisions have been interpreted through the cognitive bias perspective [42,43], where cognitive biases can be defined as a belief that hampers one's ability to make rational decisions given the facts and evidence [44]. It has been supported by various studies that cognitive biases are influential in decision making and planning [44]. Yet, little attention has been given to the potential influence of cognitive biases in GSI implementation, despite numerous studies on perceptions of various GSI stakeholder groups [45–47]. This study aims to bridge this knowledge gap.

Historically, quantitative decision support tools have been developed with the main aim to maximize GSI performance to control runoff and water pollution and to be cost-effective [48–52]. On the other hand, despite the extensive attempts made to expand the assessment work to include the social aspect of decision support [17,48,53–64], they lack a deeper understanding of the public perceptions and associated cognitive bias perspective to resolve the implementation dilemma from a bottom-up approach [65] as examined in other environmental issues [43,44]. This shortcoming can affect the expected outcomes envisioned by major decision-makers [42,66]. This study focuses on the barriers that could be linked to biased perceptions due to social factors in GSI development and implementation.

This work was conducted to examine the relevant social factors through the lens of cognitive biases, which may lead to implementation barriers during GSI adoption processes. The scope of social factors can vary significantly as they are commonly assessed in combination with factors from other dimensions, such as socio-ecological, social-cultural, socio-economic, and socio-technical factors [10,67–70]. We use a concept adapted from Gifford and Nilsson [71] to define social factors as the internal differences among people and the contextual factors that define them in this study. This study aims to understand the potential connections of cognitive biases with these barriers, and to recommend an approach to analyze and address the associated problems. Studies have been conducted to analyze cognitive biases with agent-based modeling (ABM) in various contexts [72–74]. However, no study has done a similar analysis in the context of GSI implementation. ABM is a methodology that can incorporate the autonomy, heterogeneity, and adaptability of individuals in a social system to study the resulting global patterns through a bottom-up approach [75,76]. It is also an approach that can carry exploratory simulations for a deeper understanding of the underlying adaptive behaviors and interactions that could lead to the emergence of phenomena that was previously overlooked [40]. However, the models developed solely based on social and physical science are usually fragmented in their fields, rely on qualitative analysis, or are difficult to incorporate into quantitative models [77]. This work was conducted to answer the following questions:

1. What social factors have been identified as barriers to GSI implementation?
2. How do these social factors connect to cognitive biases?
3. How can ABM accommodate these cognitive biases for better quantitative decision support?

To address these research questions, we reviewed the literature on GSI implementation barriers that arise from social aspects and on the connections between cognitive bias with these barriers. Subsequently, we reviewed the literature to show and assess the applicability of ABM in addressing the issue of social factors' hindrances to GSI adoption and implementation.

2. Materials and Methods

A literature review was conducted on two main topics in this study using a combination of platforms, including literature search engine Web of Science (WoS) and relevant referenced articles in the papers collected through the means mentioned above. Firstly, studies that were conducted to understand the restraints to wider/efficient/effective GSI adoption were examined. Reported barriers to GSI implementation that may link to social factors in the literature were identified using the search terms: 'social', 'barrier* OR challenge* OR difficult*', 'stormwater OR storm water', and 'infrastructure' as the primary screening criteria. Only peer-reviewed papers written in English published between 1900 to 2020 were considered. Seven records were first excluded prior to the screening due to lack of access to the full text. Four book chapters and 20 articles that were not directly relevant to the social barriers in GSI were eliminated. Finally, because the social context that could contribute to barriers that are dependent on local governmental regulations and governance practices [48,64,78] and socio-ecological context [64,79], the records that did not explicitly study the social barriers in the US were excluded from the final results. As a result, the search within the scope of this study yielded 34 papers in total (Figure 1). The final results are further divided into two groups, where one (20) is the collection of empirical-based studies that examined the barriers, and another (14) is the collection of studies that developed qualitative frameworks to incorporate social factors to reduce such barriers as decision support tools (the works focused solely on qualitative post-construction performance evaluations were excluded). Note that analytical simulation-based works found through this search were rearranged to the second part of the review. These barriers were reviewed through the concepts of cognitive biases proposed by Haselton, et al. [80]: Biases resulted from heuristics, artifacts, and error management.

In their article, Bukszar Jr [81] provided strong evidence that failing to address cognitive biases among decision-makers can cause strategic heuristics and biases, thus hampering the strategy's adaptability. They argued for the need for a higher capability to accommodate such cognitive biases for greater strategy success. Thus, the second part of this review was conducted using the same search platforms of records written in English and published between 1900 and 2020 to evaluate the potential applicability of ABM in addressing the issues studied in the first review topic. Due to the limited studies conducted within stormwater management, research that analyzed innovation diffusion in water infrastructure, in general, were also considered in this review. Thus, a total of 10 results were finalized (Figure 2). The key search terms used were 'agent based OR agent-based', 'infrastructure', 'perception* OR cogniti*', 'model*', and 'water'. This yielded 6 outcomes with 11 additional articles from external references. Additionally, the Institute of Electrical and Electronics Engineers (IEEE) was employed due to its particular research focus on computational simulations using a combination of key search terms of 'water', 'infrastructure', 'percept*', 'cogniti*', and 'agent-based'. It yielded 34 additional results. One record was eliminated from the WoS results because it was a conference proceeding. A total of 38 additional studies were excluded after abstract screening because they were not directly relevant to the interpretation of cognitive biases or perceptions of innovative water management strategies simulated through ABM. It was noted that all search outcomes from IEEE were not within the scope of the search objectives for this review.

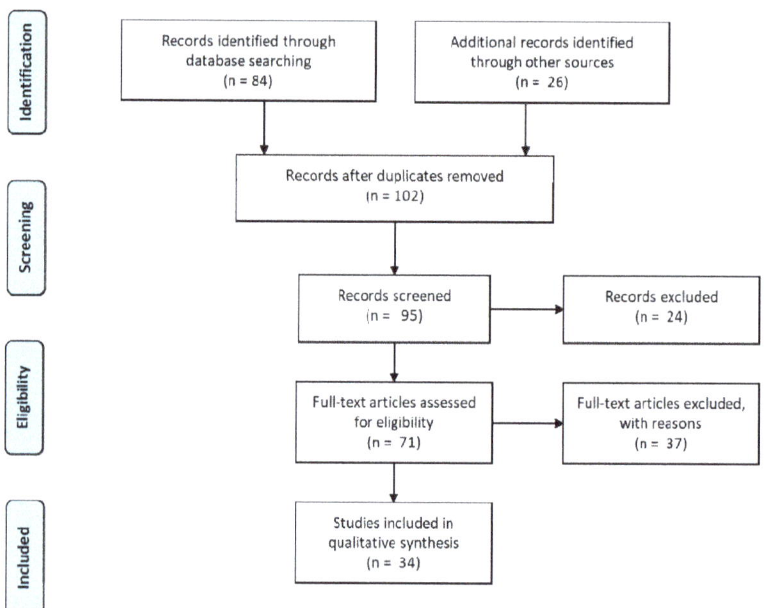

Figure 1. Flow diagram of the search results of the first topic following the Preferred Reporting Items for Systematic Reviews and Meta-Analyses (PRISMA) protocol [82].

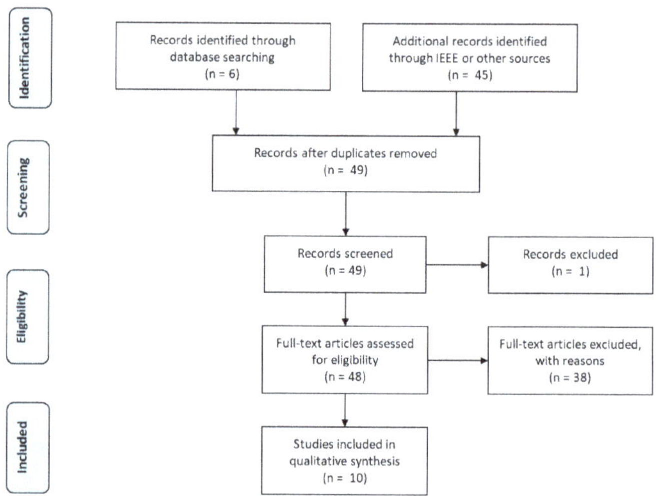

Figure 2. Flow diagram of the search results of the second topic following the PRISMA protocol [82].

3. Results and Discussion

3.1. Identified Social Barriers to GSI Implementation

The barriers to GSI have been studied by numerous international research teams, ranging from individual perceptions and attitudes, financial burdens, resource allocations, and governance rigidity to conflicts across institutions [45,67,79,83–86]. Barriers originating from social factors may be harder to address, as the values of which are usually difficult to quantify yet should not be overlooked [55,58,65]. Barriers primarily identified as associated with social factors, in terms of their potential influence on the implementation of GSI, are attributed to three main categories from the literature. They mostly cover governance discord, public participation, and demographic constraints (Table 1). Governance refers to the inconsistent strategies among or within governance entities; public participation refers to the involvement of the public in the decision-making of GSI regulations and collaborations; and demographic constraints refers to the general demographic factors, social norms, and perceived environmental concerns. However, there always is a possibility of unrecognized social factors in the published studies. For example, though not directly addressing the issues in stormwater management adaptation, a study brought forth the dilemma in regenerating historical cities of which preserving the historical cores were paramount [87]. It is thinkable that advancing GSI in such areas may encompass greater complexities than others. Additionally, the underlying interrelations across infrastructure sectors and even industries are also likely to influence sustainable decision-making in general [88,89].

Table 1. Relevant social factors that could influence the implementation of GSI in the US.

Social Barriers	Barrier Subcategories	GSI Types	Spatial Scales	Location	Stakeholder	Study Methods	Source
Demographic constraints & public engagement	Race, ownership status, relevant knowledge of GSI, knowledge dissemination platform	Rainwater harvesting, pervious paving, rain gardens, lawn depression	Sub-watershed	Two sub-watersheds in Chesapeake Bay watershed	Private landowners	Knowledge, attitude, and practice questionnaire	[90]
	Age, education, homeownership, prior experience of floods, lack of awareness, underuse of social capital	Rain barrels, rain gardens, and permeable pavement	Region	Knoxville, TN	Private landowners (households)	Survey	[91]
Governance	Limited focus on the multifactional of GSI to respond to local needs, lack of interdepartmental collaboration, and private-public partnership	Green alleys with various GSI features	Region	Various locations in the US	Government agencies, non-governmental organizations (NGOs), community groups	Narrative analysis	[34]
	Conflicting visions in hydro-social relations	GSI in general	Region	Chicago, IL, and Los Angeles, CA	Government entities, NGOs	Interviews, participant observation, literature review, survey	[92]
	Leadership in transitioning governance (informal, multiorganizational)	GSI in general	Region	Ohio	Community NGOs, environmental NGOs/land trust, federal government, local government/regional authority, university /contractor	Social network analysis survey	[93]
	Departmental silos (stakeholders' multiple and competing social perspectives)	GSI in general	Region	Chicago, IL	NGOs, governmental entities	Q-methodology	[94]
	Tensions and convergences among different management strategies	GSI in general	Region	Pittsburgh, PA	Community organizations, municipalities, advocacy groups	Interviews, participant observation	[95]

Table 1. *Cont.*

Social Barriers	Barrier Subcategories	GSI Types	Spatial Scales	Location	Stakeholder	Study Methods	Source
	Conflicting perceptions, implementation priority, limited focus on the multifunctionality during planning	GSI in general	Region	New York, NY	Agencies, city departments, national and local nonprofits, research institutions	Spatial analyses, survey, interview, participant observation	[78]
	Inequity for disadvantaged communities	GSI in general	Sub-watershed	Los Angeles, CA	Government agencies, non-profits, community organizations, and others	Statistical analyses	[96]
Public engagement	Failing to recognize the values of social capitals for long-term productivity	Rain gardens, rain barrels	Household site	Cincinnati, OH	Landowners	Experimental reverse auction	[97]
	Perception (status quo bias)	Rain gardens, bio-swales, green alleys with permeable pavement	Region	Cincinnati, OH, and Seattle, WA	Engineering graduate students	Functional near-infrared spectroscopy	[38,97]
	Ineffective information dissemination, underuse of social capital	Rain barrels, rain gardens, permeable pavement	Region	Washington DC	Homeowners	Voluntary stormwater retrofit program with statistical analyses	[98]
	Stormwater context (perception of neighborhood-level challenges, town-level stormwater regulation)	Rainwater harvesting, rain gardens, permeable pavers, infiltration trenches, and tree box filters	Cross-scale	Vermont	Residents	Statewide survey	[79]
	Depreciation of community involvement (expertise, education)	GSI in general	Region	Houston, TX	Researchers, community	Participatory action research	[99]
Governance & public engagement	Lack of awareness and responsibility for maintenance, education programs not aligned with local preferences	Stormwater ponds	Community	Southwest Florida	Homeowners, governmental entities	Survey, interviews	[100]
	Lack of awareness, ineffective regulation enforcement	Stormwater ponds	Region	Manatee County, FL	Landscape professionals, residents, government agents	Interviews, surveys, participant observation, and literature review	[101]
	Lack of awareness, understanding, and sense of responsibility; geographic disconnection between watersheds and governing entities; fragmentation of responsibility among stakeholder groups	GSI in general	Region	Cleveland, OH, and Milwaukee, WI	Practitioners (regional sewer districts, local governments, community development organizations)	Interviews	[28]

Table 1. *Cont.*

Social Barriers	Barrier Subcategories	GSI Types	Spatial Scales	Location	Stakeholder	Study Methods	Source
	Lack of awareness and adaptivity in policies to prioritize GSI measures to align with local values	Bioswales, green roofs, street trees, parks & natural areas, community gardens, and permeable playgrounds	Region	New York, NY	Residents and practitioners (individual sprofessionally engaged in the siting, design, maintenance, public engagement, and/or monitoring of GSI programs)	Preference assessment survey and semistructured interviews	[46]
	Outdated regulatory constructs, conflicted views among gray and green advocates, jurisdictional overlap, influences of social media coverage, leadership gaps or influence of lobbying	GSI in general	\	USA	Residents, governmental entities, engineers	Narrative analysis	[102]

The unclear distribution of responsibilities among stakeholders can impede the decision-making processes associated with GSI implementation. Particularly, the general public's involvement is the fundamental building block that could be influential in shaping the direction of GSI implementation [17,28,47]. Dhakal and Chevalier [83] stated in their study that, above all challenges, cognitive barriers and socio-institutional factors should be the primary issue to focus on. Furthermore, the multi-sector benefits will only be nuanced if the public is not willing to implement GSI [103]. Similarly, one study stated that sustainable GSI implementation would necessitate the need for structured public participation and local partnerships. They emphasized that, in addition to putting more reach effort onto comprehensive cost-benefit evaluations on GSI, such needed engagement would fortress the networks of non-governmental organizations, county and state agencies, municipal sewer districts, and federal research support, which could lead to a faster adaptation of GSI on larger scales [104]. Therefore, the barriers to the general public to accept GSI are crucial to dissect these aforementioned disconnections and provide practical yet effective decision support. To date, there is a limited number of conceptual frameworks that capture social factors in GSI implementation processes (Table 2). Yet there still is a need for quantitative analysis measures for better decision support for case-based GSI adoption using standardized methods that could assist in horizontal comparison and further knowledge transfer. The frameworks listed in Table 2 were categorized based on their main purpose: Classification scheme (proposed to enhance terminology clarity), planning strategy (suggesting new approaches to be adopted in current management regimes), process conceptualization (promoting a better understanding of complex socio-infrastructure systems), and framework efficacy assessment (evaluating the existing frameworks' usefulness in promoting GSI implementation).

Table 2. Conceptual frameworks that consider social factors in GSI implementation processes.

Framework Nature	Social Factors	Sub-Categories	Stakeholders	Method	Scale	Source
Classification Scheme	Governance, stakeholder engagement	Stakeholder interactions, governance, political contexts	Individuals and groups involved in rule-making processes, property owners	Social-ecological services framework	Cross-scale	[54]
	Public engagement, governance	Policy instrument assessment	Citizens	Policy instrumentations scheme	Region	[56]
	Public engagement, governance	Ownership status, political power	Governmental entities	Topology framework	Region	[64]
Planning Strategy	Governance, demographic constraints	Equitable GSI distribution, age, income, education, ownership status	Governmental entities, residents	Green infrastructure equity index	Region	[60]
	Public engagement, governance	Multifunctional strategy, multisectoral communication	All involved in decision-making processes	Millennium ecosystem assessment classification-based framework	Cross-scale	[105]
	Governance, public engagement, demographic restraints	Adaptive governance, stakeholder participation, inclusion	Governance, nongovernmental organizations, communities, academia, industry	Adaptive socio-hydrology framework	Cross-scale	[106]
	Public engagement	Interdisciplinary collaboration, university-stakeholder partnership, institutional capacity	Universities	Integrated framework combining social-ecological dynamics, knowledge to action processes, organizational innovation	Region	[63]
Process Conceptualization	Public engagement	Community participation in three themes (context, participation processes and outputs, and implementation results)	City, federal government agencies, community residents, and community NGOs	Public participation conceptual model	Watershed	[61]
	Public engagement, governance	Low stakeholder buy-in, discoordination in management objectives and goal among stakeholders, lack of awareness	Government researchers, stormwater managers, and community organizers	Adaptive management framework	Site	[62]

Table 2. *Cont.*

Framework Nature	Social Factors	Sub-Categories	Stakeholders	Method	Scale	Source
	Governance, public engagement, demographic restraints	Stakeholder interactions, governance and political contexts	All that are involved in stormwater management	Integrated structure-actor-water framework	Cross-scale	[55]
	Public engagement, governance	Hybrid governance envisioning (management and monetary responsibilities)	Regulatory agencies, residents	Multi-criteria governance framework	Cross-scale	[17]
	Public engagement, governance	Perceptions, stewardship, human-environment interactions	Residents	Coupled human and natural systems framework	Region	[58]
Existing Framework Efficacy Assessment	Governance	Governance, capacity, urbanization rate, burden of disease, education rate, political instability	Government agencies, NGOs	City Blueprint® Approach	Region	[53]
	Public engagement, governance	Community education and awareness campaign, multifunctional strategy	Residents, governmental entities	Socio-ecological framework	Watershed	[107]

3.2. Interpretations through Cognitive Biases

Kahneman and Tversky [108] pointed out that human decision making can be subjected to cognitive biases (or cognitive illusions) especially when under uncertainty, which infers that an erroneous judgment may be formed subjectively (as judgmental heuristics). It is particularly profound when forming judgment based on certainty and probability under uncertainty [109]. Over the past several decades, research efforts have been made to study cognitive biases and how they can influence decision making [41,44,66,110,111]. A deeper understanding of cognitive biases can assist in effective debiasing and re-biasing measures for better decision making [112–114]. Cognitive biases have been studied extensively in the sociological and psychological fields, yet these intellectual outputs have rarely been considered in other research domains [112], such as in the stormwater management sector. In the context of governance strategy primarily for managing complex systems, such as natural resources, hazards, and the environment, one review study pointed out that there was a need to enhance participatory processes connecting scientists with stakeholders and policy-makers to propel successful governance and policy enforcement, in which biases, beliefs, heuristics, and values were the critical influencing factors [111]. The authors believe that, despite being intrinsic to a certain extent [110], cognitive biases are shaped by surrounding contextual factors, such as social factors. Hence, this work is an early attempt to connect these two pieces in the context of GSI implementation with an envision of advancing quantitative insights on the slow progress in GSI adoption in the majority of the US territories. Only a limited number of studies have explored the social factors involved in the decision-making process of stakeholders at various levels in the context of stormwater management, and they tend to be based on simplified concepts to interpret the information transfer tarnished by cognitive biases [40,115,116].

Historically, there has been an ongoing debate on the definition and categorization of cognitive biases across different scientific domains. Furthermore, according to Caverni, et al. [117], cognitive biases is an evolving topic. Thus, this review is based on the theory developed by Haselton, Nettle, and Murray [80] based on its wide acceptance among scholars, how suitable it is to interpret social factors-related barriers to GSI implementation and its year of publication. Through a literature search of the social barriers mentioned in the literature, three are salient in the context of stormwater management that may be associated with cognitive biases (Table 1). However, the authors acknowledge the limitation on the selection of the theory due to its novelty in the context of GSI adoption, particularly the three biases chosen in this review. Furthermore, interdisciplinary discussions are encouraged to strengthen research efforts in this topic for practical decision support.

3.2.1. Uncoordinated Regulations and Governance—Biases Resulted from Heuristics

People tend to rely on rules of thumb to simplify problems at hand that may deviate from the optimum range of decisions, which can be considered heuristics [80]. The most commonly studied bias based on heuristics is the status quo bias which can be seen in regulation adaptation progresses. The status quo bias first received a greater level of scientific attention through the work of Fernandez and Rodrik [118], which can be used to explain the resistance to change within a group of people where the beneficiaries of the status quo have a stronger influence than the other group, which they referred to as the non-neutrality. This can be considered a bias due to human's insensitivity to make predictions under the influence of representative heuristics where people predict future events based on the intuition under uncertainty [119,120]. Hu and Shealy [38] conducted a study to illustrate how setting up GSI resolutions can overcome the status quo bias which limits its adoption. They demonstrated that simple public engagement strategies using factual endorsement in a municipal resolution by regulatory organizations could favor GSI over conventional practices.

Status quo bias can also be observed among the key professionals whose preferences may largely set the direction of the reform. One study identified five typical types of decision-making patterns of students in civil engineering, which include risky, social, conflicted, purchasing, and influenced by built-environment decision making [121]. By carefully examining these thinking patterns, it could contribute to overcoming potential cognitive biases among stormwater engineers. On the other hand, biases might be amplified if the role of the GSI-related implementation processes is heavily played by one stakeholder group, such as the contractor company, which takes the responsibility from the design to the construction phase. This might limit their scopes, such as potential risks or alternatives. Rather, they could distribute the workload to a third-party design company, allowing further discussions on the optimal plan. A study found that professionals who had hands-on experience favored GSI [39].

The general situation of stormwater management in the US has been depicted as lacking clear guidance and regulation [12,83]. Stormwater management was not brought into the National Pollutant Discharge Elimination System (NPDES) program until 1987 [13]. Further challenges lie in the adaption of drainage system management when facing climate change and anthropogenic stressors, which has propelled the use of GSI [122]. Attempts made through the established federal regulations often conflict with the existing rules set on state and local levels, which have more discretions on primary goals and responsibility distributions. This has resulted in the current dilemma that, even though private sources count for a greater percentage of the flow generation or have a higher potential in fortifying stormwater storage capacity, NPDES and municipalities cannot enforce regulations in these areas [13,19]. In summary, the major weaknesses and gaps in these regulation-related issues are poor coordination across institutions due to land use as private properties and not prioritizing the control and storage capacity of the discharge volume [13]. Several other studies listed in Table 1 have also observed such barriers.

3.2.2. Low Public Engagement and Inefficient Knowledge Transferring—Biases Resulted from Artifacts

Artifact biases intentionally form unrealistic conditions on which people make decisions, for instance, framing and anchoring biases [80]. It could suggest that if the information was not translated into a language that is appropriate to a specific audience, the efficiency in the transfer of such knowledge could be reduced, even causing the generation of erroneous interpretation. The framing effect occurs when a person changes their decision based on how the information is presented [123]. A study has demonstrated that the biases can be prevented in the early stage during education by using the sustainability-conscious teaching approach to assist in decision making for sustainable infrastructure like GSI, such as by using the Envision rating system [124]. On the other hand, it may lead to an anchoring effect if the parameters used in said rating systems are not properly determined [42], where a biased estimate toward the set arbitrary values will be formed even though they are far from rational estimations [125].

Even though it can bring forth multi-sector benefits, GSI implementation still faces a range of practical barriers, including the poorly perceived necessity of effective stormwater management [126]. In addition, miscommunication due to terminology confusion or ineffective knowledge transfer can also hinder the progression of GSI development to the optimum level [36,127]. These miscommunications might link to the conservative mindset about gray infrastructure, risk aversion attitude toward the related cost and performance of GSI, confusion between GSI and the gray option, and fear of taking maintenance responsibility as identified in the literature [45,79,83–85]. It was also pointed out by the U.S. Environmental Protection Agency (US EPA) that many of the barriers could be overcome if sufficient efforts were made as the policies and regulations evolved on a need basis. Given that, these aforementioned efforts need to be initiated first in order to achieve the expected outcome. The results from a study demonstrated that solely relying on GSI implementation was not adequate if public education and social learning were not enforced at the same time [85]. The authors suggested the diversity of perspectives could not be omitted to encourage the successful transitioning of this stormwater management regime. To attract more financial support to advance and accelerate research on gathering reliable GSI performance data, inadequate public (especially the major stakeholders') awareness needs to be appropriately addressed [128,129].

3.2.3. Perceived Demographic Constraints—Biases Resulted from Error Management

Error management bias occurs when people make decisions primarily to reduce consequential losses [80]. The typical bias that falls into this category is risk (or loss) aversion. As pointed out by Tversky and Kahneman [130], people tend to value any amount of loss greater than the same amount of gain, which infers that losses (or disadvantages) will be considered more than gains (or advantages). In the context of GSI implementation, one factor that hinders the decision-making process is the lack of convincing empirical data on multi-sector functionality in a life cycle [16,131,132]. This bias might emerge due to unfamiliarity with long-term GSI performance and with the demand for capital cost and maintenance fees, of which the payback has not been clearly quantified. A study found that the most salient barrier to adopting innovations is the perception of risks [39]. The authors suggested that extensive knowledge transfer in a combination of equal sharing of contractual risk through team collaborations could contribute to easing such perceptual barriers. Great progress has been made to minimize these barriers. Without enough perceived incentives, it would be difficult for any major stakeholder to bring forth the input, whereas other studies have shown some positive influence of GSI in the triple bottom line (i.e., economic, social, and environmental) [14,15,133–135].

In a study performed by Di Matteo, et al. [136], their results suggested that being able to review trade-offs among solutions can minimize biases at the decision-making stage. According to Coleman's finding [79], some private landowners favored small-scale GSI practices over community-wide alternatives, as they were more focused on addressing local issues rather than collective actions. On the other hand, some GSI practices are more likely to provide better performance if used in tandem [79,137], which could further complicate the multi-sector performance monitoring processes. Of particular note was that social performance was considered a critical factor for enhancing multi-sector funding opportunities and the adoption of GSI [68,138]. Further studies are needed on the influential social features that affect the development of GSI to resolve the knowledge gaps among the public and to elucidate major social restraints (e.g., demographics and ruling regulations). Demographic factors were regarded as the contextual background. Policy enforcement and revision according to the current GSI implementation situation were mainly the responsibility of governmental entities at federal, state, and county levels. The field experts were considered the leading personnel responsible for designs based on the built environment within the region and the outreach for knowledge diffusion. Compared to the households that prioritize individual benefits, the local community tracks the inter-connective components. Despite the efforts invested into understanding the influence of the social environment on GSI implementation, only limited research studied individual behaviors at the system level to identify the most potentially effective approach to increase social acceptance at a regional scale [139].

3.3. Applied Agent-Based Modeling in Quantitative Decision Support

Tremendous research has addressed the hydraulic and hydrological and economic uncertainties of GSI, yet social contextual factors remain under-studied given its complexity and challenges in quantitative analysis. Our work reviews and analyzes the most identified social barriers including governance inconsistency, low public participation, and demographic constraints from the consequential behavior patterns by incorporating knowledge in cognitive biases. Table 2 presents the most relevant frameworks that qualitatively assist in decision support for GSI implementation. They brought forth early attempts to solve the social dilemma identified in Table 1 through various degrees of active public engagement, collaborative governance regimes, and strengthened knowledge transfer among stakeholders. A new conceptual framework (Figure 3) was proposed to take into consideration such barriers on their potential impacts on the adoption of GSI.

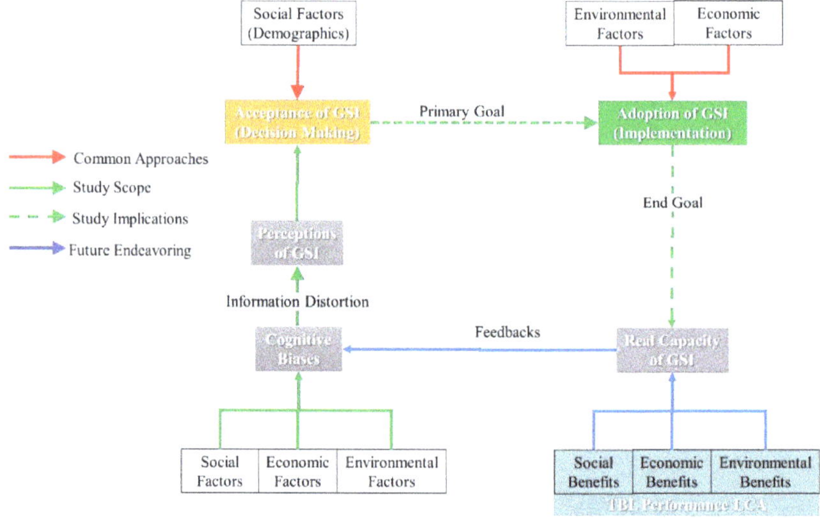

Figure 3. The conceptual framework proposed in this study.

Policymakers are usually required to make science-based decisions and actions by which they need to provide transparency in their prediction of the expected impacts of their decisions [111]. Hence, further efforts are needed to provide evidence-based quantitative analysis to gain advanced insights on practical decision support. Existing quantitative decision support tools used to simulate or evaluate GSI performance rely on the assumption of rationality, omitting the potential interference to the outcomes due to cognitive biases. For instance, several multi-criteria decision-making (MCDM) support systems made of decision support tools (DSTs or DSSs) have gradually incorporated as many relevant factors as possible [132].

Despite their capacity in being able to address multiple criteria, these decision support tools for GSI implementations have limited considerations on the potential cognitive biases, which could result in less effective strategies implementation. For instance, a study indicated that individual bias has various effects on the organization's objectivity in both positive and negative directions and distort individuals' process of creating, retaining, and transferring knowledge. Their study results suggested that for a system with high complexity, reducing individual bias may not necessarily enhance the objectivity of the organization. Thus, it is wise to examine specific social systems when developing cost-effective mediation strategies in case of simulating individual biases [74]. Psychological- and sociological-based behavioral rules have been adopted in ABM by macroeconomics since the 1960s [140]. As reviewed by Bharathy [77], the combination of an understanding of human behaviors and systems thinking is crucial for successful decision-making. Their study identified the research niche on human behavioral modeling with an emphasis on the coordination among stakeholder groups in different fields. Despite being unable to truly reflect on realistic situations, human behavioral models can still assist decision-makers in the understanding of social systems. However, the models that are developed solely based on social and physical science are usually fragmented in their fields, rely on qualitative analysis, or are difficult to incorporate into quantitative models. For models with agents to behave more realistically, one must expand their study scope to incorporate the models developed in social sciences (such as psychological and cultural studies). Limited research was able to accomplish this task [77].

In terms of complexity among available DST, ABM is more robust at detailed micro-level simulation than qualitative studies, yet less dependent on sophisticated mathematical logic than some quantitative models, such as system dynamics. This methodology can simulate global emergent patterns/social consequences by setting up only individuals' characteristics and behaviors. ABM is better at capturing the non-linear interactions between human behaviors based on various factors and the macro-environment through feedback effects and at explaining the collective outcomes resulted from a given set of interactions among individuals. So far, the primary use of ABM for policy decision support has primarily been in the fields of sociology, epidemiology, and urban planning [75,114,141–143].

The theory of innovation diffusion was developed to conceptualize innovation adoptions through communication channels over time, which are determined by individuals' personal and social characteristics in a social system, and the decision-making logic of individuals regarding the associated social changes [144]. ABM is advantageous at micro-level simulations that can account for the heterogeneity and autonomy of individuals during the innovation diffusion process to a greater extent in comparison to aggregate-level models. Kiesling, et al. [145] conducted an extensive review on ABM applied in this theory, which has been used for two main purposes: To advance the theoretical development, and to forecast outcomes for decision support using empirical data. Similar to other simulation models, ABM has its own limitations. To date, no ABM framework has been widely agreed on for innovation diffusion due to the diverse selections of sub-theories, parameters, and equations to interpret the adoption processes. The two major challenges are: The lack of capability in capturing opinion changes as models generally assume a binary decision switch from a non-adopter to an adopter with a presumption of global success as the final outcome [115]. Therefore, there is a research need to continue extending and revising the existing ABM framework to better simulate more realistic innovation diffusion, particularly water-related infrastructure due to the pressing issues highlighted in the background section of this article.

Though different in prior aims, the use of ABM to assist in decision support for diffusing innovative water-saving technologies shares similarity in the general concepts with GSI technology diffusion in terms of the simulations and behavior rules. Therefore, studies conducted on innovation diffusion of water conservation were reviewed in this section as well (Table 3).

A few studies have applied ABM to analyze isolated influences of certain demographic, household, social, and external factors on water conservation technology adoption. However, they failed to take into account the potential simultaneous influence of these attributes on agents' acceptance decision making. One empirical-data-driven study argued that ABM was favorable in simulating innovation diffusion than the Bass model and cellular automata for its greater capacity in incorporating heterogeneity of agents and explicit special relationships [146]. The statement was also supported by another study [40]. Another study discovered a research gap on the observed disagreement between the overall numbers of the households that indicated their wills to adopt certain water conservation technologies and the number of the populations that implemented said technologies. They suggested it could be due to the additional costs and motivation required to install these inventions into one's household. They used ABM to simulate the innovation diffusion process by the state transition approach as mentioned in the previous section. Their results shed light on the importance to consider various characteristics of the communities when developing intervention strategies for the effective adoption of water-saving technology by households such as income growth, water pricing structure, the cost of rebated programs compared to the affluence of the community, and social network connections [40].

One study based in Germany [146] adopted the integrated ABM approach to combine the theoretical aspects of innovation diffusion, social psychology, sociology, and decision theory to enhance the accuracy of realistic decision-making processes using an empirical study of diffusion of water-saving technologies. This model contributed to an advanced decision-making process during water-saving innovation diffusion. On a different aspect during the adoption process, few researchers have developed ABM models that are capable of incorporating the dynamics between public adoptions that are affected by changes in demands for resources and services and infrastructure expansion. A study [115] approached the issue through an ABM framework, which simulated the perception changes in risks/benefits of water reuse during the course of infrastructure expansion by incorporating the theory of risk publics to simulate the social networks. It overcomes several limitations of cognitive models and diffusion of innovation models because the risk publics theory is relatively more comprehensive in reflecting real decision making compared to other existing theories in that it assumes definitive connections among agents who held similar opinions about the risk/benefits of a technology based on a social psychology approach. This work is one of the few that applied social psychology-based ABM in innovation diffusion for water reclamation among households and has the potential to be adopted for decision support for GSI implementation.

Note that the review in this study is limited to the research works conducted solely through ABM. However, there have been several studies that used hybrid simulation models as a decision support tool in water infrastructure management. For instance, Faust, et al. [147] developed a hybrid quantitative system dynamics-ABM framework to investigate the water demand dynamics in shrinking cities. This type of hybrid model showed its advantages in capturing the sophisticated socio-technical interactions within the human-infrastructure system through feedback loops compared to using ABM. On the other hand, simulations of cognitive biases using ABM have been explored on various types of cognitive biases, such as risk aversion, confirmation bias, motivated reasoning, cognitive filtering within social science, and economy domains [148–152]. These scholarly contributions can be substantially beneficial in driving insightful decision support tools for GSI implementation that reflect realistic public opinions and actions.

Table 3. Innovative strategy diffusion in water management using ABM.

Simulation Objectives	Agents	Behavior Rules	Social Networks	Time Step	Platform	Calibration, Verification & Validation	Novelty/Advantages	Limitations	Location	Source
Water consumption behaviors	Households	Reversible stochastic diffusion of opinions, Bass' model of innovation diffusion	Random graph	Three-month (10 years)	Java	Calibration with empirical data, face validation	Integrate geographical, cultural, and socioeconomic factors with ABM for decision support in water demand	Requires exhaustive efforts into interdisciplinary empirical validation, demands advanced expertise and computation power to embed GIS into ABM	Valladolid (Spain)	[153]
Flood risk communication strategies effectiveness	Households	Protection motivation theory	Stochastic with pre-defined connection rules	Yearly (7 years)	NetLogo	Calibration with empirical data and sensitivity analysis	Simulates micro-level diffusion of information for flood risk communication	Requires sufficient empirical data to minimize uncertainty	Rotterdam-Rijnmond (Netherlands)	[154]
Adoption of water reuse measures	Households	Risk publics ABM framework	Small-World	Yearly (30 years)	Not specified	Calibration with historical data and sensitivity analysis, validation through comparing results from another model	Captures opinion dynamics and adoption decisions on water reuse innovations under various infrastructure expansion scenarios	Assumes several parameters of fixed values, simulates at the unitary household level, limited capacity in capturing opinion dynamic resulted from external factors	Town of Cary, NC	[115, 155]
Innovation processes in urban water infrastructure systems	Water supplier, water consumers, sewage system operator, technical components producer	Bounded rationality with utility functions	Simplified structured models	Yearly (50 years)	Not specified	Not specified (theoretical development only)	Captures the transition patterns of water supply infrastructure influenced by interactions of multiple stakeholder groups	Lacks of agent heterogeneity of simulated stakeholder groups, omits some relevant stakeholder groups	Not specified	[156]
Spatiotemporal emergence of GSI	Residential property owners	Probability-based GSI adoption rules	Simplified structured models	Monthly (30 years)	NetLogo	Calibration with historical data	Simulates micro-level spatiotemporal adoption rates of two GSI practices determined by physical compatibility and socio-economic factors	Requires expertise in collecting, characterizing, and modeling with the relevant data, the behavioral rules need further data collection to reflect the decisions made under various constraints and conditions	Philadelphia, PA	[139]
Effect of various factors on residential water conservation technology adoption	Households	Innovation diffusion, affordability theory, peer effect	Various (random, distance-based, ring lattice, small-world, and scale-free)	Yearly (20 years)	AnyLogic	Calibration with historical data, internal validation with sensitivity analysis, external validation through comparison with similar studies' results	Explored the influence of various social factors, social networks, and water policies on water conservation technology adoption under	Fails to capture all impactful demographic factors due to data limitations and potential feedback mechanisms through dynamic factors	City of Miami Beach, FL	[40]

Table 3. *Cont.*

Simulation Objectives	Agents	Behavior Rules	Social Networks	Time Step	Platform	Calibration, Verification & Validation	Novelty/Advantages	Limitations	Location	Source
Diffusion of water-saving innovations	Households	Innovation characteristics, Theory of Planned Behavior, lifestyles, decision theory	Small-World	Monthly (14 years)	Java	Calibration with empirical data, validation with independent empirical data	Simulates the diffusion of water conservation technology among households (heterogeneous agents) based on two decision algorithms and driven by empirical data	Sensitive to the values set to categorize households based on lifestyles, model accuracy can be improved by adding other economic factors	Southern Germany	[146]
GSI adoption optimization	Water utility, local community organizations, and property owners	Probability-based rules	\	Quarterly (30 years)	NetLogo	Calibration with historical data	Simulates multi-agent simulation of GSI adoption based on physical compatibility and socioeconomic factors with undergoing synergistic infrastructure transitioning and ownership scenarios	Relies on numerous yet reasonable assumptions	Pint Breeze, PA	[157]
Assessments of the long-term resilience of water supply infrastructure	Users, agencies, wells, stressors, wastewater treatment plant	Bounded rationality and regret aversion, stochastic processes, consequential impacts of the other two agents	\	Yearly (100 years)	AnyLogic	Internal verification through component verification assessment, external verification through tracing, calibration with empirical data, face validation	Provides insights on theoretical, computational, and practical decision support for water supply infrastructure resilience under various scenarios of sea-level rise and adaptation strategies	Omits the salinity fluctuation caused by overexploited freshwater aquafer, and other adaption solutions by households	Miami-Dade County, FL	[158, 159]

4. Conclusions and Recommendations

The burgeon urbanization and rapidly increased impervious surfaces have led to the increment of runoff volumes and peak flows casting burdens on existing stormwater management infrastructure. Conventional gray infrastructure utilizes a centralized management approach to control stormwater through treatment facilities or direct discharge into receiving water bodies bypassing the treatment process. It is environmentally inadequate in modern societies as climate change has gradually intensified its impacts worldwide. On the contrary, GSI exploits decentralized natural processes to treat stormwater runoff at its source, which also provides additional benefits to the community contributing to urban resilience and sustainability. However, it still faces various barriers to GSI implementation in the US mainly due to existing presumptions that can lead to a lack of funding allocation. Conceptual frameworks are directing tools that can be used to standardize GSI project planning. There is an urgent need for inclusive decision support tools to better evaluate the perceptions of private landowners (homeowners and renters) of GSI so as to devise effective intervention strategies for encouraging GSI implementation. This can minimize the erroneous perceptions of GSI of the stakeholders, compared to the existing gray infrastructure. This paper made the first attempt to bring forth the connections between such social barriers to GSI implementation in the US and the potentially linked cognitive biases that had hampered rational decision making, which few studies have set their research efforts on. The authors acknowledge the limitation of this review regarding the connections due to its novelty in relevant research fields applied in GSI adoption, particularly the three biases chosen in this review. Further interdisciplinary discussions are encouraged to strengthen the research efforts on this topic to drive evidence-based local data analysis in addition to systematic analyses of these cognitive biases among stakeholder groups.

On the other hand, despite their capacity in being able to address multiple criteria, the existing decision support tools omitted some common cognitive biases which could result in less effective strategy implementation as pointed out in an article [74]. Various scholarly publications reached an agreement on ABM's robusticity in simulating individual-level decision-making processes. Thus, this paper reviewed quantitative analysis for decision support to promote innovative strategies in water management for long-term resilience. Yet there have been no ABM models developed to approach the well recognized social factor-related biases in GSI adaptation using the social-psychological approach of innovation diffusion. Thus, we proposed a conceptual framework to bridge this disconnection as shown in Figure 3. In this framework, assumptions of the presence of biases could be safely made if differences are recognized between the empirical data on households' perceptions of GSI, thus the acceptance and adoption and simulated results using the common mathematic theories in a multi-agent model. To further advance the realistic simulation of socio-infrastructure systems such as GSI implementation processes, future efforts should be made to incorporate the complex opinion dynamics due to cognitive biases into advanced hybrid models to explore the interdisciplinary interactions on a broader scale that have not yet been well examined for implementing innovative strategies of water infrastructure systems.

Author Contributions: Writing—original draft preparation, J.Q.; Writing—review and editing and supervision, N.B. All authors have read and agreed to the published version of the manuscript.

Funding: The authors thank the Department of Engineering Technology and Construction Management, University of North Carolina at Charlotte for internal funding.

Conflicts of Interest: The authors declare no conflict of interest.

References

1. Ntelekos, A.A.; Oppenheimer, M.; Smith, J.A.; Miller, A.J. Urbanization, climate change and flood policy in the United States. *Clim. Chang.* **2010**, *103*, 597–616. [CrossRef]
2. Blöschl, G.; Ardoin-Bardin, S.; Bonell, M.; Dorninger, M.; Goodrich, D.; Gutknecht, D.; Matamoros, D.; Merz, B.; Shand, P.; Szolgay, J. At what scales do climate variability and land cover change impact on flooding and low flows? *Hydrol. Process.* **2007**, *21*, 1241–1247. [CrossRef]
3. Brath, A.; Montanari, A.; Moretti, G. Assessing the effect on flood frequency of land use change via hydrological simulation (with uncertainty). *J. Hydrol.* **2006**, *324*, 141–153. [CrossRef]
4. Recanatesi, F.; Petroselli, A. Land Cover Change and Flood Risk in a Peri-Urban Environment of the Metropolitan Area of Rome (Italy). *Water Resour. Manag. Int. J. Publ. Eur. Water Resour. Assoc.* **2020**, *34*, 4399–4413. [CrossRef]
5. Wang, G.; Liu, J.; Kubota, J.; Chen, L. Effects of land-use changes on hydrological processes in the middle basin of the Heihe River, northwest China. *Hydrol. Process. Int. J.* **2007**, *21*, 1370–1382. [CrossRef]
6. Barbosa, A.E.; Fernandes, J.N.; David, L.M. Key issues for sustainable urban stormwater management. *Water Res.* **2012**, *46*, 6787–6798. [CrossRef]
7. Howard, G.K.; Bowen, M.P.; Antoine, R.W. Reducing Phosphorus Contamination in Stormwater Runoff. In Proceedings of the Howard2016ReducingPC, Norfolk, VA, USA, 4 March 2016.
8. McIntyre, J.K.; Lundin, J.I.; Cameron, J.R.; Chow, M.I.; Davis, J.W.; Incardona, J.P.; Scholz, N.L. Interspecies variation in the susceptibility of adult Pacific salmon to toxic urban stormwater runoff. *Environ. Pollut.* **2018**, *238*, 196–203. [CrossRef]
9. Tsihrintzis, V.A.; Hamid, R. Modeling and management of urban stormwater runoff quality: A review. *Water Resour. Manag.* **1997**, *11*, 136–164. [CrossRef]
10. Chini, C.M.; Canning, J.F.; Schreiber, K.L.; Peschel, J.M.; Stillwell, A.S. The Green Experiment: Cities, Green Stormwater Infrastructure, and Sustainability. *Sustainability* **2017**, *9*, 105. [CrossRef]
11. Walsh, C.J.; Booth, D.B.; Burns, M.J.; Fletcher, T.D.; Hale, R.L.; Hoang, L.N.; Livingston, G.; Rippy, M.A.; Roy, A.H.; Scoggins, M.; et al. Principles for urban stormwater management to protect stream ecosystems. *Freshw. Sci.* **2016**, *35*, 398–411. [CrossRef]
12. Roy, A.H.; Wenger, S.J.; Fletcher, T.D.; Walsh, C.J.; Ladson, A.R.; Shuster, W.D.; Thurston, H.W.; Brown, R.R. Impediments and solutions to sustainable, watershed-scale urban stormwater management: Lessons from Australia and the United States. *Environ. Manag.* **2008**, *42*, 344–359. [CrossRef] [PubMed]
13. NRC. *Urban Stormwater Management in the United States*; National Academies Press: Washington, DC, USA, 2009.

14. Li, C.; Peng, C.; Chiang, P.-C.; Cai, Y.; Wang, X.; Yang, Z. Mechanisms and applications of green infrastructure practices for stormwater control: A review. *J. Hydrol.* **2019**, *568*, 626–637. [CrossRef]

15. Yang, B.; Li, S.J. Green Infrastructure Design for Stormwater Runoff and Water Quality: Empirical Evidence from Large Watershed-Scale Community Developments. *Water* **2013**, *5*, 2038–2057. [CrossRef]

16. Wise, S.; Braden, J.; Ghalayini, D.; Grant, J.; Kloss, C.; MacMullan, E.; Morse, S.; Montalto, F.; Nees, D.; Nowak, D. Integrating valuation methods to recognize green infrastructure's multiple benefits. *Low Impact Dev.* **2010**, *2010*, 1123–1143.

17. Porse, E. Stormwater governance and future cities. *Water* **2013**, *5*, 29–52. [CrossRef]

18. Cherrier, J.; Klein, Y.; Link, H.; Pillich, J.; Yonzan, N. Hybrid green infrastructure for reducing demands on urban water and energy systems: A New York City hypothetical case study. *J. Environ. Stud. Sci.* **2016**, *6*, 77–89. [CrossRef]

19. Malinowski, P.A.; Wu, J.S.; Pulugurtha, S.; Stillwell, A.S. Green Infrastructure Retrofits with Impervious Area Reduction by Property Type: Potential Improvements to Urban Stream Quality. *J. Sustain. Water Built Environ.* **2018**, *4*, 04018012. [CrossRef]

20. Golden, H.E.; Hoghooghi, N. Green infrastructure and its catchment-scale effects: An emerging science. *Wiley Interdiscip. Rev. Water* **2018**, *5*, e1254. [CrossRef]

21. Berndtsson, J.C. Green roof performance towards management of runoff water quantity and quality: A review. *Ecol. Eng.* **2010**, *36*, 351–360. [CrossRef]

22. Chui, T.F.M.; Liu, X.; Zhan, W. Assessing cost-effectiveness of specific LID practice designs in response to large storm events. *J. Hydrol.* **2016**, *533*, 353–364. [CrossRef]

23. Jennings, A.A.; Adeel, A.A.; Hopkins, A.; Litofsky, A.L.; Wellstead, S.W. Rain barrel–urban garden stormwater management performance. *J. Environ. Eng.* **2012**, *139*, 757–765. [CrossRef]

24. Liu, J.; Sample, D.J.; Bell, C.; Guan, Y. Review and Research Needs of Bioretention Used for the Treatment of Urban Stormwater. *Water* **2014**, *6*, 1069–1099. [CrossRef]

25. Saraswat, C.; Kumar, P.; Mishra, B.K. Assessment of stormwater runoff management practices and governance under climate change and urbanization: An analysis of Bangkok, Hanoi and Tokyo. *Environ. Sci. Policy* **2016**, *64*, 101–117. [CrossRef]

26. Tavakol-Davani, H.; Goharian, E.; Hansen, C.H.; Tavakol-Davani, H.; Apul, D.; Burian, S.J. How does climate change affect combined sewer overflow in a system benefiting from rainwater harvesting systems? *Sustain. Cities Soc.* **2016**, *27*, 430–438. [CrossRef]

27. Vacek, P.; Struhala, K.; Matějka, L. Life-cycle study on semi intensive green roofs. *J. Clean. Prod.* **2017**, *154*, 203–213. [CrossRef]

28. Keeley, M.; Koburger, A.; Dolowitz, D.P.; Medearis, D.; Nickel, D.; Shuster, W. Perspectives on the Use of Green Infrastructure for Stormwater Management in Cleveland and Milwaukee. *Environ. Manag.* **2013**, *51*, 1093–1108. [CrossRef]

29. Copeland, C. *Green Infrastructure and Issues in Managing Urban Stormwater*; Congressional Research Service: Washington, DC, USA, 2016.

30. Zhang, K.; Chui, T.F.M. A comprehensive review of spatial allocation of LID-BMP-GI practices: Strategies and optimization tools. *Sci. Total Environ.* **2018**, *621*, 915–929. [CrossRef]

31. Eckart, K.; McPhee, Z.; Bolisetti, T. Performance and implementation of low impact development—A review. *Sci. Total Environ.* **2017**, *607*, 413–432. [CrossRef]

32. Li, C.; Fletcher, T.D.; Duncan, H.P.; Burns, M.J. Can stormwater control measures restore altered urban flow regimes at the catchment scale? *J. Hydrol.* **2017**, *549*, 631–653. [CrossRef]

33. Gordon, B.L.; Quesnel, K.J.; Abs, R.; Ajami, N.K. A case-study based framework for assessing the multi-sector performance of green infrastructure. *J. Environ. Manag.* **2018**, *223*, 371–384. [CrossRef]

34. Newell, J.P.; Seymour, M.; Yee, T.; Renteria, J.; Longcore, T.; Wolch, J.R.; Shishkovsky, A. Green Alley Programs: Planning for a sustainable urban infrastructure? *Cities* **2013**, *31*, 144–155. [CrossRef]

35. Venkataramanan, V.; Packman, A.I.; Peters, D.R.; Lopez, D.; McCuskey, D.J.; McDonald, R.I.; Miller, W.M.; Young, S.L. A systematic review of the human health and social well-being outcomes of green infrastructure for stormwater and flood management. *J. Environ. Manag.* **2019**, *246*, 868–880. [CrossRef]

36. Fletcher, T.D.; Shuster, W.; Hunt, W.F.; Ashley, R.; Butler, D.; Arthur, S.; Trowsdale, S.; Barraud, S.; Semadeni-Davies, A.; Bertrand-Krajewski, J.-L. SUDS, LID, BMPs, WSUD and more–The evolution and application of terminology surrounding urban drainage. *Urban Water J.* **2015**, *12*, 525–542. [CrossRef]

37. Thornton, K.; Laurin, C. Soft sciences and the hard reality of lake management. *Lake Reserv. Manag.* **2005**, *21*, 203–208. [CrossRef]

38. Hu, M.; Shealy, T. Overcoming Status Quo Bias for Resilient Stormwater Infrastructure: Empirical Evidence in Neurocognition and Decision-Making. *J. Manag. Eng.* **2020**, *36*. [CrossRef]

39. Olorunkiya, J.; Fassman, E.; Wilkinson, S. Risk: A fundamental barrier to the implementation of low impact design infrastructure for urban stormwater control. *J. Sustain. Dev.* **2012**, *5*, 27. [CrossRef]

40. Rasoulkhani, K.; Logasa, B.; Presa Reyes, M.; Mostafavi, A. Understanding fundamental phenomena affecting the water conservation technology adoption of residential consumers using agent-based modeling. *Water* **2018**, *10*, 993. [CrossRef]

41. Battaglio, R.P., Jr.; Belardinelli, P.; Bellé, N.; Cantarelli, P. Behavioral public administration ad fontes: A synthesis of research on bounded rationality, cognitive biases, and nudging in public organizations. *Public Adm. Rev.* **2019**, *79*, 304–320. [CrossRef]

42. Klotz, L. Cognitive biases in energy decisions during the planning, design, and construction of commercial buildings in the United States: An analytical framework and research needs. *Energy Effic.* **2011**, *4*, 271–284. [CrossRef]

43. Zhou, Y.; Chen, H.; Xu, S.; Wu, L. How cognitive bias and information disclosure affect the willingness of urban residents to pay for green power? *J. Clean. Prod.* **2018**, *189*, 552–562. [CrossRef]

44. Acciarini, C.; Brunetta, F.; Boccardelli, P. Cognitive biases and decision-making strategies in times of change: A systematic literature review. *Manag. Decis.* **2020**. [CrossRef]

45. Barnhill, K.; Smardon, R. Gaining ground: Green infrastructure attitudes and perceptions from stakeholders in Syracuse, New York. *Environ. Pract.* **2012**, *14*, 6–16. [CrossRef]

46. Miller, S.M.; Montalto, F.A. Stakeholder perceptions of the ecosystem services provided by Green Infrastructure in New York City. *Ecosyst. Serv.* **2019**, *37*, 100928. [CrossRef]

47. O'Donnell, E.; Maskrey, S.; Everett, G.; Lamond, J. Developing the implicit association test to uncover hidden preferences for sustainable drainage systems. *Philos. Trans. R. Soc. A* **2020**, *378*, 20190207. [CrossRef] [PubMed]

48. Wu, T.; Song, H.; Wang, J.; Friedler, E. Framework, Procedure, and Tools for Comprehensive Evaluation of Sustainable Stormwater Management: A Review. *Water* **2020**, *12*, 1231. [CrossRef]

49. Van Oijstaeijen, W.; Van Passel, S.; Cools, J. Urban green infrastructure: A review on valuation toolkits from an urban planning perspective. *J. Environ. Manag.* **2020**, *267*. [CrossRef]

50. Pellicani, R.; Parisi, A.; Iemmolo, G.; Apollonio, C. Economic risk evaluation in urban flooding and instability-prone areas: The case study of San Giovanni Rotondo (Southern Italy). *Geosciences* **2018**, *8*, 112. [CrossRef]

51. Carrera, L.; Standardi, G.; Bosello, F.; Mysiak, J. Assessing direct and indirect economic impacts of a flood event through the integration of spatial and computable general equilibrium modelling. *Environ. Model. Softw.* **2015**, *63*, 109–122. [CrossRef]

52. Huizinga, J.; De Moel, H.; Szewczyk, W. *Global Flood Depth-Damage Functions: Methodology and the Database with Guidelines*; Joint Research Centre (Seville Site): Sevilla, Spain, 2017.

53. Feingold, D.; Koop, S.; van Leeuwen, K. The City Blueprint Approach: Urban Water Management and Governance in Cities in the U.S. *Environ. Manag.* **2018**, *61*, 9–23. [CrossRef]

54. Flynn, C.D.; Davidson, C.I. Adapting the social-ecological system framework for urban stormwater management: The case of green infrastructure adoption. *Ecol. Soc.* **2016**, *21*. [CrossRef]

55. Hale, R.L.; Armstrong, A.; Baker, M.A.; Bedingfield, S.; Betts, D.; Buahin, C.; Buchert, M.; Crowl, T.; Dupont, R.R.; Ehleringer, J.R.; et al. iSAW: Integrating Structure, Actors, and Water to study socio-hydro-ecological systems. *Earths Future* **2015**, *3*, 110–132. [CrossRef]

56. Lieberherr, E.; Green, O.O. Green Infrastructure through Citizen Stormwater Management: Policy Instruments, Participation and Engagement. *Sustainability* **2018**, *10*, 2099. [CrossRef]

57. Schirmer, J.; Dyer, F. A framework to diagnose factors influencing proenvironmental behaviors in water-sensitive urban design. *Proc. Natl. Acad. Sci. USA* **2018**, *115*, E7690–E7699. [CrossRef] [PubMed]

58. Shandas, V. Neighborhood change and the role of environmental stewardship: A case study of green infrastructure for stormwater in the City of Portland, Oregon, USA. *Ecol. Soc.* **2015**, *20*. [CrossRef]

59. William, R.; Garg, J.; Stillwell, A.S. A game theory analysis of green infrastructure stormwater management policies. *Water Resour. Res.* **2017**, *53*, 8003–8019. [CrossRef]

60. Heckert, M.; Rosan, C.D. Developing a green infrastructure equity index to promote equity planning. *Urban For. Urban Green.* **2016**, *19*, 263–270. [CrossRef]

61. Barclay, N.; Klotz, L. Role of community participation for green stormwater infrastructure development. *J. Environ. Manag.* **2019**, *251*. [CrossRef]

62. Chaffin, B.C.; Shuster, W.D.; Garmestani, A.S.; Furio, B.; Albro, S.L.; Gardiner, M.; Spring, M.; Green, O.O. A tale of two rain gardens: Barriers and bridges to adaptive management of urban stormwater in Cleveland, Ohio. *J. Environ. Manag.* **2016**, *183*, 431–441. [CrossRef]

63. Hart, D.D.; Bell, K.P.; Lindenfeld, L.A.; Jain, S.; Johnson, T.R.; Ranco, D.; McGill, B. Strengthening the role of universities in addressing sustainability challenges: The Mitchell Center for Sustainability Solutions as an institutional experiment. *Ecol. Soc.* **2015**, *20*. [CrossRef]

64. Young, R.; Zanders, J.; Lieberknecht, K.; Fassman-Beck, E. A comprehensive typology for mainstreaming urban green infrastructure. *J. Hydrol.* **2014**, *519*, 2571–2583. [CrossRef]

65. Baptiste, A.K.; Foley, C.; Smardon, R. Understanding urban neighborhood differences in willingness to implement green infrastructure measures: A case study of Syracuse, NY. *Landsc. Urban Plan.* **2015**, *136*, 1–12. [CrossRef]

66. Das, T.; Teng, B.S. Cognitive biases and strategic decision processes: An integrative perspective. *J. Manag. Stud.* **1999**, *36*, 757–778. [CrossRef]

67. Turner, V.K.; Jarden, K.; Jefferson, A. Resident perspectives on green infrastructure in an experimental suburban stormwater management program. *Cities Environ.* **2016**, *9*, 4.

68. Tayouga, S.J.; Gagné, S.A. The socio-ecological factors that influence the adoption of green infrastructure. *Sustainability* **2016**, *8*, 1277. [CrossRef]

69. Kati, V.; Jari, N. Bottom-up thinking—Identifying socio-cultural values of ecosystem services in local blue–green infrastructure planning in Helsinki, Finland. *Land Use Policy* **2016**, *50*, 537–547. [CrossRef]

70. Staddon, C.; Ward, S.; De Vito, L.; Zuniga-Teran, A.; Gerlak, A.K.; Schoeman, Y.; Hart, A.; Booth, G. Contributions of green infrastructure to enhancing urban resilience. *Environ. Syst. Decis.* **2018**, *38*, 330–338. [CrossRef]

71. Gifford, R.; Nilsson, A. Personal and social factors that influence pro-environmental concern and behaviour: A review. *Int. J. Psychol.* **2014**, *49*, 141–157. [CrossRef]

72. Sobkowicz, P. Opinion Dynamics Model Based on Cognitive Biases of Complex Agents. *J. Artif. Soc. Soc. Simul.* **2018**, *21*, 8 [CrossRef]

73. Chen, S.-H.; Gostoli, U. Behavioral macroeconomics and agent-based macroeconomics. In Proceedings of the Distributed Computing and Artificial Intelligence, 11th International Conference, Salamanca, Spain, 4–6 June 2014; pp. 47–54.

74. Xu, B.; Liu, R.; Liu, W. Individual bias and organizational objectivity: An agent-based simulation. *J. Artif. Soc. Soc. Simul.* **2014**, *17*, 2. [CrossRef]

75. Bruch, E.; Atwell, J. Agent-based models in empirical social research. *Sociol. Methods Res.* **2015**, *44*, 186–221. [CrossRef]

76. Gray, J.; Hilton, J.; Bijak, J. Choosing the choice: Reflections on modelling decisions and behaviour in demographic agent-based models. *Popul. Stud.* **2017**, *71*, 85–97. [CrossRef] [PubMed]

77. Bharathy, G.K. Agent Based Human Behavior Modeling: A Knowledge Engineering Based Systems Methodology for Integrating Social Science Frameworks for Modeling Agents with Cognition, Personality and Culture. Ph.D. Thesis, University of Pennsylvania, Philadelphia, PA, USA, 2006. Dissertations available from ProQuest. AAI3246140.

78. Meerow, S. The politics of multifunctional green infrastructure planning in New York City. *Cities* **2020**, *100*. [CrossRef]

79. Coleman, S.; Hurley, S.; Rizzo, D.; Koliba, C.; Zia, A. From the household to watershed: A cross-scale analysis of residential intention to adopt green stormwater infrastructure. *Landsc. Urban Plan.* **2018**, *180*, 195–206. [CrossRef]

80. Haselton, M.G.; Nettle, D.; Murray, D.R. The evolution of cognitive bias. *Handb. Evol. Psychol.* **2015**, 1–20. [CrossRef]

81. Bukszar, E., Jr. Strategic bias: The impact of cognitive biases on strategy. *Can. J. Adm. Sci./Rev. Can. Sci. Adm.* **1999**, *16*, 105–117. [CrossRef]

82. Moher, D.; Liberati, A.; Tetzlaff, J.; Altman, D.G.; Group, P. Preferred reporting items for systematic reviews and meta-analyses: The PRISMA statement. *PLoS Med.* **2009**, *6*, e1000097. [CrossRef]

83. Dhakal, K.P.; Chevalier, L.R. Managing urban stormwater for urban sustainability: Barriers and policy solutions for green infrastructure application. *J. Environ. Manag.* **2017**, *203*, 171–181. [CrossRef]

84. Qiao, X.-J.; Kristoffersson, A.; Randrup, T.B. Challenges to implementing urban sustainable stormwater management from a governance perspective: A literature review. *J. Clean. Prod.* **2018**, *196*, 943–952. [CrossRef]

85. Winz, I.; Brierley, G.; Trowsdale, S. Dominant perspectives and the shape of urban stormwater futures. *Urban Water J.* **2011**, *8*, 337–349. [CrossRef]

86. Bain, D.; Elliott, E.; Thomas, B.; Shelef, E.; River, M. *Green Infrastructure for Stormwater Management: Knowledge Gaps and Approaches*; University of Pittsburgh: Pittsburgh, PA, USA, 2019.

87. Chahardowli, M.; Sajadzadeh, H.; Aram, F.; Mosavi, A. Survey of Sustainable Regeneration of Historic and Cultural Cores of Cities. *Energies* **2020**, *13*, 2708. [CrossRef]

88. Laspidou, C.S.; Mellios, N.K.; Spyropoulou, A.E.; Kofinas, D.T.; Papadopoulou, M.P. Systems thinking on the resource nexus: Modeling and visualisation tools to identify critical interlinkages for resilient and sustainable societies and institutions. *Sci. Total Environ.* **2020**, *717*, 137264. [CrossRef] [PubMed]

89. Nosratabadi, S.; Mosavi, A.; Shamshirband, S.; Kazimieras Zavadskas, E.; Rakotonirainy, A.; Chau, K.W. Sustainable business models: A review. *Sustainability* **2019**, *11*, 1663. [CrossRef]

90. Maeda, P.K.; Chanse, V.; Rockler, A.; Montas, H.; Shirmohammadi, A.; Wilson, S.; Leisnham, P.T. Linking stormwater Best Management Practices to social factors in two suburban watersheds. *PLoS ONE* **2018**, *13*, e0202638. [CrossRef] [PubMed]

91. Mason, L.R.; Ellis, K.N.; Hathaway, J.M. Urban flooding, social equity, and "backyard" green infrastructure: An area for multidisciplinary practice. *J. Community Pract.* **2019**. [CrossRef]

92. Cousins, J.J. Structuring Hydrosocial Relations in Urban Water Governance. *Ann. Am. Assoc. Geogr.* **2017**, *107*, 1144–1161. [CrossRef]

93. Chaffin, B.C.; Floyd, T.M.; Albro, S.L. Leadership in informal stormwater governance networks. *PLoS ONE* **2019**, *14*, e0222434. [CrossRef]

94. Cousins, J.J. Infrastructure and institutions: Stakeholder perspectives of stormwater governance in Chicago. *Cities* **2017**, *66*, 44–52. [CrossRef]

95. Finewood, M.H. Green Infrastructure, Grey Epistemologies, and the Urban Political Ecology of Pittsburgh's Water Governance. *Antipode* **2016**, *48*, 1000–1021. [CrossRef]

96. Porse, E. Open data and stormwater systems in Los Angeles: Applications for equitable green infrastructure. *Local Environ.* **2018**, *23*, 505–517. [CrossRef]

97. Green, O.O.; Shuster, W.D.; Rhea, L.K.; Garmestani, A.S.; Thurston, H.W. Identification and induction of human, social, and cultural capitals through an experimental approach to stormwater management. *Sustainability* **2012**, *4*, 1669–1682. [CrossRef]

98. Lim, T.C. An empirical study of spatial-temporal growth patterns of a voluntary residential green infrastructure program. *J. Environ. Plan. Manag.* **2018**, *61*, 1363–1382. [CrossRef]

99. Meyer, M.A.; Hendricks, M.; Newman, G.D.; Masterson, J.H.; Cooper, J.T.; Sansom, G.; Gharaibeh, N.; Horney, J.; Berke, P.; van Zandt, S.; et al. Participatory action research: Tools for disaster resilience education. *Int. J. Disaster Resil. Built Environ.* **2018**, *9*, 402–419. [CrossRef] [PubMed]

100. Monaghan, P.; Hu, S.C.; Hansen, G.; Ott, E.; Nealis, C.; Morera, M. Balancing the Ecological Function of Residential Stormwater Ponds with Homeowner Landscaping Practices. *Environ. Manag.* **2016**, *58*, 843–856. [CrossRef] [PubMed]

101. Persaud, A.; Alsharifa, K.; Monaghan, P.; Akiwumi, F.; Morera, M.C.; Ott, E. Landscaping practices, community perceptions, and social indicators for stormwater nonpoint source pollution management. *Sustain. Cities Soc.* **2016**, *27*, 377–385. [CrossRef]

102. Shuster, W.D.; Garmestani, A.S. Adaptive exchange of capitals in urban water resources management: An approach to sustainability? *Clean Technol. Environ. Policy* **2015**, *17*, 1393–1400. [CrossRef]

103. Baptiste, A.K. "Experience is a great teacher": Citizens' reception of a proposal for the implementation of green infrastructure as stormwater management technology. *Community Dev.* **2014**, *45*, 337–352. [CrossRef]

104. Shuster, W.D.; Morrison, M.A.; Webb, R. Front-loading urban stormwater management for success–a perspective incorporating current studies on the implementation of retrofit low-impact development. *Cities Environ.* **2008**, *1*, 8. [CrossRef]

105. Hoover, F.-A.; Hopton, M.E. Developing a framework for stormwater management: Leveraging ancillary benefits from urban greenspace. *Urban Ecosyst.* **2019**, *22*, 1139–1148. [CrossRef]

106. Schifman, L.A.; Herrmann, D.L.; Shuster, W.D.; Ossola, A.; Garmestani, A.; Hopton, M.E. Situating Green Infrastructure in Context: A Framework for Adaptive Socio-Hydrology in Cities. *Water Resour. Res.* **2017**, *53*, 10139–10154. [CrossRef]

107. Hager, G.W.; Belt, K.T.; Stack, W.; Burgess, K.; Grove, J.M.; Caplan, B.; Hardcastle, M.; Shelley, D.; Pickett, S.T.A.; Groffman, P.M. Socioecological revitalization of an urban watershed. *Front. Ecol. Environ.* **2013**, *11*, 28–36. [CrossRef]

108. Kahneman, D.; Tversky, A. On the reality of cognitive illusions. *Psychol. Rev.* **1996**, *103*, 582–591. [CrossRef] [PubMed]

109. Tversky, A.; Kahneman, D. Causal schemas in judgments under uncertainty. *Prog. Soc. Psychol.* **2015**, *1*, 49–72.

110. Barnes, J.H., Jr. Cognitive biases and their impact on strategic planning. *Strateg. Manag. J.* **1984**, *5*, 129–137. [CrossRef]

111. Glynn, P.D.; Voinov, A.A.; Shapiro, C.D.; White, P.A. From data to decisions: Processing information, biases, and beliefs for improved management of natural resources and environments. *Earth's Future* **2017**, *5*, 356–378. [CrossRef]

112. Cantarelli, P.; Bellé, N.; Belardinelli, P. Behavioral public HR: Experimental evidence on cognitive biases and debiasing interventions. *Rev. Public Pers. Adm.* **2020**, *40*, 56–81. [CrossRef]

113. Morewedge, C.K.; Yoon, H.; Scopelliti, I.; Symborski, C.W.; Korris, J.H.; Kassam, K.S. Debiasing decisions: Improved decision making with a single training intervention. *Policy Insights Behav. Brain Sci.* **2015**, *2*, 129–140. [CrossRef]

114. Bhandari, G.; Hassanein, K. An agent-based debiasing framework for investment decision-support systems. *Behav. Inf. Technol.* **2012**, *31*, 495–507. [CrossRef]

115. Kandiah, V.K.; Berglund, E.Z.; Binder, A.R. An agent-based modeling approach to project adoption of water reuse and evaluate expansion plans within a sociotechnical water infrastructure system. *Sustain. Cities Soc.* **2019**, *46*. [CrossRef]

116. Rasoulkhani, K.; Logasa, B.; Reyes, M.P.; Mostafavi, A. Agent-based modeling framework for simulation of complex adaptive mechanisms underlying household water conservation technology adoption. In Proceedings of the 2017 Winter Simulation Conference (WSC), Las Vegas, NV, USA, 3–6 December 2017; pp. 1109–1120.

117. Caverni, J.-P.; Fabre, J.-M.; Gonzalez, M. *Cognitive Biases*; Elsevier: Amsterdam, The Netherlands, 1990.

118. Fernandez, R.; Rodrik, D. Resistance to reform: Status quo bias in the presence of individual-specific uncertainty. *Am. Econ. Rev.* **1991**, *81*, 1146–1155.

119. Tversky, A.; Kahneman, D. Judgment under uncertainty: Heuristics and biases. *science* **1974**, *185*, 1124–1131. [CrossRef]

120. Samuelson, W.; Zeckhauser, R. Status quo bias in decision making. *J. Risk Uncertain.* **1988**, *1*, 7–59. [CrossRef]

121. Hu, M.; Shealy, T. Application of functional near-infrared spectroscopy to measure engineering decision-making and design cognition: Literature review and synthesis of methods. *J. Comput. Civ. Eng.* **2019**, *33*, 04019034. [CrossRef]

122. Babovic, F.; Mijic, A.; Madani, K. Decision making under deep uncertainty for adapting urban drainage systems to change. *Urban Water J.* **2018**, *15*, 552–560. [CrossRef]

123. Gonzalez, C.; Dana, J.; Koshino, H.; Just, M. The framing effect and risky decisions: Examining cognitive functions with fMRI. *J. Econ. Psychol.* **2005**, *26*, 1–20. [CrossRef]

124. McWhirter, N.; Shealy, T. Teaching engineering students about cognitive barriers during design: A case study approach using the Envision Rating System for sustainable infrastructure. In Proceedings of the International Conference on Sustainable Infrastructure 2017, New York, NY, USA, 26–28 October 2017; pp. 418–431.

125. Strack, F.; Mussweiler, T. Explaining the enigmatic anchoring effect: Mechanisms of selective accessibility. *J. Personal. Soc. Psychol.* **1997**, *73*, 437. [CrossRef]

126. Carlson, C.; Barreteau, O.; Kirshen, P.; Foltz, K. Storm water management as a public good provision problem: Survey to understand perspectives of low-impact development for urban storm water management practices under climate change. *J. Water Resour. Plan. Manag.* **2014**, *141*, 04014080. [CrossRef]

127. Ugolini, F.; Massetti, L.; Sanesi, G.; Pearlmutter, D. Knowledge transfer between stakeholders in the field of urban forestry and green infrastructure: Results of a European survey. *Land Use Policy* **2015**, *49*, 365–381. [CrossRef]

128. Derkzen, M.L.; van Teeffelen, A.J.A.; Verburg, P.H. Green infrastructure for urban climate adaptation: How do residents' views on climate impacts and green infrastructure shape adaptation preferences? *Landsc. Urban Plan.* **2017**, *157*, 106–130. [CrossRef]

129. Giacalone, K.; Mobley, C.; Sawyer, C.; Witte, J.; Eidson, G. Survey says: Implications of a public perception survey on stormwater education programming. *J. Contemp. Water Res. Educ.* **2010**, *146*, 92–102. [CrossRef]

130. Tversky, A.; Kahneman, D. Loss aversion in riskless choice: A reference-dependent model. *Q. J. Econ.* **1991**, *106*, 1039–1061. [CrossRef]

131. Demuzere, M.; Orru, K.; Heidrich, O.; Olazabal, E.; Geneletti, D.; Orru, H.; Bhave, A.G.; Mittal, N.; Feliu, E.; Faehnle, M. Mitigating and adapting to climate change: Multi-functional and multi-scale assessment of green urban infrastructure. *J. Environ. Manag.* **2014**, *146*, 107–115. [CrossRef] [PubMed]

132. Gogate, N.G.; Kalbar, P.P.; Raval, P.M. Assessment of stormwater management options in urban contexts using Multiple Attribute Decision-Making. *J. Clean. Prod.* **2017**, *142*, 2046–2059. [CrossRef]
133. Shaver, E. *Low Impact Design Versus Conventional Development: Literature Review of Developer-Related Costs and Profit Margins*; Auckland Regional Council: Auckland, New Zealand, 2009.
134. Raucher, R.; Clements, J. A triple bottom line assessment of traditional and green infrastructure options for controlling CSO events in Philadelphia's watersheds. *Proc. Water Environ. Fed.* **2010**, *2010*, 6776–6804. [CrossRef]
135. Suppakittpaisarn, P.; Jiang, X.; Sullivan, W.C. Green infrastructure, green stormwater infrastructure, and human health: A review. *Curr. Landsc. Ecol. Rep.* **2017**, *2*, 96–110. [CrossRef]
136. Di Matteo, M.; Maier, H.R.; Dandy, G.C. Many-objective portfolio optimization approach for stormwater management project selection encouraging decision maker buy-in. *Environ. Model. Softw.* **2019**, *111*, 340–355. [CrossRef]
137. Bell, C.D.; McMillan, S.K.; Clinton, S.M.; Jefferson, A.J. Hydrologic response to stormwater control measures in urban watersheds. *J. Hydrol.* **2016**, *541*, 1488–1500. [CrossRef]
138. Carlet, F. Understanding attitudes toward adoption of green infrastructure: A case study of US municipal officials. *Environ. Sci. Policy* **2015**, *51*, 65–76. [CrossRef]
139. Montalto, F.A.; Bartrand, T.A.; Waldman, A.M.; Travaline, K.A.; Loomis, C.H.; McAfee, C.; Geldi, J.M.; Riggall, G.J.; Boles, L.M. Decentralised green infrastructure: The importance of stakeholder behaviour in determining spatial and temporal outcomes. *Struct. Infrastruct. Eng.* **2013**, *9*, 1187–1205. [CrossRef]
140. Akerlof, G.A. Behavioral macroeconomics and macroeconomic behavior. *Am. Econ. Rev.* **2002**, *92*, 411–433. [CrossRef]
141. Groeneveld, J.; Müller, B.; Buchmann, C.M.; Dressler, G.; Guo, C.; Hase, N.; Hoffmann, F.; John, F.; Klassert, C.; Lauf, T. Theoretical foundations of human decision-making in agent-based land use models—A review. *Environ. Model. Softw.* **2017**, *87*, 39–48. [CrossRef]
142. DeAngelis, D.L.; Diaz, S.G. Decision-making in agent-based modeling: A current review and future prospectus. *Front. Ecol. Evol.* **2019**, *6*, 237. [CrossRef]
143. Marsella, S.C.; Pynadath, D.V.; Read, S.J. PsychSim: Agent-based modeling of social interactions and influence. In Proceedings of the International Conference on Cognitive Modeling, Pittsburgh, PA, USA, 30 July–1 August 2004; pp. 243–248.
144. Rogers, E.M. *Diffusion of Innovations*; Simon and Schuster: New York, NY, USA, 2010.
145. Kiesling, E.; Günther, M.; Stummer, C.; Wakolbinger, L.M. Agent-based simulation of innovation diffusion: A review. *Cent. Eur. J. Oper. Res.* **2012**, *20*, 183–230. [CrossRef]
146. Schwarz, N.; Ernst, A. Agent-based modeling of the diffusion of environmental innovations—An empirical approach. *Technol. Forecast. Soc. Chang.* **2009**, *76*, 497–511. [CrossRef]
147. Faust, K.M.; Abraham, D.M.; DeLaurentis, D. Coupled human and water infrastructure systems sector interdependencies: Framework evaluating the impact of cities experiencing urban decline. *J. Water Resour. Plan. Manag.* **2017**, *143*, 04017043. [CrossRef]
148. Wang, X.; Sirianni, A.D.; Tang, S.; Zheng, Z.; Fu, F. Public discourse and social network echo chambers driven by socio-cognitive biases. *Phys. Rev. X* **2020**, *10*, 041042.
149. Abbott, R.; Hadžikadić, M. Complex adaptive systems, systems thinking, and agent-based modeling. In *Advanced Technologies, Systems, and Applications*; Springer: Berlin/Heidelberg, Germany, 2017; pp. 1–8.
150. Geschke, D.; Lorenz, J.; Holtz, P. The triple-filter bubble: Using agent-based modelling to test a meta-theoretical framework for the emergence of filter bubbles and echo chambers. *Br. J. Soc. Psychol.* **2019**, *58*, 129–149. [CrossRef] [PubMed]
151. Demary, M. Transaction taxes, greed and risk aversion in an agent-based financial market model. *J. Econ. Interact. Coord.* **2011**, *6*, 1–28. [CrossRef]
152. Cannella, S.; Di Mauro, C.; Dominguez, R.; Ancarani, A.; Schupp, F. An exploratory study of risk aversion in supply chain dynamics via human experiment and agent-based simulation. *Int. J. Prod. Res.* **2019**, *57*, 985–999. [CrossRef]
153. Galán, J.M.; López-Paredes, A.; Del Olmo, R. An agent-based model for domestic water management in Valladolid metropolitan area. *Water Resour. Res.* **2009**, *45*. [CrossRef]
154. Haer, T.; Botzen, W.W.; Aerts, J.C. The effectiveness of flood risk communication strategies and the influence of social networks—Insights from an agent-based model. *Environ. Sci. Policy* **2016**, *60*, 44–52. [CrossRef]
155. Kandiah, V.; Binder, A.R.; Berglund, E.Z. An empirical agent-based model to simulate the adoption of water reuse using the social amplification of risk framework. *Risk Anal.* **2017**, *37*, 2005–2022. [CrossRef] [PubMed]
156. Kotz, C.; Hiessl, H. Analysis of system innovation in urban water infrastructure systems: An agent-based modelling approach. *Water Sci. Technol. Water Supply* **2005**, *5*, 135–144. [CrossRef]
157. Zidar, K.; Bartrand, T.A.; Loomis, C.H.; McAfee, C.A.; Geldi, J.M.; Riggall, G.J.; Montalto, F. Maximizing Green Infrastructure in a Philadelphia Neighborhood. *Urban Plan.* **2017**, *2*, 115–132. [CrossRef]
158. Rasoulkhani, K.; Mostafavi, A. Resilience as an emergent property of human-infrastructure dynamics: A multi-agent simulation model for characterizing regime shifts and tipping point behaviors in infrastructure systems. *PLoS ONE* **2018**, *13*, e0207674. [CrossRef]
159. Rasoulkhani, K.; Mostafavi, A.; Reyes, M.P.; Batouli, M. Resilience planning in hazards-humans-infrastructure nexus: A multi-agent simulation for exploratory assessment of coastal water supply infrastructure adaptation to sea-level rise. *Environ. Model. Softw.* **2020**, *125*, 104636. [CrossRef]

Communication

Analytical Protocol to Estimate the Relative Importance of Environmental and Anthropogenic Factors in Influencing Runoff Quality in the Bumbu Watershed, Papua New Guinea

Willie Doaemo [1,2], **Lawrence Wuest** [3,*], **Shaurya Bajaj** [4], **Wan Shafrina Wan Mohd Jaafar** [5] and **Midhun Mohan** [4,6]

[1] Department of Civil Engineering, Papua New Guinea University of Technology, Lae 00411, Papua New Guinea; willie.doaemo@pnguot.ac.pg
[2] Morobe Development Foundation, Doyle Street, Trish Avenue-Eriku, Lae 00411, Papua New Guinea
[3] United Nations Volunteer, Stanley, NB E6B 2K5, Canada
[4] United Nations Volunteer, Morobe Development Foundation, Lae 00411, Papua New Guinea; shaurya.bajaj2014@vitalum.ac.in (S.B.); mid_mohan@berkeley.edu (M.M.)
[5] Earth Observation Centre, Institute of Climate Change, National University of Malaysia (UKM), Bangi 43600, Selangor, Malaysia; wanshafrina@ukm.edu.my
[6] Department of Geography, University of California, Berkeley, CA 94720, USA
* Correspondence: wuestL@nbnet.nb.ca

Received: 9 September 2020; Accepted: 12 October 2020; Published: 14 October 2020

Abstract: The wellbeing, socio-economic viability and the associated health of the inhabitant species of any ecosystem are largely dependent on the quality of its water resources. In this regard, we developed a protocol to measure the potential impact of various environmental and anthropogenic factors on runoff quality at 22 water sampling sites across the Bumbu Watershed in Papua New Guinea. For this purpose, we utilized Digital Elevation Models and several GIS techniques for delineation of stream drainage patterns, classification of the watershed based on Land Use/Land Cover, spatial interpolation of rainfall patterns and computation of the corresponding factor runoff. Our study concludes that a variety of potential challenges to surface water quality are possible such as natural geologic and geochemical inputs, runoff accumulation of precipitation and organic matter pollutants. The developed protocol can also accommodate socio-economic factors such as community and household health, sanitation and hygiene practices, pollution and waste disposal. This research lays the foundation for further development of an all-inclusive correlational analysis between the relative importance values of the factors influencing runoff and spatially distributed water quality measurements in the Bumbu basin.

Keywords: digital elevation model; maximum likelihood estimation (MLE) classification; runoff quality; social, economic and environmental (SEE) factor; Triangulated Irregular Network (TIN); urbanization; vegetation density

1. Introduction

As a result of rapid urbanization, population expansion and concomitant technological developments, substantial strain is being placed on the water systems around the world, making quality of water a matter of global significance [1]. The waters of an ecosystem have a symbiotic relationship with the health and welfare of the resident communities consuming it [2]. Study of the probable impacts of various sources on water quality has been elegantly summarized by a conceptual framework put forward by Granger et al. [3] that can arguably be described as the process of:

1. Identifying potential sources of impact;
2. Identifying the means of mobilization of the sources;
3. Assessing the impact of delivery of the sources to receiving waters.

Previous studies have revealed that immediate landscape features and land use patterns significantly influence water quality [4–11]. Nevertheless, these topographical features are susceptible to alterations and are usually examined under Land Use/Land Cover Change (LULCC) initiatives. LULCC can be defined as a convoluted process of transformation of landscape and its related patterns of utility due to environmental and human-induced interactions [12]. These interactions in turn have a corresponding effect on the physicochemical composition of water. Given these interdependencies, water systems are best identified through river basins and watersheds that comprise a uniquely integrated hydrological network through which precipitation runoff flows into a specific larger body of water such as a river, lake or ocean. The network of streams across a drainage basin is the conduit for all precipitation deposited onto the catchment area in its journey to reach the sea. In passing through the catchment area, precipitation receives the detritus of human existence as well as the accumulation of geologic minerals dissolved and granulated to such a degree as to be mobile in the air and on the earth's surface [13]. These constituents in water are indicative of the impact of various environmental and anthropogenic parameters.

With respect to the aforementioned context, the Morobe Development Foundation (MDF) has undertaken a study to understand the mutual dependency that exists between waters of the Bumbu Watershed and the residents of communities of Lae in Papua New Guinea who rely on these waters for their continued existence and sustainability. Resident communities utilize water of the Bumbu river primarily for drinking and sanitation purposes, but simultaneously, they lack access to proper toilet facilities and a treated water supply. This problem is compounded by untreated sewage and the dumping of pollutants by industries established in the vicinity. As a result, water quality has become a serious concern in the region exposing human health to various risks in the form of skin infections and waterborne diseases [14]. The study seeks to fill a gap to address the risks to human health and security created by poor water quality by analyzing influence of different factors on the Bumbu basin.

The protocol developed in this study is designed to explore the relationship between the spatial distribution of diverse factors that have the potential to influence water quality, and the spatial distribution of measured water quality. The use of spatial analysis and corresponding tools for this purpose incorporates a divergent approach in comparison to methods such as pH Redox Equilibrium modeling (phreeqc), which is designed to explore the mechanistic relationship between water composition and the actual geological and hydrological conditions affecting it. Once an exploratory study using the protocol developed in this work is completed, a more detailed approach in the latter direction can be taken up for the Bumbu Watershed. Consequently, to study this relationship, we utilized available Geographic Information System (GIS) methodologies to develop a protocol which not only measures the influence of anthropogenic and environmental factors on runoff but also has the capability to be extended to socio-economic parameters such as community health, pollution, waste disposal, crime and sanitation. The evolution of Geographic Information System (GIS) has led to the advancement of analytical tools to comprehend the interrelationship existing between land use and the corresponding quality of water, which in turn has led to commendable contributions in watershed management [8–10]. With respect to this initial research, we consider roads, streets, rainfall patterns, forested and industrialized landscapes as examples of factors that present a possible correlation with runoff water quality. Data for such features often exist in a variety of formats familiar to spatial analysts—i.e., in vectorized and/or gridded databases depending on the data source (these formats will be explained in more detail in further sections).

In pursuit of the above objective, a review of available literature, databases and analytic techniques has been conducted to gather relevant insights. Guoyu Xu et al. [4] studied the effect of multiple temporal and spatial scales on the quality of water across 32 sampling sites in the Wujiang River Watershed in China. They examined eight variables as possible indicators of quality, utilized Partial

Least Scale (PLS) regression and found that quality was influenced by landscape configuration, composition and precipitation. The levels of Dissolved Oxygen (DO) were found to be higher in dry season and higher levels of other contaminants were found during the wet season. Only landscape level metrics in periods of rainfall were found to be related to organic matter. They also concluded that watershed buffer areas involving small patches of cropland with high aggregation of forested areas lead to better quality of water. Likewise, Xiao et al. [6] investigated the relationship between water quality and landscape metrics at multiple spatial scales in different seasons. For this purpose, they took into consideration 34 sampling sites across Huzhou City. They utilized stepwise regression and found that built-up land "has a role in influencing" water quality at a smaller scale, whereas at a local scale, multiple land use categories can be expected to impose an influence. Total Nitrogen (TN) was found to be negatively correlated with the index of build-up land, whereas landscape index of forest was positively correlated with it. Moreover, Putro et al. [7] investigated the impact of land use pattern and climate on the quantity and quality of water across two urbanized catchment areas and two rural catchment areas located in the United Kingdom. Using multivariate regression models, they assessed the influence of rainfall and urbanization on the trends in the DO, runoff and temperature of the water network involved. They found that temperature and dissolved oxygen variation with respect to catchment in urban areas are not driven by climatic variables. The temperature, total runoff and DO displayed an upward trend for urban catchments, but the same was not true for undeveloped catchments. In another study conducted by Lintern et al. [5], the authors studied existing literature to understand how spatial variability of landscape characteristics and interseasonal variation lead to variations in water quality. They analyzed different correlations that exist for different landscape characteristics including land use, geology, topography, hydrology, soil type and climate through a rigorous literature review. For example, their review revealed that rainfall is positively correlated with Electrical Conductivity (EC) for developed landscape factors such as urban areas, and negatively correlated with EC for undeveloped factors such as grasslands. Similarly, topography depicted by slope/elevation for undeveloped landscape was positively related with Total Suspended Solids (TSS), Total Phosphorus (TP) and Total Kjeldahl Nitrogen (TKN), whereas slope/elevation for developed landscape factors was negatively related to the same constituents. The authors stressed the need to consider the relation existing between the numerous catchment characteristics, impact of the spatial setting of different landscape features, interannual and interseasonal variability to comprehend the relationship existing between water quality and landscape features. Their paper revealed the wide range of factors that can influence runoff and highlighted the need to take into account a wide range of environmental data.

In this preliminary study, we elaborate how we measured the relative influence of the different factors indicative of environmental and anthropogenic impact on surface runoff at the respective water sampling sites using GIS tools and techniques. We also discuss the relevance of this protocol to the ability to use the factor runoff importance values and to be able to accommodate other parameters such as water, sanitation and hygiene (WASH) practices and other socio-economic factors for our future water quality studies. With the livelihood, health and welfare of the communities dependent on the waters of the Bumbu Watershed, it has become a necessity to explore the relationship that exists between the anthropogenic and environmental factors influencing runoff, WASH conditions, physiochemical analysis of the waters of the Bumbu and the corresponding water quality. This protocol is the first major step, and a stepping-stone in this direction. In Section 2, we elucidate the methodology and the protocol involved based on the different formats and characteristics of the available factors. In Section 3, we present our results related to the computations of raw runoff and relative impact of the factors. In Section 4, we discuss our findings, and our plans to utilize the protocol and its results in our upcoming water quality studies.

2. Materials and Methods

2.1. Study Area

Our investigation is centered around the Bumbu basin, located in Morobe Province in Papua New Guinea (PNG). Figure 1 illustrates the study area at different scales. The watershed is bounded on the west by the Markham river basin, and on the east by the Busu river basin. The Bumbu River traverses through Lae, the capital of Morobe Province and the second largest city in PNG. The river originates from the Atzera Range and is relatively narrow as it flows downstream at a medium pace. However, during the extreme rainfalls of the flooding season, the rate of flow is much higher [15], resulting in rapid erosion of the sandy loam that is the main constituent of the Bumbu floodplain [16]. All coordinates in this study are based on the World Geodetic System of 1984 (WGS84) datum.

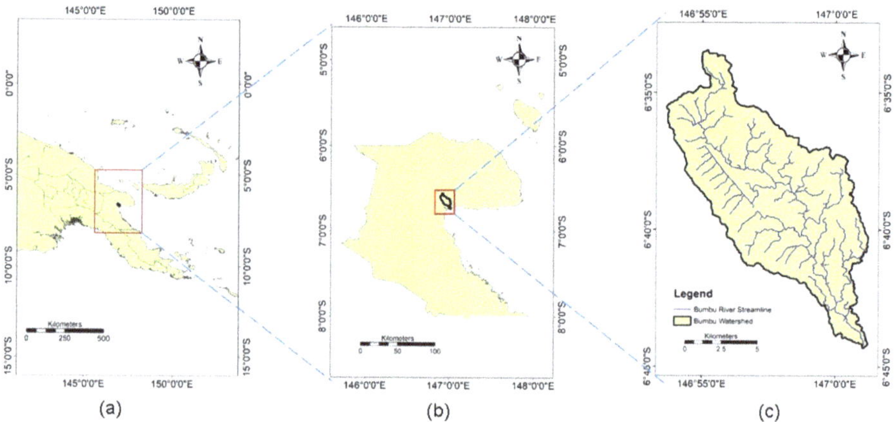

Figure 1. Depiction of the study area on different scales: (**a**) Papua New Guinea (PNG), (**b**) Morobe Province, (**c**) Bumbu basin.

In total, 22 water sampling points spread across the Bumbu river basin were chosen for this research as shown in Table 1. These sites lie at different locations and elevations, with varying levels of vegetation and urbanization. Water samples were collected at these points for further research analysis, and their respective position coordinates were captured with the help of GPS. The position of these sampling sites with respect to the Bumbu river basin are represented in Figure 2. The sampling points on the Bumbu river are divided into three main categories, namely Bumbu main channel, left hand Bumbu stream and right hand Bumbu stream sampling points. The Station IDs of these points belong to UA, UB and UC series, respectively. The captured GPS details can be found in Appendix A, Tables A1–A3.

Table 1. Water sampling points and the names of the respective sites.

Sampling Site Number	Sampling Site ID	Sampling Site Name	Sampling Site Number	Sampling Site ID	Sampling Site Name
1	UA1	Bumbu Upstream	12	UB4	Igam Creek
2	UA2	Bumbu Trench	13	UB5	Butu Stream
3	UA3	Bumbu main	14	UB6	Sukos
4	UA4	CIS Bridge	15	UB7	Butibam 1
5	UA5	Kamkumu Bridge	16	UB8	Butibam 2
6	UA6	Cassowary Road	17	UC1	Ambiun 1
7	UA7	Butibam Main	18	UC2	Ambiun 2
8	UA8	Bumbu Downstream	19	UC3	Wara Rice
9	UB1	Irom	20	UC4	Wara Misin
10	UB2	Wombong	21	UC5	Sopwara
11	UB3	Wongkos	22	UC6	Sikambu Creek

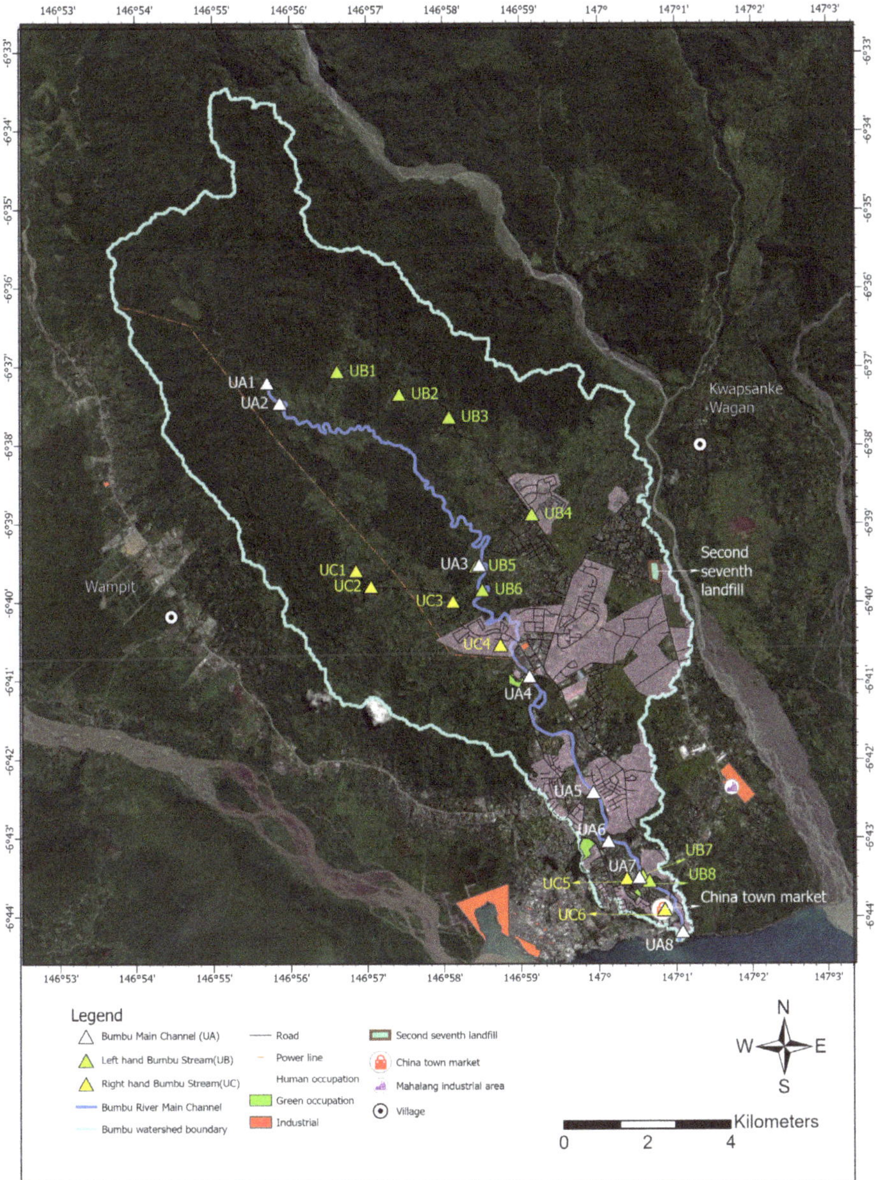

Figure 2. Position of 22 water sampling sites, Bumbu river main channel and other relevant sites of interest in the Bumbu Watershed.

2.2. Overview of the Protocol

Multiple geographically distributed factors such as rainfall, geophysical and geochemical conditions, human and animal populations, residential and commercial WASH conditions and practices, and general landscape use and conditions exist in the watershed [4–11,17–20]. These together with roads, habitations and forests potentially impact water quality and represent the social, economic and environmental (SEE) factors present. To analyze and understand the impact of these factors on water quality, it is imperative that these factors are assessed appropriately at water quality (WQ) sampling stations throughout the watershed. In practice, for almost all social, economic and environmental (SEE) variables, direct measurement of their impacts is practically impossible. Consequently, it is essential to incorporate into the analysis a systematic and consistent spatial interpolation protocol for estimating the accumulation of these factors at each Water Quality sampling station. The data for the factors taken into consideration for such studies usually exist in spatial formats such as raster layers and vectors shapefiles. Vector GIS layers can be point-, line- or polygon-based. Vector features are carried within a shapefile using geographic latitude and longitude coordinates to define points, lines and edges of geometric shapes. Individual survey points lend themselves to point characterization, while boundaries lend themselves to line characterization. Buildings and other surface structures lend themselves to polygonal characterization. Rivers and streams can be characterized by lines or polygons, but the relevant hydrologic information can be conveyed by line vectors.

As a result of varying datasets, we need to apply different techniques to extract appropriate runoff information for distinct factors involved based on how each factor can be described by line, point and raster-based GIS layers. Due to lack of WASH-related data at the time of this exploratory study, we limited the application of the protocol to the measurement of the influence of runoff of anthropogenic and environmental factors. The data related to WASH parameters and other SEE variables are usually gathered by community surveys of households, and hence are restrained to estimation at point sources. In Section 2.5, we explain how we can study point sources of information by considering the example of rainfall data in this format and demonstrating how we make use of spatial interpolation techniques. Figure 3 depicts the generalized process of runoff extraction using the different environmental and anthropogenic factors formatted in point-, line- and raster-based GIS layers. The resulting outputs include Flow Runoff, Road Runoff, Dense Forest Runoff, Green Space Runoff, Highly Urban Runoff, Habitation Runoff, Semi-Urban Runoff and Rainfall Runoff. The categorization based on environmental and anthropogenic factors is also shown. The procedures involved in extracting raw runoff are more thoroughly explained in the upcoming Sections 2.3–2.6 and procedures for compiling relative importance for the factors are discussed in Section 3.1.

2.3. Line Vector-Based GIS Layers

This section explains the methodology to analyze factors which have line vector characteristics. In this context, we delineated the stream drainage network and extracted the pattern and associated values of road network runoff.

2.3.1. Dataset

The first stage of a drainage-based analysis is to procure a Digital Elevation Model (DEM) of the study area. As currently available stream data for the study area are insufficient to directly yield the required stream drainage information, it was determined that a DEM or Digital Surface Model (DSM) of high resolution would be required to achieve the project objectives. In this initial exploratory work, a 1 arc-second DEM supplied by USGS [21] with 30 m resolution was utilized. Efforts to procure a DEM of higher resolution, though preferred, have proven difficult to obtain. The Shuttle Radar Topography Mission (SRTM) starting in February 2000 generated 1 arc second DEMs with 30 m horizontal resolution ranging from 56° south latitude to 60° North Latitude in gridded files encompassing 1° longitude by 1° latitude. The Shuttle Radar Topography Mission (SRTM) dataset has been shown to provide more accurate modelling over other datasets [22]. In many cases, a study area will include several of these files. In this study, the Lae region of Morobe Province is covered by two DEM files (S06E146.hgt.zip and S06E147.hgt.zip). After expansion and importation of the available band interleaved (BIL) formatted files into Idrisi formatted raster files (RST) in the GIS, the two rasters were merged into a single DEM. Visual examination of Google aerial photography of the study area around Lae suggested a suitable window of the expansive DEM to create a more manageable raster for further analysis.

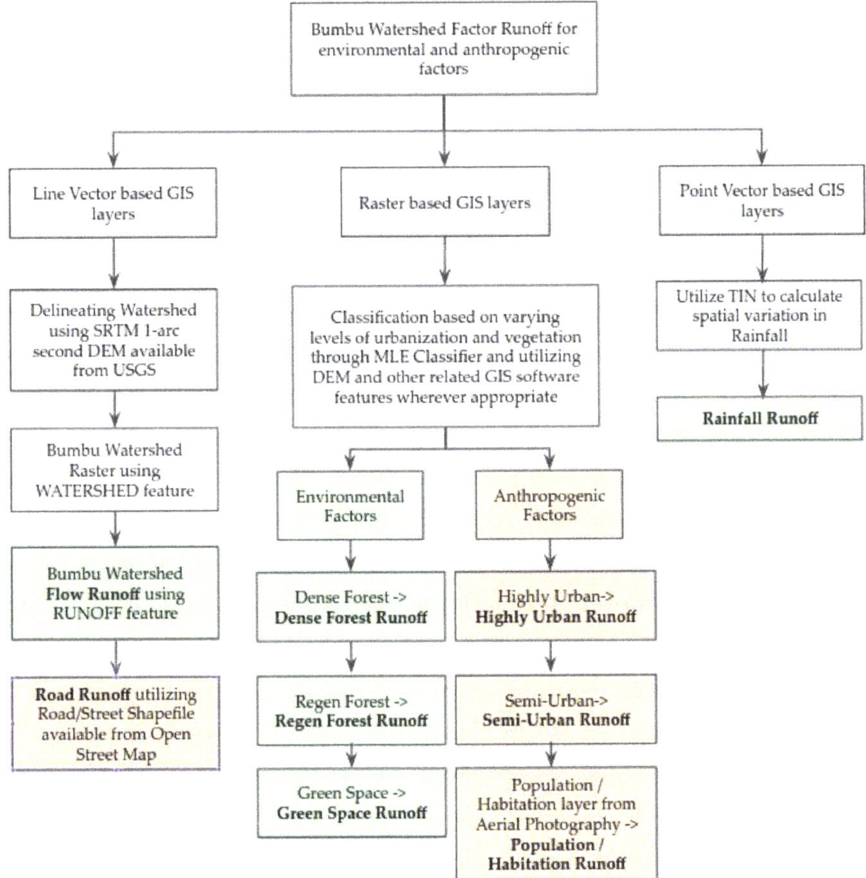

Figure 3. Flow chart depicting different anthropogenic and environmental factors influencing runoff and their characterization based on point-, line- and raster-based GIS layers.

2.3.2. Implementation

Runoff is among the components of the water cycle as surface water flows overland instead of infiltrating into the ground or evaporating. The GIS RUNOFF procedure has the ability to directly measure the amount of catchment of every pixel in a grid scene under certain assumptions of the distribution of rainfall over the study area. As a result, we utilized the "WATERSHED" and "RUNOFF" procedures of the IDRISI analysis package of TerrSet Software2 [23]. A critical element of watershed delineation is the location and vectorization of points at the lowest elevation of the watershed—i.e., at the mouth of the stream that has gathered all the streams of the catchment area into a single channel just before it empties into a larger body of water such as the sea or a more major river. In the case of the Bumbu, this point, also known as the pour point, lies at the entrance of the Bumbu Stream onto the Huon Gulf. The watershed delineation can be sensitive to the seed image provided as the lower extremity of the watershed. The CONTOUR function under "feature extraction" was found to be helpful in locating a proper seed image. Next, the steps involved for a drainage analysis based on the line vector characterization of stream data are summarized as follows:

- Obtain the SRTM 1 arc second DEM supplied by USGS with 3 0 m resolution and window the DEM to the study area. Using the DEM and the appropriate WATERSHED function of available

GIS software, delineate the [WATERSHED] raster (See Figure 4a) and convert the raster to a watershed vector polygon (Figure 4b).

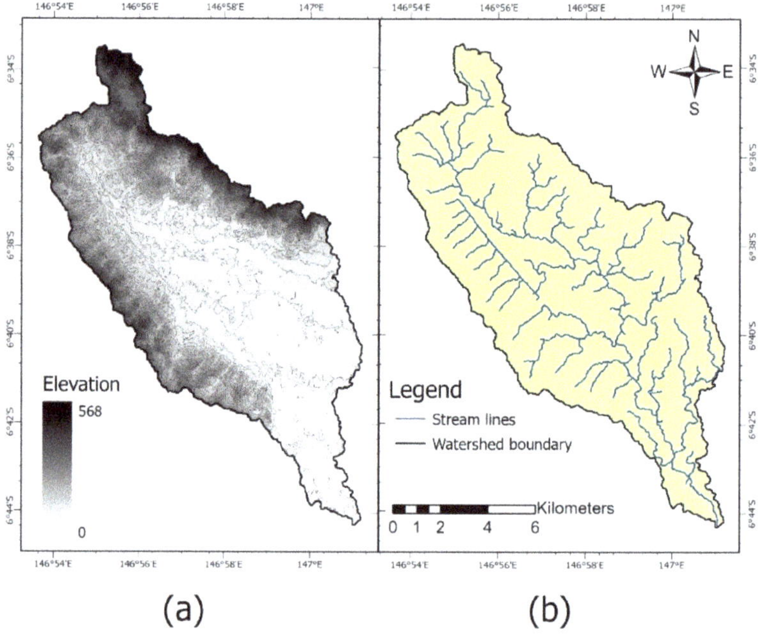

Figure 4. Bumbu Watershed Polygon overlain with (**a**) 20 m elevation contour lines and (**b**) major stream lines.

- Using the DEM bounded by the watershed polygon, apply the "RUNOFF" feature of applicable GIS Software, delineate the raster of stream channels in the watershed and reformat the raster into stream line vectors with the individual stream catchment area as each vector's feature value. Retain this raster layer and shapefile as a relative measure of the [FLOW RUNOFF] traversing each pixel in the catchment area. A basic assumption of this step is that overall, and on average, precipitation, infiltration and absorption are spatially and temporally uniform across the watershed. This is a first order assumption. Extension of the protocol to incorporate spatially variable precipitation will be considered in Section 2.5 below to address some of the shortcomings in this assumption.

- Acquire a shapefile of roads and streets in the project area from Open Street Map [24], convert the road shapefile to a raster format on a blank raster of the same location and dimensions as the DEM. RECLASS all non-zero road pixels as 1 on a 0 background.

- Using the elevation [DEM] and the "WATERSHED" function of the GIS software, with the road raster overlain as the precipitation image, again collect runoff of the catchment area. This can be retained as the [ROAD RUNOFF] layer. Again, the assumption is that all roads have an equal pollution potential per pixel.

- Utilizing the shapefile of water quality sampling stations, again create a blank raster of the same dimensions and location as the DEM. Reformat the blank by projecting the sampling station points onto this raster. Save this raster as the [SAMPLING POINT] raster.

- By overlaying the [SAMPLING POINT] raster onto the [ROAD RUNOFF] raster, a road runoff value can be assigned to each water sampling point and saved in an attribute values file for later incorporation into further analyses along with other sampling station results.

It is important to note that runoff units are all in pixels where each pixel is equivalent to 977.21 m^2. The assumption is that one unit of precipitation falls on each pixel unit of the watershed, or in the case of categories, on each pixel unit of the category. In the case of differential precipitation, fractional precipitation is assumed to fall on each pixel. The runoff procedure accumulates pixels into the stream network as a proxy for actual precipitation.

2.4. Raster Based GIS Layers

Similar techniques can be applied to capture the relative potential of other layers to correlate with water sampling station results. This section elaborates on how we rely on aerial photography and/or satellite imagery to create vegetative density and urbanization layers for similar runoff extraction at sampling stations. High-definition aerial photography and satellite imagery of the study area are available from USGS Earth Explorer [21]. Categorization of landscape elements in imagery can be accomplished using various techniques of cluster and classification analysis. In the aerial photography shown in Figure 5, it can be readily seen that the study area varies from what appears to be a mature virgin forest to a highly industrialized urban environment. A study by Doaemo et al. [25] revealed that Bumbu Watershed has undergone extensive deforestation and an increase in urbanization in the last 33 years (1987–2020). In this instance, we settled on five arbitrary but intuitively selected categories of (1) dense forest, (2) regen (regenerating) forest, (3) green space (4) semi-urban and (5) highly urban environments as relevant to the water quality study. The land-use types are largely self-explanatory with the exception of "green space". This land-use category arose as a result of aerial photo interpretation of the landscape. "Green Space" characterization was designed to differentiate between land primarily characterized by vegetation in various stages of tree growth (mature and regenerating forest) and non-vegetated land (designated urban classes). Close inspection of these vegetated areas in aerial photography revealed extensive garden cultivation of otherwise vacant land. The proximity of these garden plots to highly urban and semi-urban areas suggests these areas are used extensively for food production.

Prototypes for the various groups were envisioned. Sampling of the prototypes was accomplished by identifying points in areas assumed to be prime examples of the proposed classes. We identified thirty sample points per class, and the sample points were saved as a shapefile and then rasterized on a raster of the same dimension and location as DEM. Sample points were expanded to rectangles of 5 by 5 pixels covering approximately 2.7 hectares each and converted back to vector shapefile polygons. The distribution of signature polygons is shown in Figure 5.

These polygons constitute the sampling areas to be superimposed on the satellite imagery for the development of class signature profiles. In this study, signature profiles were developed by sampling the individual color bands of the satellite imagery of Hansen et al. (2013) [26]. The profiles/signatures that were developed were then used to hard classify the entire study area using maximum likelihood estimation (MLE) for final classification of the watershed as shown in Figure 6.

Each land-use class was individually coded as a categorical variable and mapped as a separate layer. As discussed in the example of Section 2.3.2, a Roads and Streets line vector shapefile available from Open Street Map [24] was similarly dummy coded and transformed into a categorical gridded map layer. Subsequently, a separate Population/Habitation layer was extracted from aerial photography by filtering pixels exhibiting high reflectance values >90 for all three RGB bands. The high reflectance was assumed to be the sun's reflection from metal rooftops. This layer was deemed advisable as a secondary measure of human habitation and human activity that might be missed by other urban classifications. Results of the grid transformations and categorizations are shown in Figure 7a,b.

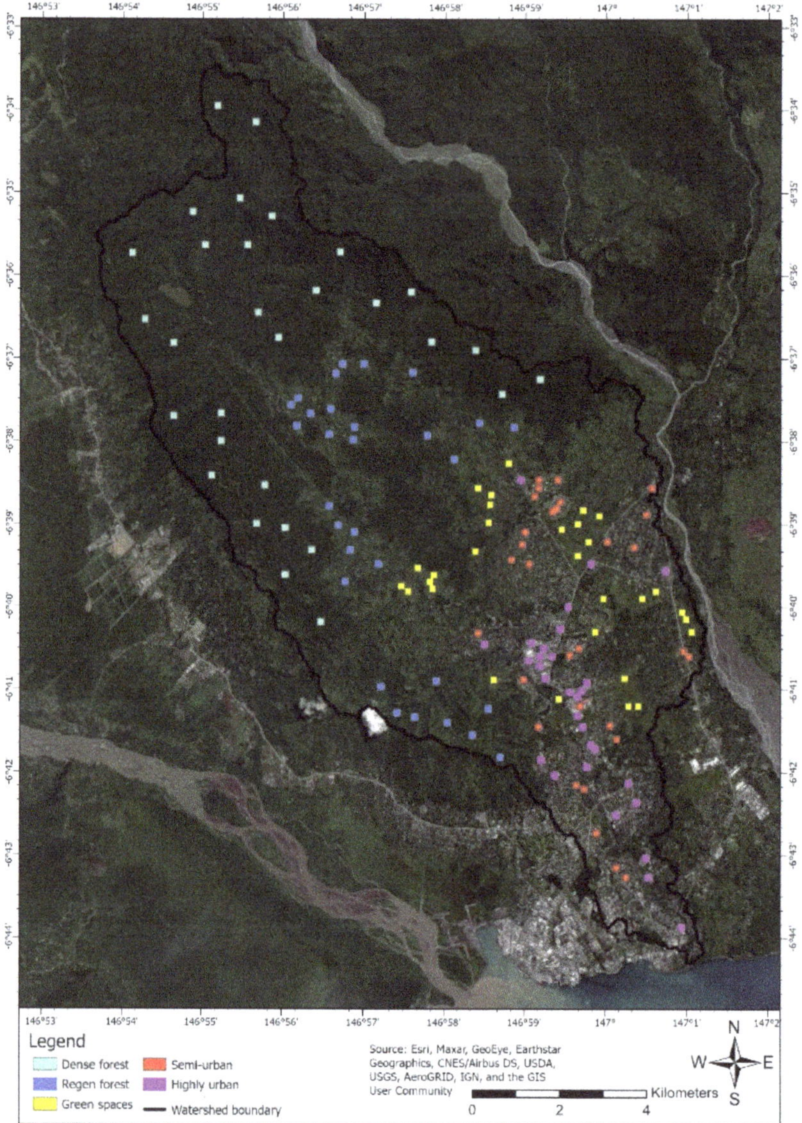

Figure 5. Signature polygons used for categorization of the Bumbu Watershed. In total, 30 sample points for each class were identified to implement the MLE clustering technique.

After categorization of the watershed, multi-category rasters are converted into individual single category feature rasters coded 1/0. The roads and population/habitation rasters are similarly recoded 1/0. The procedure from this point follows the same procedure described in previous section for roads. We utilized the DEM to accumulate [CLASS# RUNOFF] for each class. Next, by overlaying the [SAMPLING POINT] raster onto each [CLASS# RUNOFF] raster, a class# runoff value is assigned to each sampling point and saved in an attribute values file for later incorporation into correlational analyses along with other sampling station results. It is again useful to convert the rasters to [CLASS# RUNOFF] line vector shapefiles and point vector shapefiles for graphic presentation of results.

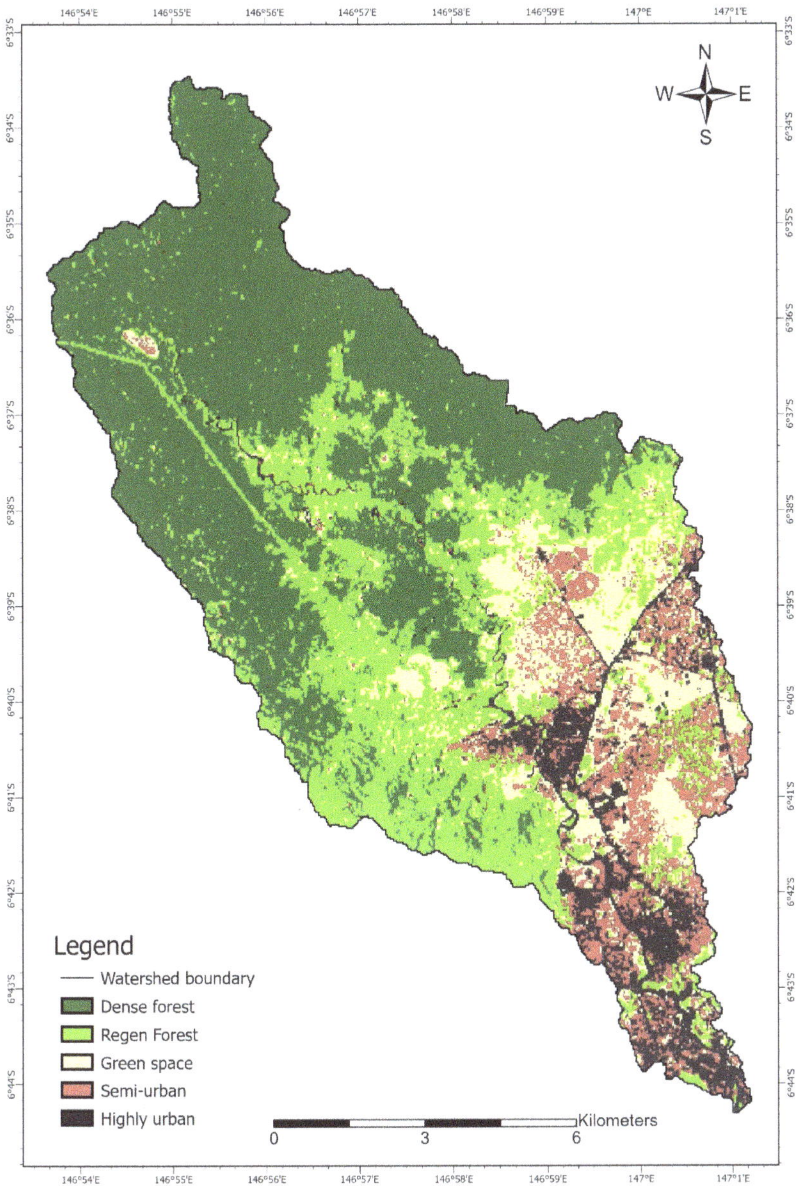

Figure 6. Categorization of the Bumbu Watershed in land use categories 1 to 5, including (1) Dense Forest, (2) Regen Forest, (3) Green Space, (4) Semi-urban and (5) Highly Urban.

2.5. Point Vector Based GIS Layers

A third scenario considers the case where only point estimates of important socio-economic or environmental variables are available. Such is the case for rainfall and other weather-related variables measured at individual sampling stations. Thus, far into the development of a protocol, only spatially and temporally uniform rainfall across the watershed was assumed. In reality, rainfall varies spatially and temporally. Such data require interpolation to landscape coverage for analysis using the methods described in Section 2 above for aerial and satellite imagery. A spatially diverse rainfall pattern is a good example for general application. Unfortunately for this study, only sparse rainfall weather station data are available for the Bumbu Watershed. Given the sparseness of available data, estimates of the spatial pattern of rainfall for the Lae area resulted in a rainfall mapping with substantial uncertainties associated with the estimated spatial pattern. For the purposes of protocol development, these large uncertainties in the estimates of the spatial and temporal distribution of rainfall will be ignored.

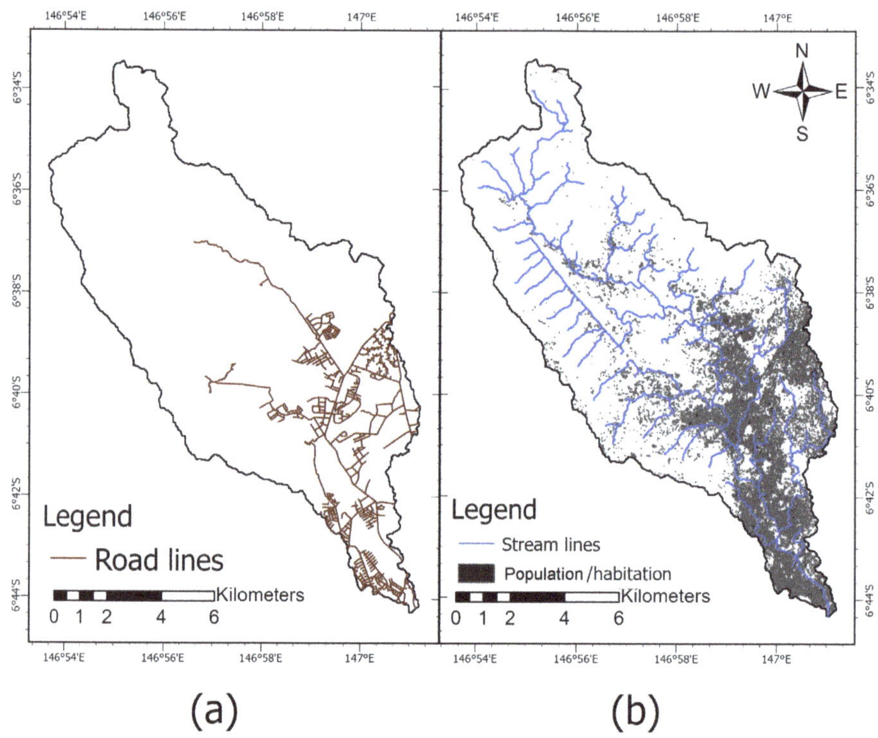

Figure 7. Shapefile input layers and result of grid transformation (**a**) road line vectors and (**b**) population/habitation polygons.

Spatial interpolation is a well-researched field of geology where point samples of geologic formations must be used for interpolation and reliable estimation of the amount and value of mineral resources. Methods of spatial interpolation include Triangulated Irregular Network (TIN) and Kriging. The sparseness of our sample points failed to satisfy Kriging requirements for a sufficient number of sample points to estimate spatial autocorrelation. Thus, for the purposes of this study, less demanding TIN methods were employed. The current study confines itself to consideration of the spatial variation of average annual rainfall. At present, current weather station data are too sparse to reliably estimate the spatial variation of rainfall across the Bumbu Watershed. Historically the situation is slightly better. McAlpine et al. (1975) [27] reported results of a 15-year study at 600 weather stations across mainland PNG and the islands. Though the McAlpine data are out of date and climate patterns are changing, the McAlpine data represent the best current available estimate of the pattern of the spatial variation of rainfall of the study area, even if absolute amounts of annual rainfall have changed.

TIN network and TIN surface estimation are a standard feature of GIS packages. Thirteen weather stations in the vicinity of Lae from the McAlpine study as shown in Figure 8a were used for this current study. Using the 13 McAlpine point estimates of average annual rainfall 1957–1972 represented in Figure 8a, TIN and TIN surfaces were compiled of the estimated pattern of spatial variation across the Bumbu Watershed study area as illustrated in Figure 8b. The rainfall surface was generated onto a grid, compatible in location and resolution with the watershed DEM used for previous RUNOFF analyses. It is convenient to convert the raw rainfall into a grid coded 0 to 1 as the fraction of maximum expected annual rainfall across the watershed. The scaled RAINFALL surface and contours are shown in Figure 8b. Other examples of spatially and temporally distributed variables, measured or estimated by point sources, are results of geo-located population surveys of water, sanitation and hygiene (WASH) practices. In upcoming studies, MDF will undertake community surveys for these variables in proximity to the same 22 water sampling points of this study of the Bumbu Watershed in order to study their relation to runoff water quality.

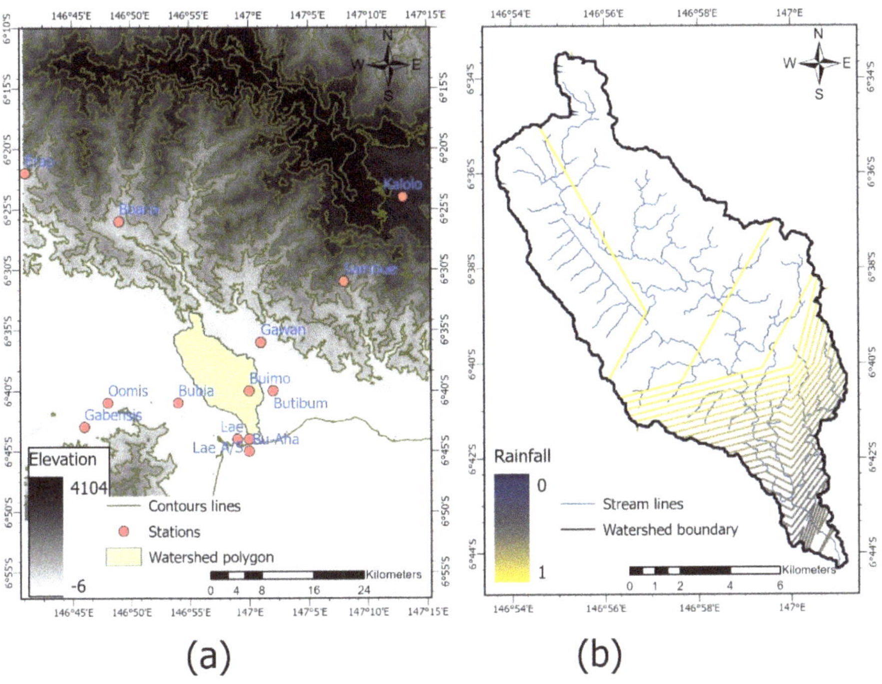

Figure 8. Rainfall analysis data sites and results: (**a**) 13 weather station locations from McAlpine (1975) used to model average annual rainfall pattern across the Bumbu Watershed and (**b**) average annual rainfall contours as estimated by Triangulated Irregular Network (TIN) scaled 1 to 100% of maximum.

2.6. Observed Limitations and Rectifications

No major impediments to using the protocol appear to exist except in the limitations of data as explained further below. The value of the protocol will emerge with application to correlation analysis of water quality measurements with the derived inputs. At this time, the imprecision of the DEM necessitated the estimation of the locations of some sampling stations on the derived stream lines. In the cases where there was a discrepancy between derived stream lines and actual streams, estimates were made of the location of sampling station points of equivalent hydrologic position. The uncertainties created by this process are unknown at this point. In Figure 9, below, the differences in positions of the actual 22 sampling sites versus their estimated "equivalent" hydrologic positions on the DEM are shown. The sites numbered 3, 12, 13, 14 and 18 required the greatest adjustment and are highlighted below.

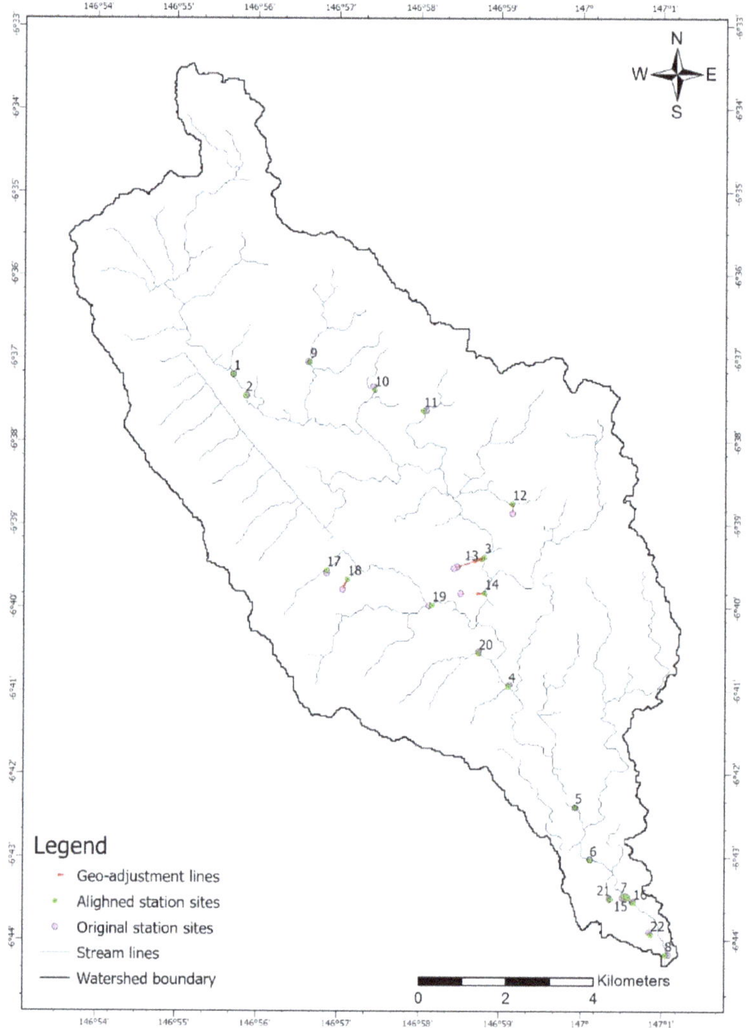

Figure 9. Geo-adjustment of sample site locations to coincide with imprecise Digital Elevation Model (DEM) stream delineation. Points of largest deviation from derived points are highlighted.

In practice, it was found that high resolution aerial photography down to 1 meter pixel resolution as portrayed on the USGS Earth Explorer [21] was superior to satellite imagery for identifying appropriate locations for signature polygons. These images are not always available for download but can nevertheless be used for informal geo-location. Landsat imagery proved superior for signature definition and more exact geo-location. There was a tendency for satellite imagery to incorrectly identify streambeds and water bodies as "highly urban" and "semi urban". Attempts to create a sixth signature and category for water were unsuccessful, but masking out of the stream layer obtained by DEM analysis partially compensated for this shortcoming. The results presented are based on the use of the 5 land-use categories for categorization of 4-band satellite imagery as compiled by Hansen et al. (2013) [25]. Refinement for any application can explore what is most appropriate in that specific study scene. For the purposes of this protocol development, no "ground truthing" other than by aerial photography verification of the categorizations was performed.

3. Results

3.1. Relative Importance of Factors

The concluding step in creating layers for incorporation into correlational analysis is the compilation of "Relative Importance Value" of the RUNOFF layers. The negative (and/or positive) influences of the runoff of SEE factors are diluted by the amount of water flowing through the same sample site. As mentioned before, runoff units are all in pixels where each pixel is equivalent to 977.21 m². The assumption is that one unit of precipitation falls on each pixel unit of the watershed, or in the case of categories, on each pixel unit of the category. After compilation of runoff, factor importance values are found by dividing the factor runoff by the rainfall runoff and multiplying by 100. The formal equation for this calculation can be given by:

$$I_f = \frac{F_f}{R_f} \times 100 \tag{1}$$

where

$I_f =$ Importance values of Factor f

$F_f =$ Factor f runoff

$R_f =$ Normalized Rain runoff

Figure 10 represents the runoff extraction protocol for different SEE factors based on point vector-, line vector- and raster-based GIS layers. Given that rainfall over the study area varies from 2751 mm per year to 4648 mm per year, the normalized rainfall surface as a percentage of maximum varies from 0.592 to 1.000, and the effect of transformation due to the rainfall pattern can have a significant effect on importance values of the runoff of anthropogenic and environmental variables. As a result of the rainfall variation in the Bumbu Watershed, the importance values fall in the range of 0 to 200. Figures 11 and 12 present the spatial distribution of the relative importance values of the factors involved.

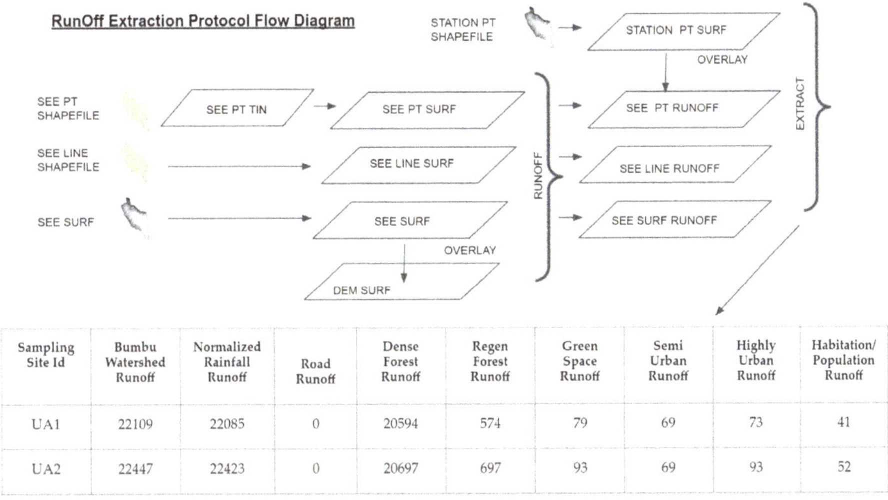

Sampling Site Id	Bumbu Watershed Runoff	Normalized Rainfall Runoff	Road Runoff	Dense Forest Runoff	Regen Forest Runoff	Green Space Runoff	Semi Urban Runoff	Highly Urban Runoff	Habitation/ Population Runoff
UA1	22109	22085	0	20594	574	79	69	73	41
UA2	22447	22423	0	20697	697	93	69	93	52

Figure 10. Flow diagram to represent the procedure to extract runoff values from different GIS layers.

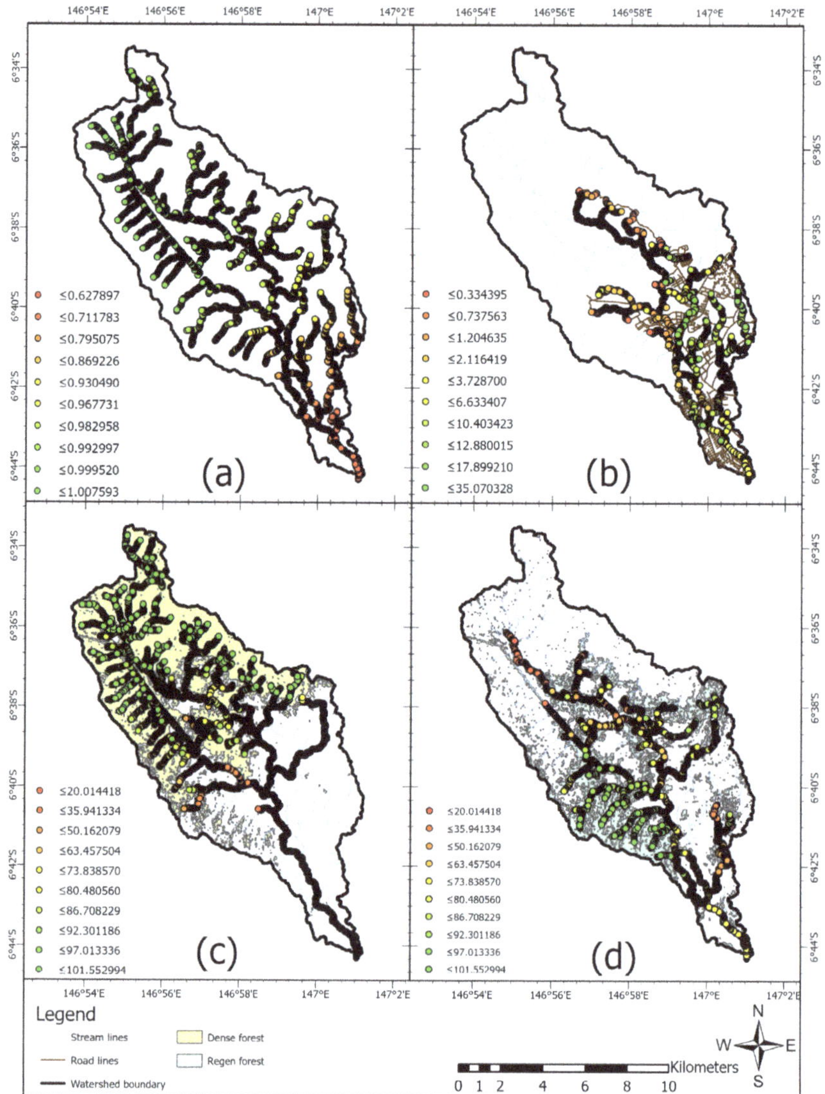

Figure 11. Distribution of (**a**) normalized rainfall pattern, (**b**) ROAD RUNOFF importance values (IV) scaled 1 to 200 overlain on roads/streets layer, (**c**) DENSE FOREST RUNOFF importance values (IV) scaled 1 to 200 overlain on dense forest polygons and (**d**) REGEN FOREST RUNOFF importance values (IV) scaled 1 to 200 overlain on regen forest polygons.

In practice, it is more practical to extract RUNOFF values from all layers in one step and to perform the "importance value" calculation in a spreadsheet. As mentioned earlier, a [SAMPLING SITE] layer was created by projecting the sampling site locations onto a raster with the same dimensions and location as the SRTM 1 arc second DEM supplied by USGS [21]. By overlaying the sampling site layer on a raster group consisting of [NORMALIZED RAINFALL], [ROAD RUNOFF], [DENSE FOREST], [REGEN FOREST], [GREEN SPACE], [SEMI URBAN], [HIGHLY URBAN] and [HABITATION/POPULATION], all the variables are captured in a spreadsheet in a single step. Tables 2 and 3 below report the results of this capture and importance value computation.

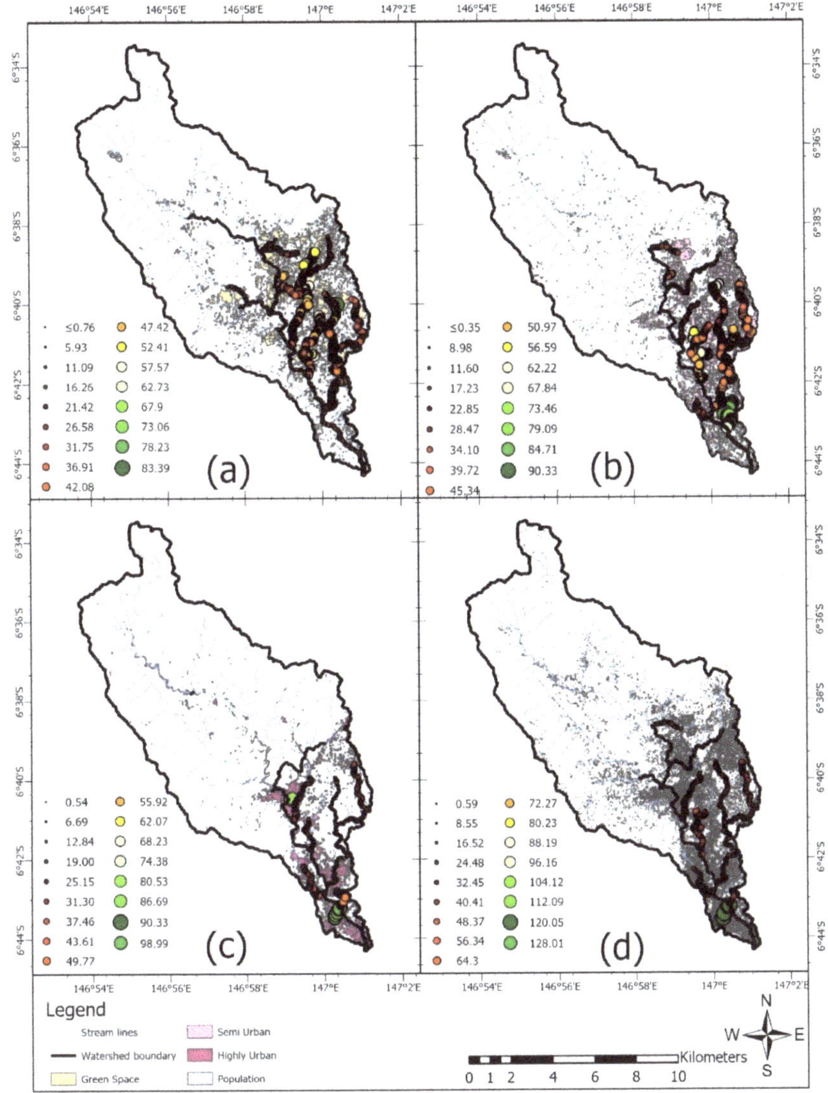

Figure 12. Distribution of RUNOFF importance values (IV) scaled 1 to 200 for (**a**) GREEN SPACE overlain on green space polygons, (**b**) SEMI-URBAN overlain on semi-urban polygons, (**c**) HIGHLY URBAN overlain on highly urban polygons and (**d**) POPULATION/HABITATION overlain on habitation/population layers.

Table 2. Raw runoff values for different factors at 22 WQ sampling sites in the Bumbu Watershed.

Site ID	Bumbu Watershed	Normalized Rainfall	Road	Dense Forest	Regen Forest	Green Space	Semi Urban	Highly Urban	Habitation/ Population
UA1	22,109	22,085	0	20,594	574	79	69	73	41
UA2	22,447	22,423	0	20,697	697	93	69	93	52
UA3	72,833	72,472	429	51,740	11,605	2985	1240	486	1269
UA4	101,605	99,732	1037	55,165	24,791	7930	3238	2042	3808
UA5	110,549	107,183	1735	55,183	25,691	10,324	5838	4088	6634
UA6	123,749	117,405	2771	55,233	27,686	13,356	10,146	6458	9749
UA7	126,752	119,108	3009	55,235	27,981	13,585	11,122	7557	10,825
UA8	128,354	120,073	3209	55,235	28,118	13,667	11,464	8395	11,550
UB1	5358	5314	3	4633	434	10	5	2	16
UB2	1856	1837	14	1317	382	15	9	0	14
UB3	2727	2691	7	2533	119	0	0	0	0
UB4	1148	1114	115	2	458	262	265	10	205
UB5	725	708	2	162	233	165	19	31	30
UB6	80,794	80,154	639	52,277	13,628	5937	2350	752	2194
UB7	13	8	0	0	3	2	5	2	1
UB8	67	41	0	0	15	16	13	10	10
UC1	1427	1426	0	681	530	8	4	1	12
UC2	359	357	13	27	296	8	3	0	15
UC3	11,479	11,193	103	2349	6601	1092	46	23	346
UC4	3501	3044	6	286	2372	174	112	29	124
UC5	123	67	27	0	0	1	35	76	62
UC6	79	47	29	0	0	0	8	66	60

Table 3. Relative importance values (IV) for different factor runoff at 22 WQ sampling sites in the Bumbu Watershed.

Site ID	Road Runoff IV	Dense Forest Runoff IV	Regen Forest Runoff IV	Green Space Runoff IV	Semi Urban Runoff IV	Highly Urban Runoff IV	Habitation/ Population Runoff IV
UA1	0.00	93.25	2.60	0.36	0.31	0.33	0.19
UA2	0.00	92.30	3.11	0.41	0.31	0.41	0.23
UA3	0.59	71.39	16.01	4.12	1.71	0.67	1.75
UA4	1.04	55.31	24.86	7.95	3.25	2.05	3.82
UA5	1.62	51.48	23.97	9.63	5.45	3.81	6.19
UA6	2.36	47.04	23.58	11.38	8.64	5.50	8.30
UA7	2.53	46.37	23.49	11.41	9.34	6.34	9.09
UA8	2.67	46.00	23.42	11.38	9.55	6.99	9.62
UB1	0.06	87.18	8.17	0.19	0.09	0.04	0.30
UB2	0.76	71.69	20.79	0.82	0.49	0.00	0.76
UB3	0.26	94.11	4.42	0.00	0.00	0.00	0.00
UB4	10.32	0.18	41.12	23.52	23.79	0.90	18.40
UB5	0.28	22.87	32.89	23.29	2.68	4.38	4.23
UB6	0.80	65.22	17.00	7.41	2.93	0.94	2.74
UB7	0.00	0.00	37.22	24.82	62.04	24.82	12.41
UB8	0.00	0.00	36.35	38.78	31.51	24.24	24.24
UC1	0.00	47.74	37.16	0.56	0.28	0.07	0.84
UC2	3.64	7.56	82.91	2.24	0.84	0.00	4.20
UC3	0.92	20.99	58.97	9.76	0.41	0.21	3.09
UC4	0.20	9.39	77.91	5.72	3.68	0.95	4.07
UC5	40.36	0.00	0.00	1.49	52.32	113.62	92.69
UC6	61.18	0.00	0.00	0.00	16.88	139.24	126.58

Table 2 represents the raw runoff at the 22 water sampling sites across the Bumbu Watershed. Herein, it becomes necessary that the relative impact of each factor is understood in the context of surface water's travel history and the varying level of influence each factor can have on runoff as it moves downstream. In most cases, it is not feasible to directly measure the exact quantity and potential impact of the anthropogenic and environmental inputs at these sites where water quality is measured. It is important to note that the potential impacts of the different factors are present in the form of several physical and chemical constituents carried by the Bumbu river. As precipitation moves across the landscape and downstream to the sea, it accumulates and transports the potential impacts of these factors in the runoff surface water and in the percolating ground water. This potential impact of various factors is diluted by the amount of water involved in their transport. The best that we can do is to infer their relative presence and concentration at the 22 spatially distributed water quality sampling stations and attempt to assess the correlation of their relative presence with accepted measures of water quality. In this regard, we utilized normalized rainfall runoff values to calculate the impact of these different factors on surface water which are enumerated in Table 3 in the form of relative importance values. The 22 water sampling sites are located across different geographical landscapes in both upstream and downstream regions of the watershed. As mentioned before, these landscapes are classified based on land use/land cover into various categories characteristic of varying levels of vegetation and urbanization. The water sampling sites of the main channel of Bumbu river are represented by the UA series. It can be observed that the class "Dense Forest" has high importance values of 93.25 and 92.30 for the sampling points UA1 and UA2 which are located in the upstream portion of the basin. As one moves the downstream (UA3–UA8), the impact of Dense Forest class is found to be continuously decreasing, accounted by a slight increase in values of other classes, namely regenerating forest, habitation and green space. The impact of the vegetation classes (Dense Forest, Regen Forest and Green Space) is usually present in surface water in the form of minerals, metallic ions, salts and several other geochemical and natural organic matter inputs.

In our future studies, we intend to verify whether the decrease in the values of the vegetation classes correlates with a simultaneous decrease in the measurement of various geochemical inputs at the water sampling sites. Road Runoff and Highly Urban runoff values are found to be insignificant when compared with other classes for the UA series. The sampling points of left hand Bumbu feeder streams are represented by the UB series, whereas those of right hand Bumbu feeder streams are represented by the UC series. Road runoff influence is negligible for all the points in the series except for UB4, UC5 and UC6 which lie in the close vicinity of residential/urban areas. The class "Green space" has a considerable influence on runoff with values 23.52, 23.29, 24.82 and 38.78 for the sampling points UB4, UB5, UB7 and UB8, respectively. This class depicts cultivated garden plots of otherwise vacant land adjacent to urban and semi-urban areas. Levels of turbidity, nitrates, nitrites and Total Phosphorous (TP) can also be examined for a possible relationship with importance values of the above classes. From the table, it can be discerned that UB1, UB2, UB3 and UB6 have high values of 87.18, 71.69, 94.11 and 65.22 for Dense Forest Runoff. The Regen Forest Runoff class has moderate relative importance values of 20.79, 41.12, 32.89, 37.22 and 36.35 for UB2, UB4, UB5, UB7 and UB8, respectively, whereas UC2, UC3 and UC4 have high values of 82.91, 58.97 and 77.91, respectively. UB7, UB8, UC5 and UC6 lie in the urban region of the city of Lae in the downstream area of the Bumbu Watershed and have high measurements for highly urban and semi-urban runoff values. The Bumbu river at these sites receives effluents from different industries located nearby, human detritus from residential areas, pollutants, waste discharge and other forms of organic matter (Figure 2) [14]. Some other physicochemical characteristics of water which can be inspected for correlation with importance values at these points for UA, UB and UC series include Total Suspended Solids (TSS), coliforms, alkalinity, pH and temperature. Thus, it can be seen that the Bumbu river experiences a wide range of influences from different landscape classes which can possibly influence its water quality as its surface water moves from the upstream regions to the downstream regions of the watershed. To accurately examine the relationship of the relative importance values with physicochemical characteristics, other social and economic parameters, a thorough correlation analysis is required, the prospects of which are further elaborated in the discussion section.

4. Discussion

The protocol developed in this study follows and expands upon the conceptual framework of Granger et al. [3] in identifying potential sources of impact on water quality, their mobilization by rainfall and their delivery across the landscape by a stream network. The methodologies presented in this study facilitate assessment of the relative importance/concentrations of different environmental and anthropogenic factors, enabling investigation of their potential impact on runoff water quality. Initially, the factors which we considered for this purpose include Road/Streets, Dense Forests, Regenerating Forests, Green Space, Highly Urban, Semi-Urban and Habitation/Population. This was followed by the application of different techniques and tools for each category characterized by line, point and raster-based GIS layers. The raw runoff values for the various factors were computed by utilizing DEM, spatial interpolation, classification methods and GIS software functionalities such as WATERSHED and RUNOFF features. The raw runoff values for each of the factors at the 22 water sampling sites are listed in Table 2. As stated in the previous section, we made use of normalized rainfall runoff values to determine the potential impact of the various factors. Figures 11 and 12 represent the spatial distribution of the relative importance values of the factors involved. Table 3 indicates the corresponding importance values at the 22 sampling sites. For instance, considering Sampling Site number 22, one can observe that the runoff importance is negligible from several factors indicative of varying vegetation density and is significantly greater from highly urban and semi-urban areas. This would imply a higher impact of urban areas on surface runoff and would indicate a greater concentration/impact of constituents contributed by urban factors. Similar observations can be made at other water sampling sites. One can observe that the runoff at each of the 22 sites involves considerable levels of varying influences from a variety of factors. This indicates the range of potential challenges possible to runoff quality in form of organic matter, coliforms, geochemical and natural ingredients to a name a few. However, to explore the exact nature of relationship existing between these factors and the waters of the watershed, a detailed correlation analysis is needed, which we discuss in more detail below.

In our upcoming studies, we plan to study how the runoff importance values for different factors are associated with the physical and chemical composition of the Bumbu waters by performing a detailed and thorough water quality study. This would include collection of water samples at the 22 sites followed by physicochemical analysis. Furthermore, a factor analysis/principal component analysis (FA/PCA) can also be performed as performed in earlier studies [28–31]. This would help us in understanding several relationships with respect to the waters of the Bumbu Watershed. For instance, whether Total Suspended Solids (TSS) and turbidity are correlated with surface runoff due to rainfall events, and to what extent and degree. Another example could be measuring the variation of DO, Electrical Conductivity (EC) and Thermotolerant Coliforms (TC), which are related to organic matter pollution. Some other examples would include evaluating how pH, alkalinity, temperature and metallic ions vary with different levels of vegetation and urbanization in the vicinity of the sampling points. Measuring the water quality index at the water sampling sites using available standards and guidelines such as those drafted by the Canadian Council of Ministers of the Environment (CCME) will also be a necessary step to comprehend how quality of water is related to the runoff importance values computed in this study [32]. We also plan to undertake a community-wide household survey to gather relevant WASH-related data such as sources of drinking water, toilet facilities, water storage and waste disposal methods used in proximity to our 22 water sampling sites. Additionally, we intend to gather community data based on parameters such as health, crime and pollution. Health-related parameters may include variables such as presence of stomach ailment, skin infections, HIV/AIDS and respiratory illness. Although crime and HIV may seem to be far-reaching candidates for correlation with water quality, the study's fundamental objective is to provide a method to explore if such non-intuitive correlations exist, and to provide a reasonable protocol for exploration, analysis and resolution of the complex questions they raise. As mentioned earlier, these types of data are constrained to be point sources of information and can be utilized easily by our protocol as implemented in the case of rainfall. All these inputs together with the anthropogenic and environmental factors form an array of socio-economic environmental (SEE) inputs. Finally, we intend to perform a correlation analysis to determine how the runoff importance values calculated from this study vary with accepted measures of water quality, e.g., the Canadian Ministers of the Environment Water Quality Index, different SEE factors as well as physicochemical parameters of water. Several multivariate analytical techniques such as Pearson and Spearman correlation, numerous regressions models and other statistical tools are available and have been used previously in this regard [4–9,28–31].

Our study is bound by data limitations such as the utilization of data collected by McAlpine et al. (1975) [27]. Although climate patterns are changing, McAlpine data represent the best current available estimate of rainfall patterns across Morobe Province to reliably perform spatial interpolation for the Bumbu basin. Another limitation is imposed by the utilization of a 30 m DEM despite our efforts to obtain a DEM of higher resolution for the watershed. Due to imprecision of the DEM, there is discrepancy between derived stream lines and observed streams, a shortcoming that we attempted to rectify as explained in Section 2.6. Despite the above limitations which we tried to overcome, the uniqueness of our approach lies in the fact that the protocol we developed not only takes into account various environmental and anthropogenic factors but also has the potential to accommodate the various socio-economic factors such as community and household health, crime and waste disposal.

5. Conclusions

Unrestricted urban expansion, burgeoning population and industrialization among various other anthropogenic factors have put significant stress on water resources all over the world [1]. In this context, the Morobe Development Foundation (MDF), a not-for-profit community-based organization located in Lae, Morobe Province, Papua New Guinea has undertaken research to understand the symbiotic relationship that exists between water systems of the Bumbu basin and the health and welfare of the resident communities. This study seeks to facilitate a capacity to link the quality of surface waters to multiple social, economic and environmental factors differentially and geographically distributed by tracing the travel history of surface waters through varying landscapes. These include urban, semi-urban, dense forests, green space and regenerating forests. For this purpose, we developed an analytical protocol to determine the potential impact the above-mentioned factors can have on surface water quality. This is essential for our future work, and a positive contribution to the study of water quality in general. It is worthwhile to mention that the protocol can be applied to many factors in addition to those mentioned in this research, based on its underlying point, line or raster-based characteristics. Although previous studies [4–11,18,20] consider a variety of factors, the uniqueness of the protocol is in its examination of the relationship between quality of water, physicochemical characteristics, WASH practices [17,19] and socio-economic parameters. The protocol will help us to understand the relationship that exists between the above parameters and the inhabitants of the Bumbu Watershed, a study that is the first of its kind for the region. Consequently, this will provide relevant insights to aid our upcoming projects which are aimed at addressing probable risks to human health created by poor water quality in the Bumbu basin [14].

We utilized a diversity of spatial analytical tools and techniques to responsibly analyze an array of environmental and anthropogenic data inputs possessing the potential to impact water quality. The study confirms the value of the conceptual framework put forward by Granger et al. [3] to identify potential sources of impact, their mobilization and their delivery to water systems. For our future studies, the practical value of the protocol will be tested in its ability to (a) interpret, interpolate and estimate appropriate values of diverse inputs of other factors associated with water quality as measured at WQI sampling stations; (b) provide the tools to estimate the spatial distribution of variables reported in household surveys; and (c) appropriately estimate the importance values of those survey parameters at the WQI sample sites. The protocol and its procedures described in this paper were generalized enough to be applicable in other geographical settings to determine the influence of runoff of a variety of factors, and to calculate the relative importance and correlation of these factors with other SEE parameters. Through this study, we intend to bring the protocol and its applications into the limelight, gather international attention of researchers and volunteers and simultaneously garner local support for our future work.

Author Contributions: Conceptualization, W.D. and L.W.; methodology, L.W.; software, L.W.; validation, L.W., W.D.; formal analysis, L.W.; investigation, W.D., L.W. and S.B.; resources, L.W. and W.D.; writing—original draft preparation, L.W., W.D. and S.B.; writing—review and editing, L.W., W.D., S.B., M.M. and W.S.W.M.J.; visualization, L.W. and S.B.; funding acquisition, M.M. and W.S.W.M.J.; supervision, W.D.; project administration, W.D. All authors have read and agreed to the published version of the manuscript.

Funding: The publication costs related to the research have been covered by the National University of Malaysia (UKM) research grant GUP-2018-132.

Acknowledgments: The research was supported and implemented with pro bono contributions by volunteers from the United Nations Online Volunteering Program. The authors would like to acknowledge Esmaeel Adrah (UN volunteer from Syria), John Kennedy Boafo (Geomatic Engineer from Ghana) and Roshan Padel (Lecturer, Acme Engineering College, Purbanchal University, Nepal) for providing GIS-related insights relevant to our research. The authors also thank Kamal Wahp (UN Volunteer from UAE) and Suwarna Shukla (Doctoral Scholar at Indian Institute of Management, Indore, India) for their suggestions and valuable edits related to the previous versions of this manuscript. The authors would also like to thank the three anonymous reviewers who helped to improve the quality of the manuscript. The authors also express their appreciation to Nancy Yang of the MDPI Remote Sensing team and to the members of the Institute of Climate Change, UKM, Malaysia.

Conflicts of Interest: The authors declare no conflict of interest.

Appendix A

Table A1. GPS survey of the 8 water sampling sites of UA series.

UA series water sampling sites									
GPS Model: Garmin GPSMAP64SC Surveyor ID: WD STD ID: 18800316									
Sampling Site No.	Sampling Site ID	Sampling Site Name	Latitude	Latitude Direction	Longitude	Longitude Direction	Elevation (m)	DEC Latitude	DEC Longitude
1	UAI	Bumbu Upstream	06° 37.192′	S	146° 55.704′	E	94	−6.6198667	146.9284000
2	UA2	Bumbu Trench	06° 37.453′	S	146° 55.867′	E	87	−6.6242167	146.9311167
3	UA3	Bumbu main	06° 39.508′	S	146° 58.465′	E	63	−6.6584667	146.9744167
4	UA4	CIS Bridge	06° 40.929′	S	146° 59.113′	E	45	−6.6821500	146.9852167
5	UA5	Kamkumu Bridge	06° 42.403′	S	146° 59.927′	E	24	−6.7067167	146.9987833
6	UA6	Cassowary road	06° 43.026′	S	147° 00.117′	E	15	−6.7171000	147.0019500
7	UA7	Butibam Main	06° 43.485′	S	147° 00.521′	E	9	−6.7247500	147.0086833
8	UA8	Bumbu Downstream	06° 44.176′	S	147° 01.078′	E	4	−6.7362667	147.0179667

Table A2. GPS survey of the 8 water sampling sites of UB series.

UB series water sampling sites									
GPS Model: Garmin GPSMAP64SC Surveyor ID: WD STD ID: 18800316									
Sampling Site No.	Sampling Site ID	Sampling Site Name	Latitude	Latitude Direction	Longitude	Longitude Direction	Elevation (m)	DEC Latitude	DEC Longitude
9	UB1	Irom	06° 37.043′	S	146° 56.618′	E	96	−6.6173833	146.9436333
10	UB2	Wombong	06° 37.340′	S	146° 57.424′	E	90	−6.6223333	146.9570667
11	UB3	Wongkos	06° 37.629′	S	146° 58.075′	E	90	−6.6271500	146.9679167
12	UB4	Igam Creek	06° 38.864′	S	146° 59.149′	E	80	−6.6477333	146.9858167
13	UB5	Butu stream	06° 39.528′	S	146° 58.423′	E	70	−6.6588000	146.9737167
14	UB6	Sukos	06° 39.829′	S	146° 58.505′	E	62	−6.6638167	146.9750833
15	UB7	Butibam 1	06° 43.468′	S	147° 00.567′	E	28	−6.7244667	147.0094500
16	UB8	Butibam 2	06° 43.531′	S	147° 00.654′	E	19	−6.7255167	147.0109000

Table A3. GPS survey of the 6 water sampling sites of UC series.

UC series water sampling sites									
GPS Model: Garmin GPSMAP64SC Surveyor ID: WD STD ID: 18800316									
Sampling Site No.	Sampling Site ID	Sampling Site Name	Latitude	Latitude Direction	Longitude	Longitude Direction	Elevation (m)	DEC Latitude	DEC Longitude
17	UC1	Ambiun 1	06° 39.583′	S	146° 56.857′	E	95	−6.6597167	146.9476167
18	UC2	Ambiun 2	06° 39.784′	S	146° 57.053′	E	85	−6.6630667	146.9508833
19	UC3	Wara Rice	06° 39.973′	S	146° 58.120′	E	72	−6.6662167	146.9686667
20	UC4	Wara Misin	06° 40.527′	S	146° 58.737′	E	54	−6.6754500	146.9789500
21	UC5	Sopwara	06° 43.498′	S	147° 00.359′	E	22	−6.7249667	147.0059833
22	UC6	Sikambu Creek	06° 43.896′	S	147° 00.849′	E	16	−6.7316000	147.0141500

References

1. Cosgrove, W.J.; Loucks, D.P. Water Management: Current and Future Challenges and Research Directions: Water Management Research Challenges. *Water Resour. Res.* **2015**, *51*, 4823–4839. [CrossRef]
2. Horwitz, P.; Finlayson, C.M. Wetlands as Settings for Human Health: Incorporating Ecosystem Services and Health Impact Assessment into Water Resource Management. *Bioscience* **2011**, *61*, 678–688. [CrossRef]
3. Granger, S.J.; Bol, R.; Anthony, S.; Owens, P.N.; White, S.M.; Haygarth, P.M. Towards a Holistic Classification of Diffuse Agricultural Water Pollution from Intensively Managed Grasslands on Heavy Soils. In *Advances in Agronomy*; Elsevier: San Diego, CA, USA, 2010; pp. 83–115.
4. Xu, G.; Ren, X.; Yang, Z.; Long, H.; Xiao, J. Influence of Landscape Structures on Water Quality at Multiple Temporal and Spatial Scales: A Case Study of Wujiang River Watershed in Guizhou. *Water* **2019**, *11*, 159. [CrossRef]
5. Lintern, A.; Webb, J.A.; Ryu, D.; Liu, S.; Bende-Michl, U.; Waters, D.; Leahy, P.; Wilson, P.; Western, A.W. Key Factors Influencing Differences in Stream Water Quality across Space: Key Factors Influencing Differences in Stream Water Quality across Space. *WIREs Water* **2018**, *5*, e1260. [CrossRef]
6. Xiao, R.; Wang, G.; Zhang, Q.; Zhang, Z. Multi-Scale Analysis of Relationship between Landscape Pattern and Urban River Water Quality in Different Seasons. *Sci. Rep.* **2016**, *6*, 25250. [CrossRef] [PubMed]
7. Putro, B.; Kjeldsen, T.R.; Hutchins, M.G.; Miller, J. An Empirical Investigation of Climate and Land-Use Effects on Water Quantity and Quality in Two Urbanising Catchments in the Southern United Kingdom. *Sci. Total Environ.* **2016**, *548–549*, 164–172. [CrossRef] [PubMed]
8. Lee, S.-W.; Hwang, S.-J.; Lee, S.-B.; Hwang, H.-S.; Sung, H.-C. Landscape Ecological Approach to the Relationships of Land Use Patterns in Watersheds to Water Quality Characteristics. *Landsc. Urban Plan.* **2009**, *92*, 80–89. [CrossRef]
9. Chen, X.; Zhou, W.; Pickett, S.; Li, W.; Han, L. Spatial-Temporal Variations of Water Quality and Its Relationship to Land Use and Land Cover in Beijing, China. *Int. J. Environ. Res. Public Health* **2016**, *13*, 449. [CrossRef] [PubMed]
10. Smart, R.P.; Soulsby, C.; Cresser, M.S.; Wade, A.J.; Townend, J.; Billett, M.F.; Langan, S. Riparian Zone Influence on Stream Water Chemistry at Different Spatial Scales: A GIS-Based Modelling Approach, an Example for the Dee, NE Scotland. *Sci. Total Environ.* **2001**, *280*, 173–193. [CrossRef]
11. Allan, J.D. Landscapes and Riverscapes: The Influence of Land Use on Stream Ecosystems. *Annu. Rev. Ecol. Evol. Syst.* **2004**, *35*, 257–284. [CrossRef]
12. Chang, Y.; Hou, K.; Li, X.; Zhang, Y.; Chen, P. Review of Land Use and Land Cover Change Research Progress. *IOP Conf. Ser. Earth Environ. Sci.* **2018**, *113*, 012087. [CrossRef]
13. Stagl, J.; Mayr, E.; Koch, H.; Hattermann, F.F.; Huang, S. Effects of Climate Change on the Hydrological Cycle in Central and Eastern Europe. In *Advances in Global Change Research*; Springer: Dordrecht, The Netherlands, 2014; pp. 31–43.
14. The National. Water Unsafe for Settlers along the Bumbu River. 2012. Available online: https://www.thenational.com.pg/water-unsafe-for-settlers-along-bumbu-river/ (accessed on 4 October 2020).
15. Jana, S.K.; Sekac, T.; Pal, D.K. *Study of Changing River Courses and Estimation of Reduction of Available Land Reserved for Development in Lae City of Papua New Guinea Using GIS and Remote Sensing Technology*; International Journal of Advance Research, IJOAR: Singapore, 2014.
16. Sekac, T.; Jana, S.K. Change detection of Busu river course in Papua New Guinea-impact on local settlements using remote sensing and GIS technology. *Int. J. Sci. Eng. Res.* **2013**, *5*, 891–899.
17. Water, Sanitation and Hygiene (WASH) Policy 2015–2030. Available online: http://extwprlegs1.fao.org/docs/pdf/png180011.pdf (accessed on 25 August 2020).
18. Ekness, P.; Randhir, T.O. Effect of Climate and Land Cover Changes on Watershed Runoff: A Multivariate Assessment for Storm Water Management. *J. Geophys. Res. Biogeosci.* **2015**, *120*, 1785–1796. [CrossRef]
19. UNICEF. Water, Sanitation and Hygiene. 2018. Available online: https://www.unicef.org/png/what-we-do/water-sanitation-and-hygiene (accessed on 25 August 2020).
20. Tanaka, M.O.; de Souza, A.L.T.; Moschini, L.E.; de Oliveira, A.K. Influence of Watershed Land Use and Riparian Characteristics on Biological Indicators of Stream Water Quality in Southeastern Brazil. *Agric. Ecosyst. Environ.* **2016**, *216*, 333–339. [CrossRef]

21. USGS EROS Earth Explorer System, USGS, USA. 2020. Available online: https://earthexplorer.usgs.gov/ (accessed on 25 August 2020).

22. Pryde, J.K.; Osorio, J.; Wolfe, M.L.; Heatwole, C.; Benham, B.; Cardenas, A. Comparison of Watershed Boundaries Derived from SRTM and ASTER Digital Elevation Datasets and from a Digitized Topographic Map. In Proceedings of the ASABE Meeting, Minneapolis, MN, USA, 17–20 June 2007.

23. Eastman, R. *TerrSet GIS; Idrisi Image Analysis*; Clark Labs, Clark University: Worcester, MA, USA, 2017.

24. OpenStreetMap Contributors; Papua New Guinea—Roads Retrieved from Humanitarian Data Exchange (HDX). 2018. Available online: https://www.openstreetmap.org (accessed on 6 September 2020).

25. Doaemo, W.; Mohan, M.; Adrah, E.; Srinivasan, S.; Dalla Corte, A.P. Exploring Forest Change Spatial Patterns in Papua New Guinea: A Pilot Study in the Bumbu River Basin. *Land* **2020**, *9*, 282. [CrossRef]

26. Hansen, M.C.; Potapov, P.V.; Moore, R.; Hancher, M.; Turubanova, S.A.; Tyukavina, A.; Thau, D.; Stehman, S.V.; Goetz, S.J.; Loveland, T.R.; et al. High-Resolution Global Maps of 21st-Century Forest Cover Change. *Science* **2013**, *342*, 850–853. [CrossRef] [PubMed]

27. McAlpine, J.R.; Keig, G.; Short, K. *Climatic Tables for Papua New Guinea*; Commonwealth Scientific and Industrial Research Organisation: Melbourne, Australia, 1975; 177p.

28. Mishra, A. Assessment of Water Quality Using Principal Component Analysis: A Case Study of the River Ganges. *J. Water Chem. Technol.* **2010**, *32*, 227–2349. [CrossRef]

29. Teixeira de Souza, A.; Carneiro, L.A.T.X.; da Silva Junior, O.P.; de Carvalho, S.L.; Américo-Pinheiro, J.H.P. Assessment of Water Quality Using Principal Component Analysis: A Case Study of the Marrecas Stream Basin in Brazil. *Environ. Technol.* **2020**. [CrossRef] [PubMed]

30. Gvozdić, V.; Brana, J.; Malatesti, N.; Roland, D. Principal Component Analysis of Surface Water Quality Data of the River Drava in Eastern Croatia (24 Year Survey). *J. Hydroinform.* **2012**, *14*, 1051–1060. [CrossRef]

31. Mahapatra, S.S.; Sahu, M.; Patel, R.K.; Panda, B.N. Prediction of Water Quality Using Principal Component Analysis. *Water Qual. Expo. Health* **2012**, *4*, 93–104. [CrossRef]

32. Canadian Council of Ministers of the Environment. Canadian Water Quality Guidelines for the Protection of Aquatic Life: CCME Water Quality Index 1.0, Technical Report. In Canadian Environmental Quality Guidelines, Canadian Council of Ministers of the Environment, 1999, Winnipeg. 2001. Available online: http://ceqg-rcqe.ccme.ca/download/en/137 (accessed on 9 September 2020).

Publisher's Note: MDPI stays neutral with regard to jurisdictional claims in published maps and institutional affiliations.

MDPI

St. Alban-Anlage 66

4052 Basel

Switzerland

Tel. +41 61 683 77 34

Fax +41 61 302 89 18

www.mdpi.com

Hydrology Editorial Office

E-mail: hydrology@mdpi.com

www.mdpi.com/journal/hydrology